Imagining Serengeti

NEW AFRICAN HISTORIES SERIES

Series editors: Jean Allman and Allen Isaacman

Imagining Serengeti

A History of Landscape Memory in Tanzania
from Earliest Times to the Present

⮌

Jan Bender Shetler

OHIO UNIVERSITY PRESS
ATHENS

Ohio University Press, Athens, Ohio 45701
© 2007 by Ohio University Press
www.ohio.edu/oupress/

Printed in the United States of America
All rights reserved

Ohio University Press books are printed on acid-free paper ⊗ ™

15 14 13 12 11 10 09 08 07   5 4 3 2 1

Library of Congress Cataloging-in-Publication Data

Shetler, Jan Bender.
  Imagining Serengeti : a history of landscape memory in Tanzania from earliest
times to the present / Jan Bender Shetler.
      p. cm. — (New African histories series)
  Includes bibliographical references and index.
  ISBN-13: 978-0-8214-1749-2 (cloth : alk. paper)
  ISBN-10: 0-8214-1749-5 (cloth : alk. paper)
  ISBN-13: 978-0-8214-1750-8 (pbk. : alk. paper)
  ISBN-10: 0-8214-1750-9 (pbk. : alk. paper)
  1. Landscape assessment—Tanzania—Serengeti National Park. 2. Landscape
changes—Tanzania—Serengeti National Park. 3. Landscape—Social aspects—
Tanzania—Serengeti National Park. 4. Geographical perception—Tanzania—
Serengeti National Park. 5. Oral tradition—Tanzania—Serengeti National Park.
I. Title.
  GF91.T36S54 2007
  967.8—dc22
                                                                    2006101502

Contents

Illustrations

TABLES

Preface

THIS BOOK IS DIVIDED into two main parts, roughly representing historical memory before and after the nineteenth century. The mid-nineteenth century is the critical chronological breaking point between these parts of the book because after that time oral traditions become more historically grounded and written sources also become available. The first section establishes the oldest and most basic ways of seeing the landscape—ways inherited from the distant past but still relevant today—and presents them in relation to the three different genres of oral tradition on which they are based. The chapters in this section explore the core spatial images in each of these genres and the appearance of these same images in other kinds of sources. These different, though intersecting, views of the landscape have ordered western Serengeti interactions with the environment, beginning with the oldest kind of oral traditions that go back to the first Bantu speakers who settled in the region about 300 CE and continuing into the most recent conversations about the park. They represent a set of strategies or approaches that western Serengeti peoples adapted in a myriad of ways to transform their society in response to changing historical circumstances.

The second section, or final three chapters of the book, then looks at how western Serengeti peoples used these three precolonial views of the landscape embedded in oral traditions in more recent social transformations. Each chapter analyzes the challenges and conflicts chronicled in historical traditions that emerged as western Serengeti peoples adapted to three major crises: the late nineteenth-century disasters, colonial rule, and the creation of Serengeti National Park. In the second section, we see how elders responded to new historical circumstances in creative ways and ultimately succeeded in adapting to change by recontextualizing older ways of seeing and using the landscape. The second section also describes new ways of seeing the landscape brought by new encounters. Although the book proceeds chronologically, in the order in which these different ways of seeing the landscape emerged, there is a constant reference back and forth in time, since spatial images cannot easily be locked into set time frames. The reader is alerted when evidence is taken from a different time period and when applications are made in the present, thus avoiding the pitfall of assuming a timeless tradition in the ethnographic present. The book ends with western Serengeti peoples' conversations about how to solve the problems they face in relation to the park that are based on a global conservationist view of the landscape.

Acknowledgments

THIS BOOK HAS BEEN ten years in the making and along the way has incurred many debts of gratitude that can never be fully repaid. I am grateful for the institutional support I received during the course of my graduate studies and research in Tanzania. The dissertation research was assisted by a grant from the Joint Committee on African Studies of the Social Science Research Council and the American Council of Learned Societies with funds provided by the Ford, Mellon, and Rockefeller Foundations. I also received a research grant from the Institute of International Education under the U.S. Fulbright Student Program (1995–96). In 2003 a generous research grant from the National Endowment for the Humanities allowed me to take a semester for additional research in Tanzania and writing. I express my sincere gratitude to the government of Tanzania for permission to do research in the country under the auspices of the Tanzania Commission for Science and Technology and the history department of the University of Dar es Salaam. Members of that department were always generous with their time and support. Special thanks to Dr. Fred J. Kaijage, Dr. Bertram Mapunda, Dr. Rugatiri D. K. Mekacha, Dr. B. Itandala, Dr. I. N. Kimambo, Dr. Nestor Luanda, and Dr. Yusufu Q. Lawi. I am thankful to the patient staff at the Tanzania National Archives, the East Africana Collection at the University of Dar es Salaam library, the Bodleian Library at Rhodes House (UK), the Public Records Office (UK), the University of Florida Africana collection, and the Goshen (Indiana) College library. Gratitude is also extended to Markus Borner of Frankfurt Zoological Society and the Serengeti Research Institute for use of their facilities while my husband Peter was working with their GIS project and for sharing mapping data.

The University of Florida, where I studied, and Goshen College, where I teach, have been excellent homes for intellectual growth. I thank the history department and the African Studies Center at the University of Florida for their support, received over my years there in the form of assistantships, travel grants, writing fellowships, and much more. At the University of Florida I experienced an atmosphere of creative interdisciplinary interaction with a community of scholars who demonstrated an unusually cooperative spirit, including Holly Hanson, Tracy Baton, Marcia Good, Edda Fields, Catherine Bogosian, Jim Ellison, Todd Leedy, Kearsley Stewart, and Kym Morrison. My deepest appreciation is extended to my mentors, especially Steve Feierman, Hunt Davis, and David Schoenbrun, who have given so much of their time and intellectual inspiration to my work. Goshen College provided ongoing financial and physical support for research through the Minninger Center and the Multicultural Affairs Office. A wonderful group of colleagues and students on Wyse third at Goshen College provided inspiration and grounding in the last phase of the research and writing. Thanks especially John D. Roth, Steve Nolt, and Lee Roy Berry in the History Department. Perhaps unknowingly, my students in the Environmental History classes of 2002 and 2004 helped me to think through many of the issues in the book. Thanks also to my wonderful research assistants, Nyangere Faini, Rose Wang'ombe Mtoka, Jessica Meyers, and Emily Hershberger. Many people read partial or complete drafts of the project and gave me comments along the way: Holly Hanson, Marcia Good, David Schoenbrun, Elizabeth Garland, Elizabeth Smucker, Kathleen Smythe, Richard Waller, and the readers and my editors at Ohio University Press, including Jean Allman and Allan Isaacman, whose suggestions made the book so much better. I am most grateful. While I was in Dar es Salaam in 1996 and 2003 I enjoyed the good conversation and camaraderie of a wider community of scholars staying at the TYCS hostel and working in the archives. Even as I finished writing my dissertation on our farm in Dove Creek, Colorado, the lovely community around me gave me the sustenance necessary to do the job. A constant source of input and ideas were my colleagues at the Tanzania Studies Association.

In Tanzania numerous people aided my work, while providing good hospitality and friendship. Thanks to Mwalimu Nyamaganda Magoto for all the time he spent going over hundreds of cultural terms in Nata; to Susana Nyabikwabe Mayani for teaching me Nata and being my friend; to Adija Sef for her friendship, quick laugh, and care of my house and children; to Mayani Magoto for his work in interviews, research on various topics, and interest in the whole process; to Goko Kimori for coming over nearly every day to find out if I had learned Nata yet; to Faini Magoto and Joseph Magoto and their families for always having a meal and a place to rest when I needed it; to Mzee Mswaga for being our community eyes and ears; to Susan Godshall for typing the Ngoreme dictionary; to Augustino Mokwe Kisigiro for lending me his Nata dictionary and tapes; to Padre James Eblin, Maryknoll Missioner from the Ishenyi mission, for his kindness and especially for twice rescuing me when the rented car broke down; to the Tanzania Mennonite Church people in Nyabange and Shirati, who always extended hospitality and support; to Bishop Solomon S. Buteng'e, Bishop Joram Mbeba, and Marehemu Bishop Naftali Birai, who were interested in the work and allowed us to rent a church car; to David and Justine Foxall for their kind hospitality in Dar es Salaam; and to Brian Farm and Bethany Woodward for their hospitality and friendship at Seronera.

I feel a deep debt of gratitude to Nyawagamba Magoto, who extended warm fellowship, invited us to become part of his family, and embraced the research as if it was his own. I am also grateful for all the work and dedication of those who acted as my colleagues to arrange interviews, interpret, make introductions, and serve as cultural translators. Without their help my work would not have been possible. They include Kinanda Sigara for Ikizu, Mayani Magoto and Nyamaganda Magoto for Nata, David Maganya Masama for Ngoreme, Pastor Wilson Shanyangi Machota for Ikoma, and Mnada Joseph Mayonga for Tatoga, all of whom also extended generous hospitality. I was also aided in arranging interviews by Rhoda Koreni, Yohana Wambura, Kennedy Sigira, Philemon Mbota, Thomas John Kazi, D. M. Sattima, and Ibrahimu Matatiro Kemuhe. Zedekia Oloo Siso is a fellow historian and gave me access to his work in the Luo-speaking area of the region. Thanks to Glen and Elin Brubaker for making that exchange possible.

Many other people facilitated aspects of the research. On the trip to Sonjo, Michael Wambura Machambire put aside his own affairs to accompany me; Ndelani Sanaya introduced me to his home village; and the chairman of the village, Emanuel G. Goroi, graciously hosted our group without prior warning. Father Ambrose Chacha helped me with the Nyegina records; G. M. Kusekwa assisted with the S. D. A. records at the high school in Ikizu; and the archivist in Musoma, Fredrick Semkiwa, helped to locate files there. In Mugumu I was always grateful for the hospitality of our friends from Missions Moving Mountains and the Mennonite Church (Daniel and Prisca Machota, and Wilson and Esta Machota). In Nyabange the door was always open from many dear friends. In Nyabasi we received welcome from Elva Landis and Elizabeth Birai; in Iramba from Father Charles Mwiguta; in Kisaka from Pastor Zakayo Jandwa and his family; in Mwanza from Juliana Matoha Magoto; in Bujora from Joseph Sungulile, Jefta Kishosha, and Mark Bessire; and in Nyegezi from Bhoke Magoto.

My extended family on both sides cheered me on with love and constant encouragement. However, my biggest debt of gratitude is to my husband, Peter, and to my children, Daniel and Paul, who suffered and rejoiced with me through it all. Their contentment and joy in living in rural Tanzania allowed me the freedom and security to do my own work, and their continued patience with my ongoing project back home helped keep things in perspective. Peter made the maps and illustrations and provided technical, logistical, and emotional support through it all. Daniel's cultural sensitivity and flexibility were always appreciated in our many travels. Paul went back with me in 2003 as my research assistant, photographer, and technology manager, as well as provided good company. I am ever grateful for that chance to work with my son and share this rich experience. I am humbly grateful for their love. The book is dedicated to my father, John M. Bender, who inspired in me a love of the environment and a curiosity about people's role in shaping it.

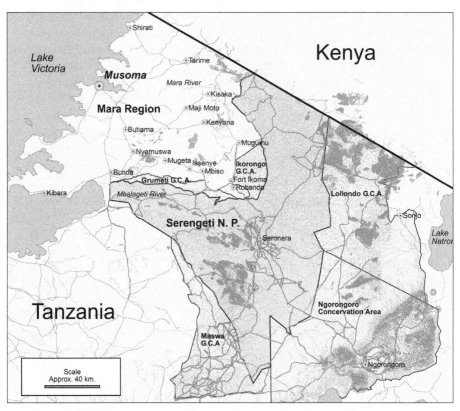

Map 1. Western Serengeti regional setting. *Map by Peter Shetler, 2005. Underlying GIS data courtesy of Frankfurt Zoological Society*

Landscapes of Memory

Standing on a rocky outcropping, one looks across the rows of low hills to Mangwesi Mountain on the far horizon. The short grass lawn is a vibrant green, dotted with well-spaced acacias (umbrella trees), beneath which graze a dozen zebras and a few Thomson's gazelles. One might see this western Serengeti landscape as nature at its finest, a last remnant of unspoiled wilderness where animals can roam free. Or one might see it as a landscape shaped by people who set fires to create openly spaced woodlands with productive grasses, tell stories about ancestors settling at Mangwesi Mountain, propitiate spirits at the nearby spring, and follow the paths of hunters, traders, and raiders that crisscross the land. This second way of seeing the landscape is that of people whose ancestors lived in the western part of the Serengeti-Mara ecosystem for the past two thousand years, including land that is now within Serengeti National Park and surrounding game reserves. Their view of the landscape has not been a part of the global conversations of other people who care about the Serengeti. Western Serengeti peoples have been dismissed as recently arrived poachers within a landscape envisioned as empty of people. Yet, for as long as we have memory, the western Serengeti has been a profoundly humanized landscape with the stories, hopes, and challenges of its people deeply embedded in its rocks and hills, pools and streams, vistas and valleys. A history of western Serengeti peoples' memory rooted in a humanized landscape introduces a new perspective to current debates about the future of African environments and the histories of people who live with them.

Serengeti National Park was founded on a view of the landscape that presents a sharp contrast to local ways of seeing.[1] When people throughout the world imagine the Serengeti, they do so through the medium of the many documentary films produced by National Geographic and others promoting it as an endangered global wildlife resource. Bernhard Grzimek, who worked for the Frankfurt Zoo in Germany, produced one of the first films of this genre, *The Serengeti Shall Not Die* in 1959.[2] The film opens with Grzimek's explanation of why he and his son, Michael, were bringing a small plane to the Serengeti—to do a count of the animals and to map the migration routes in order to aid the new park in establishing its "natural" boundaries. The beautiful images of wildlife and scenery in the Serengeti, anthropomorphized stories of animals and interesting biological facts are interspersed with the plea to save "the last refuge for the great herds of the African plains." The narrative suggests that animals can be saved only by establishing parks, aided by the efforts of people like the Grzimeks, who perform difficult and selfless acts in harnessing science and technology for the task. Even the walls of the Ngorongoro Crater are presented as enclosing "the most magnificent zoo on earth." In the Grzimeks' previous African film, *No Room for Wild Animals*, parks are described as "a forbidden land for man" where the animals know no fear of people.[3] The image of the Serengeti landscape (or any other African park) in these films is entirely wild and natural, without history or social context. They describe a landscape broken into ecological zones—plains, water holes, and hills—but devoid of names or information that would differentiate one place from another either in time or space.

Portrayals in these films of a landscape for wild animals alone is rooted in the Grzimeks' overriding compassion for and delight in the animals and their disdain for "civilization" and urbanization, which inevitably lead to ecological destruction. Local people, manufactured objects, and the colonial context in which the films were shot seldom appear at all; the park is depicted as a completely natural space that must be kept separate from people for the wildlife to survive. The film views hunter-gatherer peoples like the wild animals themselves, in danger of extinction, while the "Negroes" and other "civilized" or "mixed-race" Africans (referred to as "human hyenas") wantonly burn the grass, cut the trees, and poach with weapons that make the animals suffer and "die a lingering, senseless death." In the Serengeti film the Maasai appear briefly as proud pastoralist warriors who recklessly cut trees and brush, causing the water holes to dry up. The only mention of the western Serengeti peoples is an oblique reference in the footage where Michael Grzimek supposedly "discovers" the German Fort Ikoma as he is looking for water after a plane crash and notes that the Ikoma, who live in this area, were, during the German period, a "frontier tribe, as unruly then as they are today."[4] These potently symbolic images of the Serengeti as one of the "last nooks of paradise," a wild Africa, existing in its pristine state since the dawn of time, proved influential in creating the global perception of the Serengeti landscape.[5]

Yet other regional ways of seeing that same Serengeti landscape still exist, present in the collective memory of people who have never been included in global narratives about Serengeti National Park except as "poachers." Calling themselves

Ikoma, Nata, Ikizu, Ishenyi, and Ngoreme, these peoples now live on the western border of the park. During the summer of 2003, in the course of historical research, I traveled with Ikoma elders, Pastor Wilson Shanyangi Machota and Edward Wambura Kora from Morotonga, out to Tanzania's Ikorongo-Grumeti Game Reserves, adjacent to Serengeti National Park, to identify the abandoned settlement sites and graves of their Ikoma ancestors, now accessible only with a special permit and a village game scout as guide. We tried to get permission to go into the park to find other Ikoma sites but were unsuccessful. Although restrictions had kept these elders out of the area for over thirty years, they directed us in the car to the old settlement sites, springs, and sacred sites for propitiating the ancestors. The elders had to search the whistling thorn brush thicket for a long time before finding the sacred site at the Kumari spring, where a snake representing an ancestral spirit guards the land. No one had brought offerings here for propitiation in decades, and the old spring was dry and barely visible. The elders pointed out, and told stories about, the origin place of the first Ikoma man and woman, who pitched their camp under the Mukoma tree after arriving from Sonjo, now on the eastern border of Serengeti National Park. The elders' ability to locate the sites in the wilderness came from hunting trips long ago with their fathers and grandfathers, who told the stories of the past as they walked over the land or camped in these spots. Seeing these places brought tears of joy to their eyes. As the trip ended they expressed their gratitude for being granted an opportunity to see this magnificent land one more time. They only wished that their children and grandchildren could also visit these places where their ancestors are buried.[6]

Figure 0.1. Wilson Shanyangi Machota *(left)* and Edward Wambura Kora at dry spring in the Ikorongo Game Reserve, abandoned Ikoma settlement site. *Photo by Paul Shetler, 2003*

In contrast to Grzimek's images, the elders see a differentiated social landscape that also includes wildlife. The places we visited evoked stories about the past that represented a variety of different landscape images and social groups. The elders identified many of the abandoned settlement sites of generation-sets by springs, now dry because the people were no longer cleaning them out. Standing on the higher places, they looked across the landscape and named the areas settled by different clans, often associated with hills. They uncovered the remains of rock walls that were once fortresses to protect the people from Maasai raids in the late nineteenth century. One elder said that as a youth he used to herd cattle and play around these walls, when they were higher than his head. The walls were now almost gone because the park had used the rock for its building projects. Because they knew this land as hunters they also knew the water holes, campsites, and paths that connected them with other communities in the region. They had walked these paths as migrant laborers going to Nairobi to find work, stopping and spending the night among friends in Sonjo, to the east, or as traders to take wildebeest tails to barter for goats and sheep in Sukuma, to the south. In these later, more historically identifiable stories the landscape visions of other peoples also became apparent—a Maasai view of the land as a pastoralist domain, a British view of the land as a resource for economic development, and a global conservationist view of the land as wilderness to be kept apart from people. In response western Serengeti peoples told new kinds of stories about these events. Although western Serengeti peoples incorporated these newer landscape visions that fundamentally altered their ways of living on the land, they continued to tell the older stories and visit the places that kept earlier landscape memories alive.

The Ikoma elders' ways of seeing the landscape, as well as the contrasting film images, are all imaginative constructions: interpretations influenced by historical experiences, social identity, and political power, rather than by objective visions of the physical land. The title of this book, *Imagining Serengeti*, captures a broad definition of landscape as an "imaginative construction of the environment."[7] David William Cohen and E. S. Atieno Odhiambo refer to landscape as "encompassing the physical land, the people on it, and the culture through which people work out the possibilities of the land," while Simon Schama writes that "landscape is the work of the mind. . . . built up as much from strata of memory as from layers of rock."[8] Thomas R. Dunlap, in his work on the British settler colonies, describes landscape as "the picture of the land people see as having significance for the nation and their culture."[9] Benedict Anderson's influential book *Imagined Communities* describes the formation of European nationalism as shifting concepts of time and technologies like the printing press and the newspaper enabled Europeans to imagine themselves as members of "nations." Similarly, western Serengeti peoples conceive of their own social identities through the "imagined landscapes" embedded in oral tradition.[10] The power of a group of people to shape the landscape is dependent on how they imagine the landscape, which, in turn, is reproduced on the landscape. However, both Grzimek and the Ikoma elders take their view of the landscape for granted as natural or objective reality and do not consciously see it as a means to assert power. It therefore takes careful analysis and comparison with other sources to unpack and "denaturalize" the meanings they attach to physical features of the landscape and to place them within particular historical contexts and contests of power.[11]

The problem for historical analysis is that the sources for making local landscape visions from the past visible and meaningful are difficult, inadequate, and not easily accessible. No written documents for this region, except the ecological or economic, exist before the beginning of the twentieth century, nor has much historical research taken place in this region. Archaeological and historical linguistic sources can be applied only at a rough regional scale, and it is problematic to project ethnographic information from recent societies onto the past. Oral traditions remain one of the few available sources, and those are fraught with inconsistencies since they have changed as they are transmitted over time, are expressed in local cultural idioms, and represent the views of only a certain segment of society. The historian struggles to find meaning in a list of place-names or the route of a generation-set walk presented by the elders in their narration of oral tradition. In a heterarchical society without chiefs or kings there is no dynastic tradition remembered by court griots or one master narrative about the past. While oral traditions seem to retain spatial images as they are transmitted over time, they lose connection to temporal sequences or to the historical context to which they first referred. Many Africanist historians use oral tradition to reconstruct nineteenth-century precolonial histories, but most have been unable to support the evidence for earlier histories without written sources.

This book addresses the problem of oral traditions as reliable sources with a new methodology for tracing a history of memory. Historical changes in ways of seeing the landscape are reconstructed by identifying core spatial images in oral traditions that can then be reinserted into historical contexts identified by other kinds of sources. The starting point for this methodology is a spatial analysis of oral traditions, based on the durability of spatial memory that is linked to social identity. I use accepted methodologies to reconstruct the basic historical contexts from archaeological, historical linguistic, ecological, ethnographic, and archival sources. Through a process of identifying congruency and logical patterns, the core spatial images are then recontextualized into historical periods or time frames. For the later periods the analysis also includes the profound material and ideological effects of introducing other ways of seeing the landscape from other regional societies, the colonial government, and global conservationists. This study relies on the interdisciplinary work of environmental and social historians, in Africa and elsewhere, who have identified key issues for the study of human communities in relation to their environment—including landscape, space, and memory—to provide the theoretical tools for analysis. With this methodology I am then able to reconstruct a long sweeping history of western Serengeti peoples as they interacted with their environment over the past two millennia. This methodology could be similarly applied for reconstructing environmental history in other places and times, especially where few historical sources exist apart from oral tradition.

New areas of inquiry open up as one incorporates, but moves beyond, a history of the environment to a history of memory connected to the environment. This lens allows us to see not only how people physically changed the environment by their presence but also how landscape memory shaped their societies and how this memory changed over time in response to new contexts. Using this critical theoretical insight, people become actors, rather than victims of environment, making environmental decisions rooted in continuity with the past while innovating as they adapted to new circumstances. Changes in ways of seeing the landscape over time indicated shifts in the physical way that people related to their environment. The two are inextricably connected and reciprocally interactive. Ways of seeing determined ways of using the land that, in turn, influenced memory as these landscapes became part of oral tradition in the core spatial images. When new contexts introduced new ways of seeing, and thus using, the landscape elders elaborated new oral traditions, while continuing to tell older traditions that retained the core spatial images from previous ways of seeing the landscape. These different ways of seeing coexisted, as they do today in the memories of elders who tell various kinds of stories about the past. But as oral, rather than written, traditions these memories depend on a physical connection to the landscapes in which they are embedded. The same physical space can be seen, and thus remembered, in a variety of ways with profound consequences for how people live in it. Paying attention to spatial patterns provides a key for recovering the historical meaning of oral tradition. It is thus through an investigation of the history of landscape memory that a long term history of people in relation to the environment can now be reconstructed.

This analysis asserts that the environment will be preserved, changed, or destroyed based on the memories imbued in it by specific groups of people. The tragedy of setting apart wild spaces that people can no longer visit is that these places cannot sustain social memories but rather become abstract, generic wild places consumed in a global marketplace. Deep social connections to specific landscapes may be more effective for protecting sustainable ecologies than an appreciation for interchangeable natural places often used to justify the destruction of the land in domesticated places. In an older western Serengeti tradition the bush is left to grow up undisturbed around the sacred sites of ancestral spirits of the land while the grass is burned in areas of habitation to create open parkland, healthy for both people and wild animals. But those human decisions depend on a historical memory connected to particular places and people. Seeing the landscape as either wild or domesticated is not the only way of creating memories that honor and preserve the land. *Imagining Serengeti*, through a varied and contested history of memories embedded in peopled landscapes, adds both a rich, new dimension to existing conversations about preserving African environments and a new methodological approach to precolonial African history.

WESTERN SERENGETI PEOPLES AND SOURCES OF EVIDENCE

The complex mix of languages, economies, and cultures making up the western Serengeti presents a challenge for historical analysis. Western Serengeti peoples are East Nyanza Bantu-speaking agropastoralists known as the Ikoma, Nata, Ishenyi, Ikizu, and Ngoreme ethnic groups who now occupy Serengeti and Bunda districts in the southeastern portion of the Mara region of Tanzania. Each of these ethnic groups claims its own unique identity and history, and no pan-ethnic identity developed here in the colonial period as it did among their Sukuma, Luo, or Maasai neighbors. Without a tradition of chiefs or hierarchical leadership, the Mara region also differed from the Great Lakes kingdoms, where dynastic history often overshadowed commoner or clan histories; thus, no centralized narrative tradition exists here.[12] The adaptation of agropastoralists to the ecology of this region and their ongoing prosperity depended on interaction with other peoples in the region, such as the Tatoga (Dadog-speaking pastoralists, including Rotigenga and Isimajek Tatoga) and Asi hunter-gatherers.[13] While I interviewed many Tatoga elders I could not identify any Asi descendants who knew their traditions other than those now integrated into agropastoralist communities. No local designation exists for this western Serengeti group of Bantu-speaking agropastoralists as a whole except *Rogoro* (the people of the east), yet even the area to which this designation refers varies relative to the location of the speaker. My research was concerned with these five ethnic groups, forming a coherent unit, and within the limitations of field research, but logically could have expanded to include other groups such as Zanaki, Sizaki, and, at a larger scale still, Kuria or the Mara region as a whole. Kuria moved into the Serengeti District during the 1950s. Limited interviews among these neighboring groups allowed for a regional comparison.

For the purposes of this analysis the western Serengeti is treated as an integral region, in the sense that it encompasses the geographical boundaries of an intercommunicating, interacting set of people. *Region* is not defined here as a homogeneous cultural or social unit, as the economic relations of exchange or as a formalized marketing system, as has been the trend in much of the recent regional analysis. Rather, regions are treated as historical products constantly in negotiation and transforming as different peoples interact in changing ways over time.[14] Even the most rigidly conceived regional boundaries of western Serengeti with the Southern Nilotic–speaking Maasai herders to the east or the Sukuma

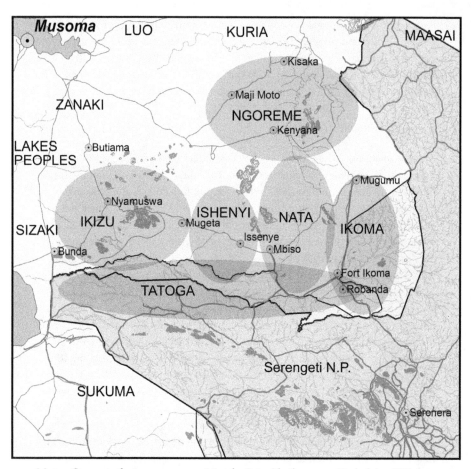

Map 2. Current ethnic group areas. *Map by Peter Shetler, 2005. Underlying GIS data courtesy of Frankfurt Zoological Society*

farmers of another Bantu-speaking family to the south were frequently crossed through trade, marriage, prophecy, or refuge. This region, both past and present, has functioned based on its linguistic, cultural, and economic diversity, even as western Serengeti peoples feel a diffuse sense of collective identity due to their common historical background, shared cultural assumptions, and proximity to each other. I have chosen to refer to this region as the western Serengeti, rather than eastern South Mara, because during the late nineteenth century, when the most significant social transformations took place, its people were oriented toward the Serengeti. From the colonial period on, the people of the western Serengeti began to see themselves as part of the Mara region, or Musoma District as it was then known. Serenget is a Maasai word, referring to a historical Maasai section and meaning wide-open spaces. The Serengeti has widespread recognition and puts this story in the current context of debates over Serengeti National Park. The people of this region do not use the name western Serengeti to refer to themselves but would recognize their common story within the historical region.[15] The western Serengeti region is bounded on the east by the great Serengeti plains and the Maasai, on the north by the Mara River and the Kuria, on the west by a gradual shift without any natural division toward the peoples of Lake Victoria, and on the south by the Mbalageti River and the Sukuma. These boundaries also correspond to an ecologically unified area, the western woodlands of the Serengeti-Mara ecosystem.

Ignored in ongoing debates about Serengeti National Park and the surrounding ecosystem, western Serengeti peoples have also been ignored as a subject of scholarship. No academic work exists beyond this study to establish even the most basic chronological framework for the region.[16] Written sources are few, including a handful of early travelers' accounts and ethnographies that are based on visits of short duration and primarily in neighboring ethnic groups within the larger region.[17] The written sources on this region, housed in the Tanzania National Archives and the East Africana Collection of the University of Dar es Salaam, are scarce, since the German records in the country were largely destroyed when they left the country during wartime. Only an incomplete set of the Musoma District files from the British period have survived—forcing reliance on papers from the provincial or territorial rather than the district level. Archival sources for the independence period are also problematic. Additional archival data from mission or British government sources is also scant, since a marginal region such as this received little attention in the home offices.[18] The archival sources that do exist were quite useful for reconstructing the historical periods documented in the last three chapters of this book, but as for the precolonial period, the historian must find other sources for reconstructing the historical context.

Archaeology and historical linguistics remain the time-tested tools for African historians interested in the distant past. Yet no archaeologists have worked in this region on eras after the earliest domestication of plants and animals, nor have researchers been interested specifically in the ancestors of western Serengeti peoples. Therefore the archaeological evidence must be used carefully to describe wider regional patterns that seem to have some bearing on developments in the western Serengeti and to establish a basic chronology. I relied on the existing reports of archeological research in the wider region.[19] Historical linguistics can tell us generally when and from what direction languages were introduced in the region. Comparing languages spoken today to identify changes in words and sound patterns over time provides a sense of how the languages are related to one another and how contact with other languages and local language innovations influenced the development of a language over time. Although dating through glottochronology has been somewhat controversial, it does provide a rough sense of when people speaking a particular language lived in the area. Tracing words and their meanings throughout the region and back in time also helps identify concepts and their variations that have been fundamental to the culture for a long time, while loanwords provide evidence for cross-cultural contacts. The historian must be careful, however, not to assume that languages are synonymous with communities of people, since languages can spread without human migration.

Historical linguistics, like archaeology, identifies older changes more accurately than it does the recent. While relying on the existing linguistic work of David Schoenbrun and Christopher Ehret as well as locally published dictionaries, my research also included collecting core vocabularies of one hundred words in Ngoreme, Ikizu, Nata, Ikoma, Ishenyi, Sonjo, and Dadog and assembling a nearly complete 1,563-word cultural vocabulary list in Nata. The word lists were used to figure out how closely the languages were related to one another and the cultural vocabulary to see how meanings of words reflected their Great Lakes Bantu roots or borrowings from other language groups.[20]

Both ecological and ethnographic sources assisted in reconstructing earlier historical contexts for oral tradition. Since this area is located in the Serengeti-Mara ecosystem and near Serengeti National Park, abundant ecological data is available.[21] While I do not claim to be an ecologist (or archaeologist, ethnographer, or linguist), I have read the works of ecologists looking for patterns that might explain or corroborate the human imagined landscapes evoked in the elders' stories. When I walked with elders out to the historical sites, I located them on a geographical grid using a Geographical Positioning System and the technology of GIS (Geographical Information Systems). Peter Shetler constructed the maps in this book using this information as well as data from the Serengeti Research Institute GIS project under the Frankfurt Zoological Society, with which he was involved in 1995 and 1996. These maps, which appear at the beginning of each chapter, allow us to see the landscape from a bird's-eye view, a vantage for visualizing the information that the elders transmit in oral traditions but see in a different way. Ecological data helps to place some evidence from oral traditions and ethnography in the time frame of slow and long-term ecological changes. I have read the existing ethnographies from the region, but I also did my own informal village survey and have learned one local language as part of my method. A friend in the village met with me regularly to learn and practice Nata, although I was never proficient and functioned mostly in Swahili. Participant observation was also an important method, as my family and I lived in the rural Nata village of Bugerera for eighteen months and were incorporated into the extended family of Magoto Mossi Magoto. I made the habit of visiting women in neighboring homesteads in the late afternoon and sometimes helped with their daily tasks. However, I had to use ethnographic data carefully in relation to other sources, since one cannot assume an unchanging "traditional" past from which these practices were transmitted.

Oral traditions, although also problematic, formed the most important source used in this study. Historical narratives in the western Serengeti, like those of many other noncentralized societies in Africa, are nonformal and loosely structured. They appear more in the form of conversation than as epic poetry in set verse.[22] No particular word exists in local languages for this genre of oral tradition except as *amang'ana ga kare* (matters of the past). No formal experts control this knowledge although some are considered more knowledgeable than others. Those who know more about "matters of the past" acquired their knowledge through personal desire or aptitude, rather than purely as a function of their social position. Some people have a gift for it, given by the ancestors. Elders attain legitimacy as narrators of "matters of the past" and specialize in particular kinds of knowledge through a combination of ability, respect, role, interest, experience, and the sanction of the ancestors—all of which are manifested in the effectiveness of their tales.[23] The people most often recommended to me as those who knew about these "matters of the past" were men more than sixty years old who occupied positions of authority or respect in the "traditional" structures of society and were often consulted for their wisdom.[24] Men who had education and held political offices were considered especially valued intermediaries for an outsider like myself. On the other hand, many felt that educated people disparaged "traditional" knowledge. Material wealth was not a particular criterion for recommendation. Almost all were born after colonial rule and had once worked as migrant laborers, but the most knowledgeable elders had spent much of their lives at home in the Mara region. Because the kinds of stories that men told were influenced by their roles, experiences, and interests, historical interpretation of oral traditions is inherently problematic.

Women's stories have not been part of the corpus of historical knowledge, thereby confirming the problem that oral traditions represent the experiences of only certain segments of society. Women possessed entirely distinct forms of knowledge about the past. When I asked to speak with those who knew about history, local colleagues, men and women alike, agreed that men of this generation were the keepers of historical knowledge. When I insisted on talking to women, I found that most women did not know the larger ethnic accounts of origin, migrations, clans, ritual, and battle, which made up the spontaneous content of interviews with men. At first I thought that women were just reluctant to give me their versions of the past, but I later became convinced that women possessed not just another version but wholly different kinds of knowledge about the past.[25] Because people learn about the past in the gendered spaces of the male courtyard beer party or the female cooking fire, men and women share neither styles of oral narration nor types of knowledge about the past. Men and women occupy separate spheres of interaction in their daily routines, sharing the same world but participating in different, though intersecting, sets of discourses about that world.[26] They keep and transmit historical knowledge by the paths they walk each day and the positions they occupy in the imagined male and female spaces that permeate their world. Women may learn some of men's knowledge about the past, but they do not transmit those stories in the narrative style of men or in the formal setting of men's courtyard meetings. Their knowledge about community relationships and genealogies is, however, critical to understanding the imagined landscapes shared with men, even though it does not appear in the formal narratives. A gendered analysis of oral tradition is necessary for finding its historical meaning.

I had access to these stories about the past as a young American woman with a husband and two children who had been associated with the Tanzania Mennonite Church. While I carried out formal research in the Mara region at various times over the past decade, I have worked and lived in this region over the past two decades: from 1985 to 1991 as a development worker with the Mennonite Central Committee, for eighteen months in 1995–96 as part of my Ph.D. dissertation research, during a brief follow-up visit in 2001, and for three months of research in 2003. Regional context also comes from living and working in Ethiopia from 1980 to 1983 and again in 2005. During my research I also collected locally written histories from groups of elders in the region that were published as *Telling Our Own Stories: Local Histories from South Mara, Tanzania.*[27] In interviews with over two hundred Nata, Ikoma, Ishenyi, Ngoreme, Ikizu, Isimajek, Kuria, Rotigenga, Ruri, Sizaki, Sonjo, and Zanaki informants I asked open-ended questions, trying to explore the range of historical knowledge.[28] When I put these accounts side by side, a common regional history emerged in many unique versions. I gained access to knowledgeable elders through the mediation and recommendation of trusted local people who introduced me in the various communities and helped with the interviews, which often lasted more than four hours at a sitting. Although I learned some Nata, the interviews that started in Swahili often turned to any of the other, often mutually recognizable, languages for which I needed interpretation. Those friends who helped me were part of social networks established during earlier work in the region, and they were themselves committed to preserving local history. The oral research depended on this network of friends, family, and colleagues. The extended family of Nyawagamba Magoto hosted our family in Nata Bugerera during the main research period, and he is responsible in large part for making the research possible. The oral histories, however, cannot stand on their own as unmediated accounts of the past. They must be reconnected to the social groups and historical contexts from which they were transmitted. This context is reconstructed by placing the evidence from other disciplines (archaeology, historical linguistics, ecology, ethnography) within a framework built by using the theoretical tools from a wide variety of scholars.

Figure 0.2. (top) Author
(left) in an interview
with Mechara Masauta,
Robanda, Ikoma. *Photo
by Paul Shetler, 2003*

Figure 0.3. (right)
Nyawagamba Magoto,
Mbiso, Nata. *Photo by
Paul Shetler, 2003*

Environmental historians provide the tools for understanding how the environment has been created through interaction with human society and human society through interaction with the environment.[29] Because the popular image of a wild and natural Serengeti has so strongly influenced conservation policy, it is important to demonstrate the argument long made by environmental historians that virtually all landscapes are created by human intervention and that the division between nature and culture is artificial.[30] On the other hand, this book further draws on the work of historians who acknowledge the biological limits to social creation of the environment and look at environmental features as historical actors in their own right. For this book, the particular kind of soil, grass species, or rainfall pattern matters in charting historical patterns.[31] Environmental historians are careful not to err too far on the side either of environmental determinism or of romanticizing the harmonious relationship of precolonial people with the environment. For example, Emmanuel Akyeampong's "eco-social" history of southeastern Ghana emphasizes the mutualism between people and their environment as a dynamic interaction.[32] These tools allow us to see that western Serengeti peoples are neither natural conservationists nor destroyers, nor can one characterize their historical interactions with the environment homogeneously.

Recent studies in African social history using oral history also assist in the analysis of pan-African concepts of social organization found in their particular form in the western Serengeti.[33] Many Africanist scholars have noticed the clear identification of people in their oral traditions with ecological zones (cattle people or hill people, for example) and the congruence of some of these categories with what we would now call ethnicity.[34] Others have documented how extensive social networks of reciprocity were mobilized in times of famine, how communities used descent as an idiom for thinking about relationships, how the multiple meanings of ritual and of sacred sites changed over time, or how economic strategies concentrated on minimizing risk.[35] The literature also provides a framework for understanding African resistance within a framework of the moral economy and local agenda.[36] All these insights and more are used in each chapter of the book as tools for analyzing the data from a variety of sources. However, few of these social histories based on oral sources extend the analysis to a precolonial narrative beyond the nineteenth century.[37] Through the new methodology introduced in this book, the tools of social historians can assist in reconstructing various time frames that are then brought together in one synthetic account of imagined landscapes moving through time. This chronology derived from local understandings disrupts and blurs the Eurocentric time periods of pre- and postcolonial.[38] For the western Serengeti, interactions with Maasai power in the second half of the nineteenth century was at least as important for social and environmental changes as colonial incursions. This book builds specifically on the work of scholars in East Africa, and especially Tanzania, who have produced a wealth of recent literature combining social and environmental history, including James Giblin, on patronage as an older way of protecting the environment; Steven Feierman, on local ways of evaluating leadership in terms of its efficacy in healing or harming the land; and the collection compiled by Greg Maddox, James Giblin, and Isaria Kimambo on the intersection of social history and environmental change.[39]

Environmental historians also provide tools for dealing with the question of the ongoing destruction of the environment. Depending on their perspective, some scholars have blamed land degradation on poor or "primitive" African techniques and management; others, on the advent of Western capitalism, with its global trade and colonialism. William Cronon argues that the narrative forms and plot lines we use to tell environmental history lead to very different conclusions about the trajectory of environmental change.[40] African historians debate degradation narratives that point to the collapse of indigenous environmental control in the late nineteenth century, emphasizing previously successful African ways of management.[41] Newer studies have cautioned against a monolithic analysis of colonial discourse and environmental science, bringing out the conflicting and ambivalent ways that colonial officers both denied and championed or learned from local knowledge.[42] Archival sources from the Musoma District, the administrative authority of the western Serengeti, show colonial officers both reprimanding chiefs for not growing cassava or cotton and investigating resistance in local populations to sleeping sickness. Some historians applied an ecological model of disequilibirum ecosystems to rethink apparent environmental degradation.[43] Examined with these tools, the late nineteenth century, a time of ecological disaster in the western Serengeti, becomes an important turning point in the narrative. However, it does not mark the destruction of the environment but rather the beginning of the end of human occupation and the development of a protected reserve for wild animals.

A growing body of literature serves as an aid for analyzing the conservation movement and the development of parks in Africa in relation to the communities that surround them.[44] African parks, following the U.S. Yellowstone model, were based on a European way of seeing the landscape that separated productive space from leisure space or civilization from nature, while at the same time legitimizing state control over land and natural resources. Jane Carruthers demonstrates how the Kruger National Park in South Africa became a symbol of Afrikaner identity, as it preserved a "remnant" of the vacant wilderness that the pioneers conquered.[45] Because Europeans assumed that their view was universal, they thought that the way to solve the problem of local opposition to the parks and poaching was to "educate" the people to appreciate nature, essentially to make them see the landscape in the same way.[46] Grzimek's film, released a few years before Tanzanian independence, said that "our Serengeti" could not be secure until the "natives" ("who believed that the animals were only being preserved so that Europeans could come and shoot them") were "won over" to protecting the animals as the "property of all mankind."[47] Although park policy is largely enforced by coercive violence, beginning in the 1980s there has been a movement throughout Africa toward community conservation, or giving the people who live near the park benefits from a stake in preserving it. While these programs have clearly improved western Serengeti peoples' attitude toward the park, community conservation efforts may actually be increasing government control over land and resources and not substantially reducing violations of park rules.[48]

Significant to this study are the tools of analysis for making the critical connections between landscape and memory. Landscape as an analytical concept developed among an interdisciplinary mix of geographers, ecologists, regional historians, and art historians in Europe, especially during the interwar years. The concept revived with a postmodern turn in the 1980s and 1990s to look at the "socially constructed and politically contested" ways of seeing the environment as an ongoing process.[49] Landscape as it was first conceived was both a genre of European seventeenth- to nineteenth-century painting and the new class-specific way that these paintings presented the environment. Because European landscapes separated people from nature, by viewing the environment from an outside and distant gaze, many scholars wondered whether the idea of landscape was entirely a European phenomenon. Scholars were quick to identify the concept of landscape in other cultures and times but also to recognize the particular connection of landscape to imperialism.[50] For example, Deborah Bird Rose finds that Australian Aborigines view the "country," or the land, as a living being with a history and a people who care for it.[51] Candace Slater compares competing visions of the Amazon rainforest, contrasting the "gigantic" images of outsiders, who project their own fears and desires on the Amazon to the "shapeshifter" images of local people, who see mystery and hope for transformation in the forest.[52] One of the first African historical landscape studies, Cohen and Odhiambo's *Siaya*, shows how Luo concepts of landscape and identity shifted over the last hundred years.[53]

Africanists, too, have used the tools of landscape analysis in historical work, concentrating on colonial conflicts over landscape perceptions and particular features of the land.[54] Tamara Giles-Vernick's study of a forest community in Central Africa explores Mpiemu people's perceptions of the environment through a category of environmental and historical knowledge called *doli*. She shows how these ways of seeing and interpreting the landscape changed in the twentieth century and how local people used doli to engage in conflicts with contemporary conservation efforts.[55] Michele Wagner investigates history as it is embedded in mental maps of the Baragane (Burundi) environment.[56] Some of these studies also demonstrate how landscape is gendered, allowing us to look critically at oral sources that have been controlled by men's knowledge and to see how men and women have used the landscape in different ways.[57] Landscape studies, both in the West and in Africa, often focus on the stories surrounding particular environmental features such as forests, rivers, or mountains, while much of the work from Africa concentrates on sacred places.[58] For example, Sandra E. Greene looks at the effects of profound shifts in understandings of sacred places during the colonial era and how that affected and was affected by new kinds of social identities.[59] In an example from the Zambezi River, JoAnn McGregor contrasts a European way of imagining the landscape that emphasizes the generic sites of scenic visual value with African accounts that feature sites of specific historic value associated with ancestors and past events.[60] Terence Ranger studied the shifting meanings associated with the sacred site of the Matopos hills of Zimbabwe that were particularly evident at moments of conflict.[61] These Africanist landscape understandings can be extended beyond specific sacred sites and into the precolonial era.

Literature on the development of Anglo-American and specifically British ideas about landscape and nature provide the tools for understanding colonial and conservationist perceptions of the landscape. Roderick Neumann describes a "nature aesthetic" that developed from artistic representations of "picturesque" idyllic or awe-inspiring landscapes, without evidence of human work and poverty, that positioned the observer outside the landscape. Elite landlords and bourgeois capitalists promoted this ideal view of the landscape separated into productive (practical, rational) and consumptive (aesthetic, recreational) spheres in the context of transition to industrialization as a way to legitimize control over land and resources.[62] Others have shown how this view of the landscape was also evident in the popular craze for natural history among settlers in the colonies, who used nature both to conquer and transform the new environment as well as to critique the problems of urban industrial society and the destructive environmental policies of the colonial government. Ecological science, and eventually environmentalism, developed out of these imagined landscapes and more specifically in relation to management of the colonies, as well as in interaction with the colonized.[63] Both hunting and appreciating nature became a mark of class distinction as social access to hunting in Britain and the colonies was gradually restricted to "sportsmanship" rather than subsistence. The idea of the game preserve was based on the model of the nineteenth-century private shooting estates that took common land to create the "planned wilderness" of paintings and literature.[64] Africans, first seen as part of nature in their primitive, innocent state, became impediments to elite consumption of nature.[65]

The way that memory and identity is "inscribed on land" as explored in the landscape literature provides a critical perspective for analysis.[66] Much of this scholarship draws on Maurice Halbwachs's and Paul Connerton's work on collective memory, which demonstrates that memory is socially constructed and structured by group identity.[67] James Fentress and Chris Wickham, interested in the social meaning of memory rather than recovering historical facts, look at the history of memory, or how the way we remember changes over time in relation to context. Memories are always connected to a context, involving a particular task and reflecting the collective experience of a particular social group when they are formed. In addition, memories must be simplified and conventionalized or transformed into conceptual images in order to be socially meaningful and thus remembered. However, the images of social memory rapidly lose connection with the original context; they are decontextualized and reinterpreted when they are transmitted as oral traditions.[68]

THE METHODOLOGY OF A SPATIAL ANALYSIS OF ORAL TRADITION

These insights from the study of social memory provide the framework for the central contribution of this study in presenting a methodology for unlocking the historical meaning of oral traditions in the reconstruction of a history of memory. I argue that through a spatial analysis of oral tradition, in conjunction with other kinds of evidence, historians can *re*contextualize landscape memory or reconnect images from oral traditions with the older contexts through which they were transmitted in the past. Historians using oral tradition as their principal source of evidence have been confounded by the central problem of the reliability of this source in accessing an objective rather than a mythical past. It has been demonstrated many times that the content of oral tradition is not stable and that it changes from performance to performance over time and in relation to the various historical contexts in which the traditions are told.[69] Different social groups tell different stories about the past and in different ways to legitimize a particular social order. When the social context changes, features of the oral tradition that no longer have meaning drop out or change to reflect new meanings.[70] Not only does the present influence the narration of the past but knowledge of the past most surely influences our experience of the present.[71] The historian's analysis of any one tradition must take into consideration the present context in which narrators tell it, as well as all the other historical contexts through which it has passed in transmission.[72] Because of these difficulties, many have despaired of finding any verifiable historical content in oral traditions.[73]

One way to assess the historical content of oral tradition is through an understanding of its narrative form. Studies of oral memory have shown that narrators construct (rather than reproduce) oral traditions in performance through the use of mnemonic systems, the central elements of which scholars of oral tradition call core images or clichés. Fentress and Wickham contrast the visual character of these images that embody knowledge ("memory of things"), common among oral societies throughout the world, to a semantic knowledge ("memory of words"), characteristic of literate societies.[74] By recalling these core images narrators improvise the entire narrative as they tell it. In the Nata origin story the core images are a hunter following his prey from the wilderness and a woman at her cave by the spring. Narrators elaborate details of how they met and what they said around these core images to form episodes or narrative units that they string together to create the larger story anew in each performance.[75] Historians of oral tradition have long postulated that it is these core images that hold the key to historical interpretation. Jan Vansina proposed guidelines for interpreting the "implicit meaning" of these core images, or "clichés," such as comparison with other traditions and other cultural expressions.[76] Joseph Miller later suggested that since core images serve as the mnemonic device for recalling the story, people pass on these images from generation to generation, even if they no longer understand the original meaning. Facts are lost during transmission not just because they are gradually generalized and forgotten but also when they no longer apply or have meaning in a new context. Miller postulated that the core images held the best possibility of bearing "information from and about the past."[77] While the reconceptualization of these core images often erases historical fact as the images are dislocated from the original context, the process also ensures an incredible stability of core images and shared meanings.[78]

Some of the most important core images found in African traditions are visual and, more specifically, spatial images of landscape, place, or topography. Work on oral traditions and memory over the past fifty years shows that while oral traditions are notoriously deficient in usable historical chronologies, they are amazingly consistent in their accurate representations of specific places and ways of understanding space, suggesting that memories themselves are attached to place and spatial representation. As Isabel Hofmeyr, writing about South Africa, puts it, "oral memory has a close mnemonic relationship with place and location, and in a variety of societies people often bank information in the landscape." She questions whether people can sustain memory if they lose touch with the places and landscapes of the core images.[79] Many of the oral traditions that I have collected over the past ten years in the western Serengeti are little more than a decontextualized string of place names in a clan migration narrative or the stories of miraculous events in specific places. Elders wanted to take me out to see the places themselves and to walk over the landscapes as they told the stories, providing more information about people, events, and ideas associated with the places. However, the recognition that core images often appear as spatial images does not solve the problem of their interpretation. The first generation of historians to interpret oral traditions in Africa accepted the literal meaning of place names in migration or clan origin traditions, which resulted in untenable reconstructions of the movement of large and discrete groups of people over long distances.[80] The internal meaning of spatial images is not always explicit and the original contexts not always clear. The central focus of my investigation was how to interpret these many references to space and place in the memories of elders.

The historian best approaches the interpretation of spatial images by understanding how and why our minds spatialize memory. Studies of memory have shown that people store the recollections of their past as spatial rather than temporal images.[81] We remember events and people by locating them in particular places, landscapes, and organizations of space rather than by reference to time or date. Thus memories appear to us as a sequence of places rather than as the orderly passage of time. For example, we cannot conceive of or tell our own family histories without memories of the succession of family homes to anchor and order the stories in time. In his exploration of the "poetics of space," Gaston Bachelard writes, "Memories are motionless, and the more securely they are fixed in space, the sounder they are."[82] This insight has profound implications for the historian using the evidence of memory as a primary source. As Bachelard notes, "to localize a memory in time is merely a matter for the biographer and only corresponds to a sort of external history, for external use, to be communicated to others."[83] The job of the historian, like the biographer, is to fix memories within a chronological sequence in order to understand change over time and its possible causes. Yet if Bachelard is right, memory cannot provide the historian with precise temporal sequences or duration.[84] Clearly, the historian cannot reconstruct the temporal framework of oral memory without paying particular attention to the indigenous conceptual frameworks that govern the use of time and space in oral narratives.

If the spatial elements of oral tradition are part of this mnemonic system of core spatial images, then the historian can use them as "evidence in spite of themselves" that provide tangible information about the past.[85] The spatial elements of oral tradition—references to place-names, landscapes, topographical features, and the social organization of space—are crucial elements in the historical reconstruction of this region, rather than geographical background.[86] While historians have often disregarded these elements as useless details they provide bits of evidence from the past, transmitted to the present because of their function in oral memory. Imagined landscapes, embedded in oral traditions as core images, are artifacts from the past that, although people might understand their meaning differently or lose their meaning altogether in different time periods, remain tenacious fragments of past social worlds transmitted in oral memory.[87] Like the ceramic artifacts that an archaeologist unearths, a particular shard may have been used in subsequent generations as a shallow water container for chicks in the yard or later picked up off the refuse heap by a child to be made into a toy. Still, the archaeologist can sometimes reconstruct its original use and historical context through careful comparison with similar shards found in other places, other kinds of artifacts found nearby, and contemporary pottery forms and their uses.[88] These encoded fragments yield information about the past only as historians interpret them within their cultural context and alongside other kinds of evidence.

Instead of preserving oral narratives for their esoteric or archaic value, oral traditions are transmitted to the next generation by and through specific social groups to communicate knowledge critical to maintaining group identity and relationships with others. Individuals preserve memories as members of a group and those memories are situated within the socially specific spatial framework provided by that group.[89] Memories are not only *spatially* located but also *socially* located within particular groups. We can identify each kind of oral tradition with the history of a particular social unit. Different social groups located in one place may preserve radically different memories about the same time period because each builds on its own "mental map."[90] Just how spatial organization relates to social organization has been the subject of much scholarly debate, beginning with geographers who argued that landscape functions not as a neutral backdrop to the events of history but shapes and is shaped by human action.[91] A classic argument, first articulated by Emile Durkheim and Marcel Mauss in 1903, holds that the built spaces we inhabit represent the social structure. Scholars have demonstrated this mainly in the layout of homestead and village as well as the interior design of houses.[92] I extend this observation to hypothesize that oral traditions encode social relationships and identities by employing a spatial imagery that includes landscapes and topography. How people order their memories within a particular spatial construct or landscape, then, depends on their own social identity—their socially shared, situational definition of self in relation to others—that both shapes and is shaped by the landscape. As social identity changes over time so do oral traditions and ways of seeing the landscape. Thus, as powerful symbols of collective identity that both drive and reflect historical change, landscapes are always social and political.

If this is to be a *history* of landscape memory, then the next step in solving the problem of interpreting oral traditions is to find a way of determining the time frame or chronology of these different spatial images that have long since been separated from their historical context. Some have described oral traditions as resembling palimpsests, or tablets that various people have written over with the older writing just barely visible beneath.[93] In the search for historical meaning, the core spatial images for each set of oral traditions must be analyzed and recontextualized. One can hypothesize and test the relative age of these images by interpreting their cultural meaning in connection with other forms of evidence and in comparison with other traditions. Through this process the historian can arrive at a tentative understanding of relative time depth and how the different elements of a single tradition were reinterpreted at different time periods. For example, references to practices associated with the core spatial images in oral traditions as well as current ethnography for which we find similar evidence in the ecology of the area—like plant adaptations or species selection that take a long time—might indicate a similarly old time depth. Likewise core spatial images in oral traditions that describe practices or settlement patterns also in the ethnography seem to be old when they are congruent with proto-Bantu words or loanwords from now extinct languages or archeological finds that can be dated to an early period.[94] These congruities of evidence can never be conclusive proof that the practices were not adopted fairly recently or at any other time in between. But with enough evidence from a variety of different kinds of sources one can build up a logical case that identifies the historical context that seems to fit the core spatial images of the oral traditions.

More problematic yet for dating is when one must deal with the material that represents continuities in social patterns over an indeterminate period. For example, chapter 2 connects the spatial images of diversification, inclusion, and distribution from clan traditions to economic strategies, homestead patterns, leadership roles, and descent systems, for which the memories of elders living today talking about the "traditional way of doing things" provides most of the evidence. Ecological evidence of climatic, demographic, and soil data does provide some evidence for the economic possibilities of early settlers while kinship and leadership terminology can be traced back in time through the methodologies of historical linguistics. Again the congruence is only circumstantial and holds up only as it presents a logical reconstruction of long-term social patterns that are reflected in this particular view of the landscape. One cannot know with certainty that western Serengeti peoples have used regional social networks established through clan and friendship to gain access to food during famine times for the past millennium. But if the core spatial images of clan stories that depict a landscape of social networks are indeed an artifact from the past, then all the other evidence that is consistent with these images might logically identify long-term regional patterns. Although the nature of the sources does not allow us to say any more than that, if or until other evidence becomes available, further confirmation comes from a study of different kinds of temporalities.

Another problem in the identifying historical sequences in oral traditions is that they cannot be read as a straightforward account of linear time because they employ different concepts of time or temporalities. African historians have proposed at least three kinds of traditions, representing three different indigenous time frames: (1) the origin traditions of clans and ethnic groups employ *mythical time*, (2) scarce and cryptic information in the form of lists of settlement sites or place-names related to descent groups, sometimes called the *floating gap*, refers to the long middle period of *social process* or *cyclical time*, and (3) stories about the more recent past that can be dated by genealogies or age-sets draw on *historical time*. These temporalities of indigenous periodization represent relative chronologies that become more disconnected and mythical the further back in time they go. Although the historian must consider that oral traditions of the first two temporalities are also used as "social charters" to justify present power arrangements, their careful analysis can still provide important evidence about the past.[95] Similarly, Fernand Braudel's classic history of the Mediterranean is structured around three different temporalities: geographical time (history of imperceptible changes in the relationships of people to their environment), social time (the slow but perceptible rhythms of social process), and individual time (the short-term political time of remembered history). His model allows for each type of analysis to simultaneously supply a different kind of historical information through its own time frame, corresponding to a different spatial scale and social unit.[96] Local concepts of both time and space must be combined in the analysis of oral traditions, although firm dates are not always possible. These concepts of time or temporalities are used as the basic organizing principle of this book's chronology.

The landscapes derived from oral traditions presented in the first three chapters represent long-term continuities in basic social processes such as economic strategies, kinship, and generational relations that may date back two thousand years but still have ongoing relevance today. These core spatial images, also evident in other kinds of sources, represent underlying themes or generative principles; they generate a range of strategies and options for social organization that change form over time while retaining the foundational approaches to the recurring problems that have confronted western Serengeti communities in relation to their environment.[97] Oral tradition rationalizes these generative principles that govern the daily elaboration of social practice into a static "official" version of what is a dynamic process to preserve the existing social order. When we understand lineage as a strategy rather than a structure, we see people's everyday actions as significant because they are making choices rather than following a script. The early-period traditions present social organization in a rigid and timeless, "traditional" form. Although this is the given spatial text, people have read it in countless ways over time.[98] The exclusive narration of the "official version" has now silenced some of these various interpretations. The historian must resist the trap of the "ethnographic present" inherent in the telling of oral tradition by using them alongside other sources and other versions that give them historical specificity.

We can be reasonably sure that oral traditions do contain information about the past because of the remarkable congruence between historical reconstructions based on the core spatial images of oral tradition and the contexts reconstructed through the evidence of historical linguistics, comparative ethnography, archaeology, or written sources. If those who tell oral traditions cannot have known about this other evidence, how else could they tell such similar stories concerning social processes in the distant past? Oral traditions provide a culturally grounded expression of historical processes about which we know from other kinds of evidence. The historians must also accept the limitations of oral traditions. The genres of oral tradition corresponding to "mythical time" and "social process time" cannot by themselves show change over time. Only the comparison with other sources and with various versions of the same types of traditions throughout the region or among different social groups can accomplish this result.[99] Even then these lead only to tenuous hypotheses, usually confirmed by written sources. However, even once the oral traditions can be confirmed by written sources in the late nineteenth century, one is still interested in how these landscape memories change within that historical context.

Ultimately, my goal as a socially engaged historian, like the western Serengeti narrators themselves, is to produce an account of the past that speaks to current debates and the concerns of common people's lives in the western Serengeti today. This book aims to insert a new historical voice, an internal perspective from the communities surrounding Serengeti National Park, into the discussions of those thinking about the future of African environments. Concerns about the environment and its degradation, including the question of how to preserve the continent's prolific wildlife, are among the most controversial and pressing issues facing Africa today.[100] The various approaches of governments, scientists, conservationists, NGOs, historians, and local people to these problems are a direct result of how they imagine those same African landscapes. While the park's managers have been successful at preserving the wildlife, they are ultimately fighting a losing battle on the borders against poachers, woodcutters, and grazers. Even though ecologists and historians alike have given more attention to indigenous environmental knowledge and championed a movement toward community involvement in and benefit from parks, after two decades of rather spotty success with community conservation they still know little about how local people who live in those environments have imagined and humanized the landscape.[101] An appreciation of local landscape memory is the first step in working toward rethinking present practice. Acquiring that environmental knowledge will surely add a rich, new dimension to present understandings of the Serengeti-Mara ecosystem and provide a methodology for accessing similar knowledge about other African landscapes.

At the same time this book presents a new methodology that speaks to the concerns of African social historians who have recognized the importance of writing history from an African perspective and from sources generated in Africa. This book models the application of a spatial analysis of oral traditions by presenting a complex and historically grounded understanding of how the people who live in the Serengeti ecosystem have viewed the landscape over time and how those views have affected their interactions with the environment. The methodology combines many of the tools of analysis and sources of evidence on which African historians have long relied. Yet it combines them in a way that allows for historical claims about the precolonial past even where sources are scarce and problematic. While recent theoretical debates among historians have cast further doubt on the use of oral traditions, my hope is that this new methodology will reinvigorate the study of precolonial African societies and the discussion about oral tradition as history.[102] The contrasting visions of the landscape represented by the Grzimeks and the Ikoma elders concern more than a conflict over control of the Serengeti's natural resources; they are also about a struggle over the symbolic meaning of the landscape and its relevance to a people's history.[103] Parks conceived of as natural spaces that exclude people and keep them from seeing and interacting with the landscape separate the people from the "texts" of their history and ultimately erase that history. Although a seemingly esoteric enterprise, taking time to see the landscape from the perspective of others in an interconnected world matters profoundly for those involved both in the study of the African past and in setting policy for the African future.

PART I

Past Ways of Seeing and
Using the Landscape

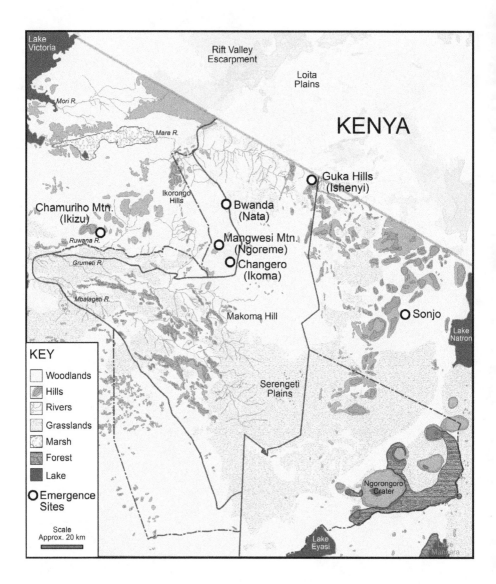

Map 3. Ecological zones and emergence sites. *Map by Peter Shetler, 2005. Land cover and park boundaries adapted from map of Serengeti National Park and the surrounding area by T. M. Caro, Serengeti Research Institute*

1 ⤳ Ecological Landscapes

Settling Frontier Environments (Asimoka), ca. 300 CE to Present

IN A HISTORY of western Serengeti memory, the first, and perhaps oldest, way of seeing the landscape is accessible in origin or emergence traditions of today's ethnic groups. This view places people practicing different economic subsistence strategies in the ecological niches of the woodlands, grasslands, and hills. In this ecological view of the landscape social identity is connected to ways of making a living by adapting to particular environments and forming interdependent relations with people in other environments. The core spatial images of interdependent ecological zones are embedded in the origin traditions about farmers, herders, and hunters brought together to form a new people. Elders from the ethnic groups of Nata or Ishenyi or Ikoma elders tell different versions of a story with the same core spatial images as the unique story of their beginnings as a people when the land was uninhabited. Even though these core spatial images are now recontextualized as ethnic origin stories, they predate ethnic groups as we know them today and refer to the social identities and interactions of farmers, hunters, and herders in the distant past. The generative principles inherent in these core spatial images of environmental management and interdependence continue to be used as the basic strategies for human interaction with the environment. We can see them repeated in the way that people talk about and manage the land. These strategies have helped create the open parkland appearance of the Serengeti and the ways that power is expressed through dominance over particular ecological zones and other people.

Origin traditions employ *mythical time*, or the geographical time of almost imperceptible changes in the relationships of people to their environment with ongoing relevance today. However, because the core spatial images of this landscape are congruent with the evidence from archaeology, historical linguistics, and ecology dating back to around 300 CE, this landscape is best understood in the historical context of the period when the first Bantu speakers began to move into the dry interior and interact with the grain farmers and pastoralists who were there before them. These core spatial images, however, have been recontextualized many times as they lost connection with their original social context, and by the nineteenth century they were attached to ethnic identity as people used them to cope with famine, disease, and raiding. While we cannot date any of these stories precisely because they have changed form over time, the core spatial images represent consistent and stable generative principles that seem to go back nearly two millennia and are still operative today. The evidence for this part of the story demonstrates the consistency between the core spatial images of the stories, the way the generative principles are discussed and practiced by elders today, and the ancient context established through archaeological, linguistic, and ecological sources.

The basic emergence story, told with many variations throughout the region, is that of first man, the hunter, meeting first woman, the farmer. This is a Nata version told by Mgoye Rotegenga Megasa:

> They came from the east, the man was Nyamunywa and the woman was Nyasigonko. He was an Asi, a hunter of zebra. The animal he was hunting fell on the rise and died, he took out the arrow and saw that there was a person in front of him. He came closer and saw that it was a woman. He went to the door and asked who was there. The answer came, "the person of this house." To which he replied, "[I am] a person of the wilderness." They greeted each other in Nata. She invited him inside.[1]

This gendered story goes on to describe how first woman, the farmer, showed first man, the hunter, how to eat millet while he showed her how to eat meat. They stayed together and gave birth to the Nata people. The ecological landscape in this basic story, and its core spatial images of a hunter from the wilderness coming to live with a farmer from domesticated space, is also found in other narratives and other forms of evidence, presenting a landscape of historical depth in the region. This is a human landscape that has been formed by the interaction of farmers, hunters, and herders and expressed in the stories they tell about their origins in this land.

Ecologists, however, often describe the Serengeti landscape without humans, or at least as one put it, as "a natural landscape as yet relatively unspoiled by man."[2] Yet it is in this mixed ecosystem of plains, hills, and woodlands that the peoples of the western Serengeti—today known as the Ngoreme, Ikoma, Nata, Ishenyi, Ikizu, and Tatoga—have lived at least for the past two millennia. The boundaries of Serengeti National Park and the Ikorongo-Grumeti Game Reserves arbitrarily divide lands once inhabited by the ancestors of these peoples and still known to them as places where their ancestors lived and were buried. Although the park claims that western Serengeti peoples are recent immigrants, their ancestors have been part of this spectacular landscape for a very long time and have helped to create the "natural" landscapes that tourists enjoy today. Ecological, archaeological, linguistic, and ethnographic sources also provide evidence for human involvement, integral to the total ecosystem for as far back as ancient people have lived in these landscapes. The Serengeti is a profoundly humanized landscape.

This view of an ecological landscape in which people have been an integral and creative part of the environment from earliest times. Ecological evidence set alongside other narratives of subsistence practices demonstrates conscious human control over the environment. The settlement of the early frontier provides the historical context for emergence stories. Today most western Serengeti peoples are farmers who speak East Nyanza Bantu languages, but their heritage draws from a rich variety of other peoples, languages, and economies with whom they interacted in the distant past. The linguistic, archaeological, ethnographic, and ecological sources also address these interactions. Elders continue to tell stories embedded in these landscape images because they provide a way to order relationships among peoples practicing different economic subsistence strategies as well as men and women controlling different economic spheres. The western Serengeti peoples are all part of a larger system in which they are integrally related and dependent on one another. This view of the landscape allowed western Serengeti peoples to make claims about authority over different ecological zones and over other peoples. An ecological view of the landscape establishes ancient claims to living on and preserving landscapes that are now understood as pristine natural wilderness.

HUMANIZED ECOLOGIES

The ecology itself provides evidence for long-term human interaction and management. The Serengeti-Mara ecosystem reaches from the hills of Sonjo, now in Maasailand in the east, to the hills of Ikizu on the western corridor, extending all the way to Lake Victoria and covering about twenty-five thousand square kilometers. Wildlife populations thrive in this varied ecosystem, producing the largest herds of grazing mammals in the world, with twenty-eight species of ungulates (some 2.4 million total) and thirteen species of large carnivores. The Serengeti-Mara ecosystem is defined as that area influenced by the migratory wildebeest population, the "keystone species" that indicates and helps determine the health of the entire system.[3] The route of the annual migration begins in the southeast and in the Ngorongoro Crater in December, when the plain is teeming with vast herds of wildebeests (about 1.3 million at last count), mixed with zebras and gazelles. These herds come to the open plains in the rainy season to give birth to their calves on pastures rich in minerals, finding water in pockets left by rain. The long view available to grazing animals on the almost treeless plains of short grass also provides better protection from predators while the calves are still vulnerable.[4]

The nearly flat Serengeti plain is a vast, high-plateau, short-grass savanna that covers nearly approximately twenty-five hundred square kilometers. Trees appear only along the rivers and among the rocky granite outcroppings known as kopjes. Elsewhere sedimentary soils of volcanic ash or limestone overlay the rock to a depth of up to 150 meters. A nearly impenetrable hardpan under much of the volcanic soils of the plains make it possible for only shallow-rooted *Cynodon* and *Digitaria*[5] grasses to flourish until they dry up when the water evaporates after the rains. These soils, though easily eroded, are quite fertile, making the Serengeti one of the most productive grasslands in the world. Yet because of the high evaporation rate of water and the paucity of nonalkaline water sources on the plains, the animals can only stay during the rainy season when the rainwater collects in pools on the rocks.[6]

As the plains dry up from May to June, the herds move north and west into the woodland hills of ancient granitoid rocks, looking for permanent water sources and fresh grass.[7] The acacia woodland northwest of the plains, drained into Lake Victoria by the Grumeti, Mbalageti, and Duma river systems, is the place where the migration splits and where some herds remain resident all year. Here the volcanic soils become a fine-grained sedimentary soil, the rainfall increases and taller grasses of different species along with bush begin to thrive, merging into woodland. The alkaline ash soils demarcate the limits of the plain. The soils of the woodlands, to the north and west of the great plains, are formed from granite or quartzite parent rock, creating an acacia thorn-tree scrub and woodland. The small whistling thorn trees (*Acacia drepanolobium*) dominate in the woodlands with poorly drained soils.[8] The woodlands and plains finally break up among the ranges of hills in the northwest nearer to the lake, creating a unique mosaic of many distinct ecotones in close proximity — savanna woodland intermingled with open grasslands and hills. Rivers form valleys between the hills, resulting in fertile but swampy low-lying areas of woodland. Directly to the south of the western hills the grassland plains again dominate, running parallel to the rivers as far west as the lake, in what is now the western corridor of Serengeti National Park. It is in this western part of the larger Serengeti-Mara ecosystem that our story takes place — where Nyamunywa, the hunter, first met Nyasigonko, the farmer.

In spite of popular images to the contrary, it should not be surprising to learn that this is a humanized landscape. After all, Olduvai Gorge, located on the eastern side of Serengeti National Park in the Ngorongoro Conservation Area, is the site where archaeologists Mary and Louis Leakey discovered one of the oldest human ancestors, the hominid *Australopithecus*, which flourished in East Africa between four million and one million years ago, using the earliest stone tool technology.[9] *Australopithecus* coexisted with *Homo erectus* for about a million years before disappearing. By that time *Homo erectus* had already moved out of Africa and across the Red Sea to the Arabian peninsula and beyond, with improved tools, intelligence, language, and the ability to control fire. *Homo sapiens* appear in the archaeological record about two hundred thousand years ago. Archaeologists still consider East Africa to be the heartland where humans first developed and then spread out around the world; all of us can ultimately trace our roots back to this dry Serengeti savanna.[10] Some have even hypothesized that hominids developed bipedalism because they needed to carry their babies as they followed the migrating wildebeest herds across the East African plains.[11] It is no wonder that this landscape bears the mark of these early inhabitants.

Visible evidence of the long-term interaction of humans with the Serengeti environment can also be found in the very presence of the large mammal herds that tourists come to see. Humans have coevolved with the wildlife over millennia and have therefore learned to coexist. In other places around the world where humans came later, such as the Americas, Australia, or Asia, the large mammal population suffered serious decline. In North America the extinction of large mammals took place between 14,000 and 11,000 because of the introduction of human predators, against which the mammals had not acquired defense mechanisms. Africa, on the other hand, is unique among the continents in that its megafauna did not experience the so-called Pleistocene overkill.[12] The same species of wildebeest that inhabits the Serengeti today was there alongside our ancient ancestors and followed similar migration routes 1.5 million years ago.[13] Large mammals remain in East Africa, not because the people were necessarily natural conservationists, but because of this long history of developing side by side.

Ecological evidence demonstrates that humans have had a profound affect in both creating and maintaining the unique Serengeti ecosystems, largely through the deliberate and controlled use of fire. Although lightening may sometimes ignite fires, now and in the past, they have most often been started by humans.[14] The woodlands of the northwest zone of what is now Serengeti National Park (closest to the area historically inhabited by western Serengeti peoples) contain species of trees making up the dominant *Acacia-Commiphora-Balanites* woodlands that not only tolerate but thrive on regular fires.[15] Experiments conducted with Serengeti tree seedlings demonstrated that burning and browsing together actually stimulated increased stem productivity. Seedlings were equipped to survive cool burns, but the rates dropped significantly after repeated hot burns that result from a buildup of dry plant material.[16] The dominant grasses of the western Serengeti, *Themeda triandra*, *Pennisetum*, and associated grasses, are also fire-adapted vegetation whose presence indicates the evolution of these ecologies in conjunction with annual burning. Burning eliminates all the tall dead grass that has almost no nutritional value for either wild animals or cattle and encourages the growth of the more nutritious grasses, such as red oat grass (*Themeda triandra*) at the expense of the coarser grasses.[17]

Serengeti ecologists have challenged the negative assumptions about human-induced change to the environment, in which old-growth wooded areas were assumed to be natural and any change meant a disturbance of the "climax" state of the ecosystem. They have discovered that the Serengeti is a "multi-stable state" or "disequilibrium ecosystem," meaning that there is no one climax state that must be maintained but rather an oscillation between woodlands and grasslands is the "natural" state of the ecosystem.[18] Ecologist Holly Dublin found that the Serengeti savanna ecosystem has oscillated between phases where more grasslands or more woodlands existed depending on large-scale perturbations in the system (some of which were human induced), such as disease, drought, hunting, elephant damage, or fire. Dublin documented three distinct vegetative shifts in the twentieth century. European visitors to the Serengeti around the turn of the twentieth century who noted the parkland appearance of the western Serengeti witnessed its dominant grasslands phase. But in response to a series of disasters, including the 1890 rinderpest panzootic, in which 95 percent of the wildebeest, buffalo, and cattle died, the grasses grew taller due to less grazing, people migrated out or died, and the woodlands and bush began to encroach. By the 1930s the tsetse fly (the vector for sleeping sickness), finding suitable habitat in the bush, prevented people and livestock from returning to these areas and further encouraged the woodlands phase. The colonial government set aside the Serengeti as a wildlife reserve at this time because people could no longer live in tsetse-infested areas. By the 1950s and 1960s, however, higher rainfall and restrictions on burning led to long grass and more bush, which provided more dry fuel for hotter and larger fires, destroyed trees, and led again to the spread of grasslands. With the successful vaccination of cattle against rinderpest, the wildebeest population increased fivefold from a quarter million to approximately 1.3 million by 1977, with few changes since in overall wildlife density.[19]

Rather than searching for ways to prevent large-scale perturbations, ecologists concluded that these extreme disturbances of the Serengeti ecosystem were, in fact, responsible for maintaining the diversity, productivity, and resilience of the system.[20] On observing the degraded and denuded appearance of the Serengeti during one season, many experts in the past predicted with alarm the destruction of the Serengeti ecosystem. Yet a few seasons later, with improved rains, the grasses returned. A North Mara colonial officer expressed amazement at the quick recovery of nearly bare "overgrazed" land where bad grasses had encroached around the European houses. "Cattle were excluded in early November 1949," he wrote. "[But] by the end of February 1950 with short rains a fine stand of grass over a good part and fair proportion of better pasture grass began."[21]

Ecologist A. R. E. Sinclair contends that the fears of "overgrazing" due to the high numbers of wildebeest on the plains were groundless. The grasslands showed no detectable change toward the dominance of less palatable species as a result of increased grazing.[22] That is not to say that the ecosystem can never be destroyed, only that it is much more resilient than models based on temperate forest ecologies might suggest.[23]

Various long-term human interactions, particularly burning, account for the appearance of the distinctive Serengeti woodland-grassland landscapes. Within weeks of a burn a new carpet of brilliant green grass covers the land with very few other ground plants, making it appear orchardlike, as a carefully mowed lawn under scattered acacia trees.[24] One of the first European hunting parties in the Serengeti described "gentle low-sloping hills as green as emeralds, beneath trees spaced as in a park," where the grass was "short and thick" of the "greenest green," as if it had been "cut and rolled."[25] Another visitor said, "a great deal of this virgin country resembled a cultivated park, making it hard to realize that we were in the wilds of Africa." He described the landscape after a burn as, "Gently rolling hills of velvet-smooth grass, with clumps of small trees dotted here and there swept toward distant mountains of purple and blue."[26] Many visitors and ecologists, including the Grzimeks, assumed that humans had created this orchardlike parkland by the more recent destruction of preexisting forests through fire rather than as the result of long-term patterns of interaction and coevolution.[27]

Although we do not know for sure when western Serengeti peoples began to practice these strategies, they seem to be long-term patterns embedded in the

Figure 1.1. Grassland plains, from Kemegesi, Ngoreme, looking toward Kyweigega Hill. *Photo by Paul Shetler, 2003*

Figure 1.2. Parkland landscape with greenflush under acacia trees after the first rains, Robanda, Ikoma. *Photo by Paul Shetler 2003*

generative principles represented by core spatial images in oral tradition and consistent with ecological realities. Farmers describing the practice passed on to them by their fathers said that after the harvest in July and August they set the dry grass on fire, burning extensive areas, but in a cool burn before the dry season progressed too far. Elders maintained that in the past the community carefully controlled which areas were burned and at what time, so that there would always be grass for livestock before the new grass sprouted. Leaders also directed the people to cut down undesirable trees and bushes when they sprouted up and instead to encourage the growth of other useful trees and grasses, such as *Themeda triandra* (called *ambirisi* in Nata, from *korisi*, to herd), acknowledged as one of the best pasture grasses and as a valuable material for thatch.[28] In local languages the month of August or September when the first rains appear is called *ekinyariri*, or green lands, referring to the greenflush of grass emerging right after a burn that is especially appealing to cattle and wild animals.[29] Some say that the wildebeest follow the smoke to find the greenflush. Elders today confirm that in their youth many of the areas that are now woodlands used to be open grasslands, which were valued because they were considered healthier for wildlife, livestock, and people alike.[30]

Using the generative principles of ecological landscapes in historical times, people modified the environment specifically to control for disease. Western Serengeti elders reported burning the grass to destroy ticks and other diseases carried by the wildlife and stock.[31] Colonial officers reported that for several years Maasai living in the Serengeti took their cattle off of grasslands infected with

Figure 1.3. View of the Serengeti plain from Sasakwa Hill, looking toward Kichelaga Hill and Butamtam Hill. *Photo by Paul Shetler, 2003*

ticks that harbored East Coast fever, burning the grass and then letting the sheep and goats graze there to finally rid the area of ticks. Maasai also allowed their cattle to be infected with mild rinderpest for inoculation if it appeared during good times, when water and grazing were abundant.[32] Western Serengeti elders said that pastures sometimes became infected by a leguminous grass called *ahuroro* (*Crotalaria burkeana*) that made the cattle's hooves so painful they would die because they could not graze or get to water to drink. Burning was also used to control this infestation.[33]

In the past East African people used a strategy of control, rather than eradication, of disease vectors, particularly in the case of endemic trypanosomiasis in cattle, or sleeping sickness in humans, spread by the tsetse fly.[34] While sleeping sickness seems to have been endemic in East Africa for a very long time, it did not reach epidemic proportions until the early colonial period, when the Germans identified it in Musoma District, where two thousand people had died by 1905.[35] Until that time sleeping sickness seems to have been kept at bay, rather than eradicated, by local practices that reduced the bush harboring the tsetse fly and increased people's immunity by limited contact.[36] Western Serengeti peoples avoided bush areas with cattle except in times of drought and found more open areas or springs for watering the livestock.[37] Elders testified that, until the colonial period, western Serengeti farmers kept mainly goats and sheep, which developed more resistance than cattle to trypanosomiasis. Goats browse further into the bush, coming into limited but regular contact with tsetse fly habitats and thus increasing their immunity.[38] People had regular but limited contact with the tsetse fly by farming in fields located near the wilderness boundaries, hunting, collecting firewood, or traveling. Not recognizing this strategy for immunization, the colonial tsetse fly officers in the 1930s disapproved of the practice of farming away from the homesteads and into the bush.[39] Yet together these practices allowed the farmers to gradually push back the cleared areas and maintain a controlled zone of regular contact with the tsetse fly.[40]

Although sleeping sickness did break out in the colonial years, the reports of British officers indicate the persistence of these older patterns of control. A British veterinary officer reported in 1928 that "the fly belt is vaguely demarcated by the natives who seem to know where they can safely graze but keep dangerously close," adding that "goats seem to thrive there."[41] In the first reports of sleeping sickness in the Ikoma area in 1927, the medical officer was surprised to find that livestock were healthy, although he found cases of trypanosomiasis among them and vast tracts of tsetse-infested bush surrounding the settlements. The same report also found, in the human case, that the incidence of sleeping sickness was only one percent, with no tendency to epidemic spread.[42] Western Serengeti communities, whether consciously or not, seem to have learned how to control their environment as much as they were controlled by it.

People created a favorable environment by keeping the grasslands open and also by maintaining watering points. In Ikoma the main river systems of the Grumeti, Ruwana, and Orangi rivers flow only in the rainy season, with isolated, often alkaline pools remaining in the dry season where the bottom-feeders like the catfish burrow into the mud until the rains come.[43] People found fresh water by digging out springs or pools in the riverbed, as drainage collected on the surface of the rock underlying the streambed.[44] Ikoma elders said that the wild animals also used these pools regularly.[45] When I took elders into the Ikorongo-Grumeti Game Reserves in 2003 to locate old settlement sites, we found that the springs identified with these sites were filled in and dry because no one had cleaned them out for the past thirty years, since they have been prohibited from entering this area (see figure 0.1).[46] Sukuma village elders, just south of the corridor, were responsible for ensuring dry-season water by organizing labor to clean out or enlarge water holes in the river bed or protected springs, build up the banks of catchment ponds, and dig shallow wells. Families who herded their livestock together would dig pools in the riverbed for water.[47] These watering points were used by wild animals as well as people.

Wild animals were part of the humanized landscape envisioned in the emergence stories and in the specific rules, prohibitions, and relationships western Serengeti hunters observed.[48] People identified with particular animals through the animal symbols of their clan, such as elephant, lion, python, zebra, hyena, or leopard and could not kill or harm the clan totem (see chapter 2).[49] Hunters were also prohibited from killing animals in circumstances that had parallels with human relations. For example, a lion that was eating grass could not be killed because he was like a person who pulled up grass in both hands and put it between the teeth as the regional symbol of begging for mercy. If an animal ran into a homestead or corral while being pursued by another animal or a hunter, it must not be killed since this was parallel to a person escaping to someone's home for sanctuary.[50] Hunters preferred adult male animals; they freed pregnant or immature animals caught in traps and took home young animals who had lost their mothers to be nursed by livestock.[51] If infractions of any of these and other prohibitions were discovered, elders fined the man and required that he undergo purification rituals. Without this atonement the man's entire household was at risk and would not prosper.[52]

This vision of a peopled landscape characterized by intimate interactions with wild animals imagines leaders as those who control the environment in order to gain authority. In the Ikizu origin story Muriho became the founder of a new people because he demonstrated authority over both the wild animals and the Hengere, the hunter people who already occupied the land. In this story the wild animals (bees, snakes, buffalos, ants), rather than the armies of men, acted as his weapons to defeat the enemy.

> Muriho tried first to send wild buffalo to bother them day and night in their homes, but they did not leave. He sent snakes to bite them and scare them, but they did not leave. Then he used ants who would bite them at night, but they did not feel it. He said, "What shall I do with these people, the moon is nearly gone." Then he used bees on them, and the very day they stung them the Hengere moved away. The bees drove them off.[53]

Knowing the critical role that fire has played in creating the Serengeti environment, it is not surprising that in the Nata story Nyamunywa demonstrated mastery over fire to establish his authority as founder of a new community. In some versions of the Ishenyi, Ikoma, and Ngoreme stories, a prophet defeats drought and disease in order to gain his authority and bring the people to a new land. In one Ikoma version the people came from Sonjo because their father, a prophet, gave them wild animals as livestock to herd. They arrived in Ikoma following their herd.[54]

None of this evidence necessarily means that western Serengeti peoples were natural conservationists who never had an adverse effect on the environment. Their purpose was to use the land's resources for their own benefit rather than for the sake of the land itself. The precolonial past was not a perfect paradise and circumstances, most significantly population growth, have certainly changed dramatically in the present. Yet the conservation of wildlife in the past has not depended on either the absence of humans or maintaining an unchanging "natural" environment. The generative principles behind long-term strategies for dealing with environmental problems have more often been to accommodate than eradicate. The Serengeti environment has proved to be incredibly resilient even in the face of dire predictions of its demise. Although humans may have destroyed the environment at times in the past, the larger picture is one of coevolution and interdependence. The wild animals profited from the control of bush areas that harbored disease vectors and preferred the open landscapes interspersed with water holes that had been deliberately shaped over the millennia by human hands.

ASIMOKA AND THE EARLY SETTLEMENT OF
FARMERS, HERDERS, AND HUNTERS

This ancient humanized western Serengeti landscape is the context that produced the core spatial images of the emergence traditions depicting the ecologies of the hills, woodlands, and grasslands, each associated with a different community of people living in separate but interdependent communities. In the Nata story Nyamunywa the hunter meets Nyasigonko the farmer and together they give birth to a new people. Although origin myths are often analyzed for their ideological content, parallel evidence in archaeology and linguistics suggests that they can be contextualized in historical processes of the distant past when the first farmers and pastoralists came to dominate the land controlled by autochthonous hunters. People developed economic specializations in relation to the particular ecological niches they inhabited. Even today the Dadog-speaking Tatoga pastoralists live on the grassland plains near the Serengeti corridor; the few Asi hunter-gatherers that remain live in the woodlands, now closer to Maasailand; and the dominant Bantu-speaking farmers of Ikizu, Ikoma, Nata, Ishenyi, and Ngoreme traditionally live in the hills. The Bantu-speaking western Serengeti agropastoralists represent a synthesis that combines many elements of this rich heritage from many different language families and economic strategies. The core spatial images of emergence stories are congruent with these ancient processes and thus represent the most likely historical context for their origin.

An analysis of *asimoka*, the local word for ethnic origin or emergence stories of the first man and first woman, provides a model for understanding the origin of western Serengeti communities that includes diverse places on the landscape. In Nata, the verb -*sisimoka* means to spring up, as in waking up from a sleep or as in small sprouts popping up out of the ground.[55] In the related language of Kuria the root is -*semoka*, meaning "to originate from, derive from, rise (of a river)."[56] Social theorists have suggested a rhizome, instead of a tree, as a model for origins that recognizes multiplicity without reduction to a single genealogical beginning.[57] Rhizomatous grasses in the Serengeti appear distinct on the surface while a network of underground stems connects them together. All the surface growth may be burned off, only to sprout up again from rhizomes beneath the ground. Asimoka stories tell of the emergence of new identities out of the old tangled underground network of rhizomes, without simple, primordial origins.[58]

The core spatial images of the emergence stories are similar for each of the ethnic groups, with important variations in the details of who was involved and where. Another version of the Nata asimoka story, told by Jackson Mang'oha Maginga, narrates the diverse beginnings when hunter and farmer, male and female, who spoke different languages, from different ecological zones, met for the first time.[59]

> Our parents, of Nata are—Nyamunywa, he was a man—and Nyasi-gonko, was a woman. Nyamunywa was a hunter—Nyasigonko was a farmer, the woman. They met—this man, Nyamunywa, shot an animal, which fell near to the field of the woman, Nyasigonko. The man, Nyamunywa, was thirsty. When he got to where the animal had fallen he saw some green grass, which is a sign of water, so he went there to look for water. When he got near, he saw there was a person coming out from that place. It was a human, like him, and the woman saw him too. He went to her house in the cave. They could only speak in signs because they did not know the same language. The man asked for water to drink. She got him some from the spring in a gourd [ekebucho]. She then took some millet from her field and brought it to him in an elongated gourd [akena ya oburwe]. She put it in his hand, and he chewed it. It was mixed with sesame. She asked him, "And what do you eat?" He showed her the animal and skinned it. The man went outside in the bush and made a fire by twirling a stick into a board using an ekengeita and ororende . . . shweeeeee. She got wood, and they roasted the meat. They ate it. They took the meat home and lived in the cave of the woman. Basi [so finally], it became their home. The man followed the woman. They gave birth to the Nata, Nyamunywa and Nyasigonko.[60]

This story can be interpreted on a number of different levels. Although it has many mythic elements related to present-day social and ideological concerns, it also uses the core spatial images to symbolize in a succinct and effective way a very long historical process where hunting peoples assimilated into farming communities (that we know about from other kinds of sources).[61] With early frontier dynamics in mind we must suppose that a woman called Nyasigonko (from the root -gonka, to suck at the breast) or a man called Nyamunywa (from the root omunywa, mouth) probably never existed. But, over many generations, countless farming women may have met hunting men, whose families decided that their cooperation would be mutually beneficial. It is significant too that each learned skills from the other that were critical to the subsistence of the new family.

The Ikizu asimoka story about how Muriho established a new people also uses the core spatial images of farmers and hunters joined together to form a new society. In this case, however, the hunter-gatherer people, the Hengere, are driven off the land rather than assimilated into the community. The following story, part of which appeared earlier in this chapter, symbolically reflects the reality of the marginalization of hunter-gatherer societies in the frontier process. But it also shows the diverse origins of the society that the interactions, friendships, and intermarriages of different people produced.

> Muriho came from the west, the lake, Nyanza, through the north to Kisu, but he did not go through Gorogosi. He was a healer and had medicine. When he left Kisii, Muriho went first out to Ngoreme, to Mangwesi Mountain. Mangwesi was a hunter and invited Muriho to be his guest. Mangwesi was a Ngoreme. Mangwesi told Muriho, "You are my friend, why not stay here in my country and build with me?" Muriho said, "I am going over there to the mountain, the big one that I dreamed about, Chamuriho."
>
> So Muriho left and came to his land and was welcomed by a man named Nyamwarati. He learned that the Hengere were bothering those people. After living with Nyamwarati for a while, Muriho asked him, "Who is the big man of the country?" Nyamwarati answered, "I am, but I am defeated by the terrible people here called the Hengere." Muriho said, "If I drive them out what will you give me?" Nyamwarati answered, "If you drive these people out of my country I will give you my daughter to marry." Muriho tried first to send wild buffalo to bother them day and night in their homes, but they did not leave. He sent snakes to bite them and scare them, but they did not leave. Then he used ants who would bite them at night, but they did not feel it. He said, "What shall I do with these people, the moon is nearly gone." Then he used bees on them, and the very day that they stung them the Hengere moved away. The bees drove them off. . . . The Hengere are the people of the Congo, the pygmies. When the Hengere were chased out, Muriho encircled the mountain and the land with protection medicine to make it safe so they would not return. The first wife of Muriho was the daughter of Nyamwarati— Wanzita. She gave birth to Mughabo. We say, "We are the Ikizu, the people of Wanzita and Mughabo."[62]

The particular location of Ikizu origins, revolving around the two most prominent mountains in the region, Mangwesi (Bangwesi) and Chamuriho, identifies these early settlers as hill farmers (see map 3 for location of emergence sites). Although the narrator describes the Hengere as pygmies, this is shorthand for all autochthonous hunting peoples, and it is not likely that Batwa people from the Congo basin lived here. Although Muriho came from the west, near the lake, the significant relationships are formed locally, with an Ngoreme man at Mangwesi Mountain and an indigenous man at Chamuriho Mountain.

Just as in the Nata story, he marries a local woman in order to start the new community.

Ikoma, Ishenyi, and Ngoreme emergence stories also use the core spatial images of farmer and hunter from different ecological zones to tell about the creation of new communities on the frontier. However, they identify as their founder a hunter from Sonjo, a community now across the park boundaries in Maasailand. In the Ikoma story this hunter marries women from the local farming community who become the founding mothers of the Ikoma, Ishenyi, and Ngoreme. As in the other stories the hunter man is incorporated into the farming community. The story also symbolically explains the relationship between the ethnic groups who claim origins in Sonjo, and differentiates them from the Nata and Ikizu.

> A Msonjo [a man from Sonjo] came from Sonjo to hunt. He got lost and went farther to the west and rested under an *omokoma* tree. His name became Mwikoma. He came with his bow and arrows and when he got lost he slept under the huge tree that was in the bush. The limbs spread out like a house, providing shelter inside. This was at the place called Chengero. He killed an animal, skinned it, made a fire and ate it under the tree. This then became his house and his camp. He would go out to hunt and return here at night. After a while he became aware that other people lived in the area. He went to their camps to talk with them but they could not understand each other because they spoke different languages. He invited a woman to his camp. She only ate grains or porridge and he gave her meat to eat. She was amazed and thought how she had only had porridge by itself and how good this was. Thus, they began to get to know each other. He said, "I am Mwikoma." They began to live together and then got married, settling among those people who were already there.
>
> He went on and married a second wife, they had children. Then he married a third wife. Soon his children had grown up and were adults themselves. Each went off in a different direction, but Mzee Mwikoma stayed back in Chengero with his first wife Nyabaikoma. His second wife's name was Nyabangoreme, and they lived around Bangwesi Mountain. The third wife's name was Nyabaishenyi, and they lived around Paori. They separated from each other and multiplied. Thus, today the Ikoma, Ngoreme, and Ishenyi are one group, one thing, they are from one family. The Nata on the other hand come from the Ikizu, and the Ikizu come from Sukuma.[63]

The locally specific nature of the story names identifiable places like Chengero and Paori, located within the park and the game reserves today (see the location of Chengero in map 3 and also on the map of abandoned settlement and emisambwa sites in map 6). Some suggest that Makoma Hill, south of Banagi in the park, might be a possible site for Ikoma origins as well. Bangwesi (Mangwesi) Mountain was depicted as the origin place of the Ngoreme, in the previous story from Ikizu.

Both the Ishenyi and Ngoreme versions of the origin story, though abbreviated, retain the cores spatial images of the meeting of people from different zones and subsistence strategies. The Ishenyi story names the people: first man Mugunyi and first woman Iyancha (meaning from the lake), and their place of origin, located in woodlands hunter-gatherer territory of the northern extension of Serengeti National Park, near Klein's camp (see map 3).

> The origins of the Ishenyi are at Guka, in the eastern part of Serengeti District. In the Ishenyi language (Rogoro) the ancestors of the Ishenyi people are Mugunyi (the man) and Iyancha (the woman). They lived at Guka as the first Ishenyi people. After that they moved to Hukano. It is well known that they moved to Rebaka and finally to Nyiberekira, at which point they had become a large group.[64]

The Ngoreme story is a cryptic praise song that gives the genealogy of their founding parent, mentioning the places that they came from near Sonjo and the ancestor that Mangwesi Mountain was named after. Elders say that the phrase about "water that floods down" refers to the irrigation agriculture practiced in Sonjo:[65] "The Ngoreme son of Sabayaya son of Wandira son of Mangwesi—from Regata—and Manyare—the people that dip out water that floods down from on high."[66]

If the landscapes of the asimoka stories tell us about anything more than present-day social structure and ideology, there must be some basis for placing them in the context of ancient frontier patterns. The underlying similarity of Nata, Ikizu, Ikoma, Ishenyi, and Ngoreme core spatial images in their origin stories, as well their location-specific variations, suggests that they date from the time before these ethnic groups, as we know them today, existed or before they differentiated linguistically over the past five hundred years. This line of reasoning is confirmed by looking at origin stories from a little further out—stories from Kuria, to the north, or Jita and Kwaya, toward the lake—which contain a different set of core spatial images and underlying structure. Western Serengeti origin stories seem to have been adopted and recontextualized as the particular origin stories of newly forming ethnic groups, if not before, during the profound late-nineteenth-century transformations in social identity that resulted from the ecological disasters described in chapter 4. The variations within the western Serengeti narratives, then, would be the result of the particular events that occurred in each group in the new context.[67] In fact, the Ikizu story narrated later in this chapter is another tale of the meeting of first man and first woman that describes their connection to rainmakers in Sukuma during this period of disasters. But it has been joined together with Muriho's story, told earlier, without apparent discomfort about this inconsistency. The remarkable congruence of these older stories with processes that are recoverable through the evidence of historical linguistics and archaeology also seems to place the original stories in this older time frame. Rather than each ethnic group entering the region separately, with its own history and identity already formed, these emergence stories seem to refer to the sense of group identity that developed locally as new kinds of farmer communities sprouted up from the rhizomatous networks that preceded them. The core spatial image of a hunter from the wilderness coming to establish a new community is ubiquitous across Africa in many different forms.[68] This is not surprising since emerging food-producing communities all across Africa faced the process of coming to terms with, or differentiating themselves from, preexisting hunter communities in the distant past. Nevertheless, the western Serengeti core spatial images appear in a form that is specific to this context and history.

The fact that the named origin places are all located within the western Serengeti and in the hill ecologies also suggests that this new common identity of farmers was formed internally rather than brought intact by immigrants. The Nata story describes its origins in situ, without reference to ancestral migrations from anywhere else. It tells of the springing up, awakening, of a new people right where they are today, in Nata territory. Although most elders have only a vague sense of where to find these places, others claim to know the very cave at Bwanda for the Nata. The stories name known mountains and other regional landmarks like the Guka hills, Mangwesi Mountain, and Chamuriho Mountain. All of the origin places in these stories are located on the hill ecologies, to the east of where western Serengeti peoples now live, just at the western edge of the Serengeti plain, within what is Serengeti National Park today, and in the woodland wilderness areas attributed to Asi hunters. Hill settlements, located to take advantage of other ecologies and of the annual wildebeest migration, provided a space in which farmers, like Nyasigonko, consistently interacted with hunter-gatherers, like Nyamunywa, in the woodlands.

Archaeological and linguistic sources provide evidence within an ancient historical context that tells a story amazingly similar to that available through oral traditions. The earliest peoples living in the western Serengeti were hunter-gatherers, probably speaking a Khoisan-related language. Hunter-gatherers in East Africa adopted livestock that had been domesticated in the Sahara from the fifth through the third millennium before the present. Pastoralism gradually came to occupy a more important place in local economies as pastoralist people gradually moving south with better-adapted cattle breeds learned how to deal with the particular climates and ecologies of East Africa. Diane Gifford-Gonzalez argues that early pastoralist spread was delayed by cattle disease, forcing pastoralists to integrate into hunter-gatherer communities already in the area until they learned how to manage the diseases. With the end of the late Holocene drying trend, about twenty-five hundred years ago, the typical annual wet-and-dry-season pattern with the same flora and fauna found today was established. These ecological patterns then encouraged the interdependent interaction of farming, hunting, and herding economies.[69] Archaeologist Stanley Ambrose further refines this reconstruction of Neolithic history by identifying the coexistence of three distinct communities that occupied diverse ecological habitats in the Rift Valley: hunter-gatherers, who were eventually confined to the montane forests, agropastoralists, in the forest-savanna ecotone, and pastoralists, in the open, lightly wooded savanna grasslands. Competition among these three groups must have been minimal because each had adapted to a different ecological zone.[70] However, farmers and pastoralists eventually restricted the ecological niche that hunter-gatherers occupied. Hunter-gatherers compensated for their loss of resources by developing relations of trade and ritual mediation with the incoming populations.[71] In historic times hunter-gatherers in East Africa nearly always lived in close interdependent, if not subservient, relationship with agriculturalists and pastoralists, adopting the many different languages of their patrons.[72]

The first agropastoralist peoples in the western Serengeti spoke either a Southern Cushitic or a Rub Eastern Sahelian language and developed an association with Khoisan-speaking hunter-gatherers that were already there.[73] The ancestors of those who now dominate the region were root-crop farmers who came from around the Lake Victoria shorelands about 300–400 CE, speaking what historical linguists now call the East Nyanza Bantu languages. At about the same time, Mara Southern Nilotic speakers also began moving from the north into the interior of what is now the Mara region. As they moved into the drier interior, East Nyanza Bantu speakers had to learn how to expand their lakeshore root-crop, fishing, small-stock, and hunting economy to adopt grain-crop farming (eleusine millet and sorghum), and how to increase their hunting and herding expertise. Another group of Southern Nilotic speakers, the Dadog-speaking Tatoga herders, later absorbed the Mara Southern Nilotic speakers and also came into long-term and continuing interaction with East Nyanza Bantu speakers, who used common age-sets and the comradeship of peers to gain access to livestock expertise and to develop new kinds of homesteads built around the livestock corral.[74] Although East Nyanza Bantu speakers came to dominate the region by about 1000 CE, a distinct western Serengeti culture developed out of the interaction of peoples speaking languages from four major language groups, with diverse economic and cultural practices, each occupying a separate community corresponding to a different ecological niche.[75] Eventually East Nyanza Bantu speakers developed a lifestyle that exploited each of the other's economic subsistence patterns.

By or before 1500 CE, the present linguistic and cultural foundations of this region were well established. Each of the East Nyanza Bantu languages that is now distinct had differentiated itself. Those who stayed near the lakeshore came to speak Suguti languages (Jita, Ruri, Regi, Kwaya), and those who went inland came to speak the Mara languages, growing distinct from each other about 500 CE. As the Mara-speaking communities spread into new lands, those who crossed the Mara River formed the language communities of North Mara—Kuria and Gusii. In South Mara they differentiated themselves into three groups, probably becoming distinct about three to five hundred years ago—Ngoreme, eastern South Mara (Nata, Ikoma, Ishenyi), and western South Mara (Ikizu, Zanaki, Shashi or Sizaki).[76] The percentage of shared core vocabulary between each of the East Nyanza languages is depicted in figure 1.4. A higher percentage indicates a shorter time span since their separation. Each of the Bantu-speaking ethnic groups today insists on its own unique culture and tradition, yet the linguistic sources provide evidence for their common and rhizomatous roots.

The references to Sonjo origins in the Ikoma, Ngoreme, and Ishenyi stories are indications that the imagined landscape of the emergence stories reached to

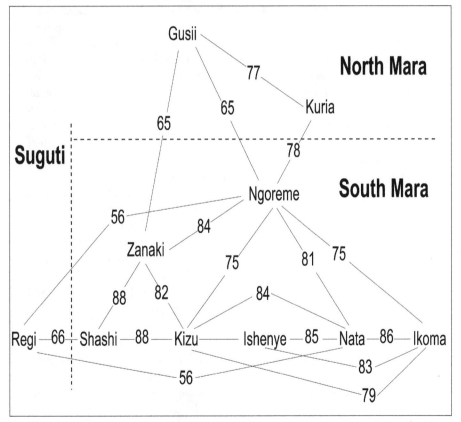

Figure 1.4. Dialect chaining of East Nyanza languages. Figures are percentages of shared core vocabulary. *Chart by Peter Shetler, 1998*

distantly related people, who nevertheless shared their ecological space. The Sonjo speak a Bantu language of an entirely different branch of the family tree, probably coming from central Kenya and leaving Gikuyu by the middle of the first millennium CE. However, they seem to have had long-term contact with western Serengeti peoples speaking East Nyanza languages since they share a high percentage of their vocabulary.[77] Sonjo also shows the influence of and interaction with Dadog and other Southern Nilotic languages, as do the East Nyanza languages.[78] Although Sonjo and western Serengeti do not seem to be genetically or even culturally and socially "related" in the way that the origin stories claim, linguistic analysis going back to the period when the first Bantu speakers arrived in this region demonstrates a relationship of long-term neighborly interaction.[79] Although we cannot date the beginning of Sonjo and western Serengeti interactions, linguistic evidence indicates that it could have started as early as 500 CE, when both groups were establishing themselves in their present homes, within a context of interdependent interactions with others.

The fact that farming settlements in Sonjo and the western Serengeti seem to have had significant contact and yet maintained separate identities may have been the result of the interaction of hunters who traveled across the woodlands and plains in search of game.[80] Traveling west, the hills of Sonjo merge into woodlands and plains until hills once more begin in Ikoma, eighty kilometers away (see map 3). Because of the ecological patterns of the wildebeest migration, hunters from Sonjo who followed the herds seasonally would have had occasion to ask for hospitality in Ikoma, while Ikoma hunters would have found themselves near Sonjo at the end of the dry season. Sonjo traditions also tell of a hunter father who brought fire and a farmer mother who controlled water. The same Sagati name is used for a Sonjo hunter clan and western Serengeti Ishenyi and Ikoma clans alike.[81] In Sonjo, the Sagati blessed bows and arrows and kept the secret for making arrow poison and preparing wild herbs for medicines. Sonjo elders said that these hunters maintained relations of trade between Ikoma and Sonjo, meeting in the wilderness where Sagati hunters gathered honey and hunted.[82] Traditions say that they came from Jaleti and Ngrumega (a transliteration of the Mbalageti and Grumeti rivers in the western Serengeti).[83] Perhaps as a way to establish fictive kinship or blood brothership of hunters, Ikoma and Sonjo both scar their right breast with the *ntemi*, a sign that they are children of "one womb."[84]

Oral traditions also suggest that farming communities from the western Serengeti might once have extended as far east and north as Sonjo. Both communities are located in the hill ecologies, although, unlike western Serengeti peoples, the Sonjo practice irrigated agriculture and may have connections to the ancient irrigation complexes at Engaruka, which seems to have flourished sometime between the fifteenth and eighteenth centuries.[85] Ngoreme praise names for Sonjo ancestors describe them as "those who irrigate."[86] Once hunters had established relationships, people might have gone to settle in similar environments, where they could practice the same kind of farming.[87] Evidence of ritual relationships across the Serengeti also exist. Some Sonjo elders say that the great prophet Khambageu came from a large mountain near Ikoma in the west where, until recently, Sonjo people returned annually to propitiate his spirit.[88] If both Ikoma and Sonjo formed their identity as hill farmers within an intercultural environment of plains herders and woodland hunter-gatherers in the distant past, then they may have felt an affinity that they explained by common origins. Although they retained their own distinct identities in spite of settling near one another or meeting on the hunt, they were "brothers" within the regional understandings of economically based identities. The fact that the Sonjo adopted more vocabulary from East Nyanza languages than vice versa would suggest that the East Nyanza communities were larger and more prestigious than the Sonjo. However, the fact that western Serengeti oral traditions recall Sonjo origins might also mean that the power equation shifted with the rising dominance of the Maasai in the nineteenth century.[89] "Coming from Sonjo," in the origin traditions, may then mean coming from settlements in the land of the Sonjo rather than from Sonjo people. Whatever the particular connection of western Serengeti peoples to Sonjo in the early period, and whenever it began in the distant past, the Serengeti plain was clearly a zone of interaction rather than a barrier, with frequent crossings from both sides and settlements in much closer proximity than previously assumed.[90] Although the origin stories of the Ikoma, Ngoreme, and Ishenyi were recontextualized in the context of the late-nineteenth-century disasters there is evidence for a much longer relationship that resulted in some shared vocabulary and tradition, with neighborly interaction but the maintenance of separate identities, as the generative principle of interdependence dictated.

Historical linguistics provides evidence for the interactions among farmers, herders, and hunters that reaches back more than two millennia. Yet, as the latest innovations in vocabulary attest, this frontier process of interaction lasted right up through the last six hundred years. At that point, Bantu speakers had become dominant by diversifying and adopting the expertise of their neighbors. Yet the larger system depended on interdependent interactions between each of these communities. Hunters became part of farming or herding communities when they gained stock, and when times were bad, farmers and herders became part of hunting communities. Even today people claim kinship across these boundaries. Cultural elements of the East Nyanza Bantu–speaking farming community draw on this rich heritage from many other language groups in the region over the past two millennia. It is difficult, if not impossible, to make a judgment about who represents the "indigenous" people in the region. Nor can the specializations of farming, hunting, and herding be isolated from one another, since they were dependent on each other for the most basic needs. Specializations were part of a larger economic system invoked by core spatial images of the ecological landscape that depended on the functioning of each of its parts.[91]

ECOLOGICAL NICHES AND CLAIMS TO AUTHORITY

The profoundly humanized landscapes evoked in the asimoka stories identified each ecological niche with a different people and subsistence pattern but further established authority over the land and over other people. This way of seeing the landscape was a device employed to legitimize power in a variety of arenas. Farmers legitimized their control over the land as they increasingly marginalized or assimilated Asi hunters, while giving some authority to and valuing both hunters and herders in particular spheres. At the same time, men legitimized their control over the operation of the homestead while giving independence and value to women in certain spheres. Both these examples demonstrate that this way of seeing the landscape was not a neutral, disinterested view but one that perpetuated relations of unequal power. Although western Serengeti peoples adopted the economic strategies of their neighbors, their source of authority and legitimacy ultimately came from maintaining distinct communities in separate ecological niches. These dynamics are evident in the way people recontextualized the core spatial images of the emergence stories in the past and still use them today.

Ecological evidence confirms placement in the emergence traditions of the origin sites in the hill ecologies. Because of social, soil, and climatic patterns, the hills are the ecological niche that farmers would have, of necessity, inhabited. Linguistic evidence shows that these agricultural settlers were following a very old pattern of their Great Lakes Bantu predecessors, who commonly inhabited the hill ridges.[92] The staple crop of the early western Serengeti farmer was eleusine, or finger, millet (oburwe) can be grown only on the lower slopes of the hills. Finger millet requires a fertile and free-draining sandy loam, since it cannot tolerate waterlogging.[93] In the western Serengeti such conditions are found only in the red-brown soils of the hills and rises.[94] Many elders described a wooden digging stick (akoromo in Ikoma) as the original farm implement, used in loose, sandy soils.[95] While a visitor to Ngoreme in 1913 observed blacksmiths forging iron hoes, he also noted that "the usual style" of soil cultivation involved the use of "sharp sticks."[96] These implements could have been used only on the hill soils, since the plains largely consist of a heavy, claylike black-cotton soil (known in colonial circles as mbuga) that becomes waterlogged and swampy in the wet season. Although, given the right rainfall conditions, the black-cotton soil is fertile and good for sorghum, western Serengeti peoples have fully exploited it only during the colonial period because it is so difficult to dig.[97] Because millet could be easily stored for a long time and is not susceptible to pest damage, it was an ideal crop for food security.[98] Historian Kirsten Kjerland identified sixteen va-

Figure 1.5. Mbuga landscape with whistling thorn trees, Robanda, Ikoma. *Photo by Paul Shetler, 2003*

rieties of finger millet used among western Serengeti's immediate neighbors, the Kuria, which they differentiated according to soil tolerances, color, and use.[99] This diversity and adaptability of native varieties indicates a long history of cultivation in the hill ecologies.

Sustained farming was not possible either in the acacia woodlands or on the short-grass plains. Although the average rainfall would allow for farming in the woodlands, the poorly drained soils prohibited it, except on the lower hillsides. A colonial resource survey of the region describes these areas of woodlands as "famine" land of heavy black clay soils, covered by acacia thorn bush.[100] The grasslands have never supported permanent farming communities because of the hardpan soils and lack of permanent water sources. An early colonial report states that "on the nine days' track Ikoma to Ngorongoro through the Zerengeti [*sic*] there are only two perennial watering-places."[101] Ikoma farmers lived as far east as Seronera, in what is now the park, but only in few numbers because of the alkalinity of the water there.[102]

The Serengeti grasslands were the dry-season grazing lands for transhumant pastoralist peoples like the Maasai, who moved into the western Serengeti, at least as far as Seronera and the Moru kopjes, during the colonial years.[103] But the pastoralists with a much longer history in the area are the Tatoga, who now graze the plains of the park's western corridor, though they once dominated the Serengeti plains as far as the Ngorongoro Crater. They call themselves the Rotigenga Tatoga, known by the farmers as the Tatiro. The other Dadog-speaking group in the area, the Isimajek, were traditionally hunters and fishermen, marginalized as dependents by the dominant Rotigenga.[104] When asked about their relationship to the Ikizu farmers, Tatoga elders expressed an older understanding of the land-scape: "We are the people of the plains, they are the people of the hills—when we go over there to the hills we say, we are going to Ikizu; when they come over here to the plains they say, we are going to Tatiro."[105] These differences are re-flected in the ways in which they "map" and name the same landscapes. The Tatoga have names for the gaps *between* the hills and the plains, while the west-ern Serengeti farmers name the rises and hills.

Oral traditions of both the Ikoma and Tatoga connect the Tatoga to the creation of wild animals and livestock, legitimize the authority of the Tatoga over the grass-lands, and thus assert their status as original guardians of the grassland plains. In Ikoma Robanda there is a rock with natural depressions that resembles a *bao* board game, in which participants place seeds as counters in parallel sets of holes. Ikoma elders claim that this was the bao of Masuche, with which he tricked God, the sun. The Ikoma story told at the site went like this:

> They call this Masuche's Bao [*agoreshi e Masuche*]. Masuche played
> bao with God, the Sun [Irioba], here. Masuche's cattle went out
> and grazed themselves and came back at night because God was
> close to Masuche. One day they quarreled and the Sun went home
> in anger. Only some cattle had come home to the corral by then,
> when it got dark, because the Sun left. So the ones who were left
> out became the wild animals—Masuche named them zebra, gazelle,
> topi, impala, and all the others. That is how the wild animals came
> to be. They are Masuche's cattle.[106]

Tatoga elders told the story of Masuje, one of the most important Tatoga prophets, in this way:

> Giriweshi was born of a woman and was the son of God. His son
> was Masuje, who tricked God, the Sun, in a game of bao because
> he knew how to make the stones revolve endlessly without coming
> to an empty hole. Because the game never ended the Sun never set
> and Masuje's cattle could graze far from home. The Sun became
> angry and retreated into the sky, taking Giriweshi with him.[107]

The fact that the Ikoma tell this story as their own demonstrates both their ac-ceptance of Tatoga dominance on the grassland plains and their close and inter-dependent relationship with them.

At least in historical times and consistent with the older generative principles, we know that Tatoga pastoralist authority was integrated in a regional system of interdependence with farmers and hunters. Although Bantu-speaking farmers also herded and the Dadog-speaking herders also farmed, their occupation of different ecologies opened options for trade in livestock and grain that were mutually beneficial. The two communities established cattle-trustee relationships to protect their herds and cooperated in the chase whenever a herd was raided. Yet Tatoga distinctiveness has been responsible, in part, for Tatoga ability to play a highly influential role among their neighbors as gifted prophets with ritual authority in both Ikoma and Ishenyi generation-set ceremonies.[108] In fact, colonial investigations of local politics concluded that the Tatoga once held a dominant position from Lake Victoria up to the Mara River and across the Serengeti plains to Ngorongoro Crater.[109] Although Bantu-speaking farmers were certainly dominant numerically and geographically by at least 1000 CE, their position was dependent on others serving their function within an integrated regional system in which the Tatoga herders had authority over the grasslands and in specific ritual roles.

The core spatial images of the emergence traditions are consistent with other evidence demonstrating that while in the distant past Bantu-speaking farmers in the hills and the Tatoga herders on the plains each became dominant in their own ecological niche, the woodlands became the space where the autochthonous Asi or Ndorobo hunter-gatherers were marginalized as lower-class dependents of farmers and herders. The earliest colonial maps and reports from the area, as well as oral tradition, identify the woodlands as the territory of Asi hunter-gatherers. One colonial report commented that they wander "in the uninhabited hinterland of Musoma," where they "form small camps in the forest and are constantly on the move."[110] The woodlands bush, with its more acidic soils was in fact the area best suited for gathering edible plants and unsuited for either grazing or farming. Its low soil fertility and low rainfall encourages larger tubers and woody plants that provide edible foods. The migrating herds of larger animals from the plains move through here in the dry season in search of the permanent water pools along the tributaries of the Grumeti and Mara rivers. The hunter-gatherer lifestyle, both from current ethnographic and ancient archaeological evidence, includes an unusually high dependence on meat, which makes up 80 percent of the Asi diet, and is supplemented by honey and fat. These populations were quite sedentary, occupying the zone between grasslands and woodlands to exploit both ecologies. They mainly hunted resident ungulates of the woodlands with traps and snares and also kept some small stock.[111] The woodlands, with a supply of edible plants, a resident game population, and a seasonal arrival of larger game, sustained a hunter lifestyle. Linguists and archeologists hypothesize that although the hunter-gatherers found what they needed in the woodlands, they had occupied the hill area because its location between grasslands and woodlands gave them access to both areas. The Asi hunted in the woodlands but also grazed small herds of livestock, probably sheep and goats, on the grasslands and found useful plants in each of the areas.[112] However, as farmers came to dominate the hill ecologies, they increasingly relegated the hunters to the woodland ecologies located to the north and east of western Serengeti settlement today. Their confinement to the marginal woodland areas meant that they had to depend on hill farmers and grassland herders for livestock and grain in return for products like meat, skins, tusks, and honey.[113]

Origin traditions are consistent with this historical context reconstructed from other sources that document the competition over hill zones between early farmers and hunter-gatherers and the eventual marginalization of hunters to the woodlands. Elders today can locate the places named as farmer origin sites in areas on the hills within what is known as Asi territory, now in the "wilderness" woodlands located to the north and east of present-day western Serengeti settlement. The Nata origin site of Bwanda (near where the town of Mugumu now stands) was in an area known traditionally as Materego ya Abaasi (the wilderness of the Asi).[114] Oral testimony also identifies the area north of Nyichoka in Nata territory, contiguous with the Ngoreme origin site of the Ikorongo hills and Mangwesi Mountain, as Asi country. The Ishenyi origin site at Guka is well into the woodland territory but also located on the hills. The earliest German maps (1910) show these same areas as Ndorobo territory, described as "undulating country of open thorn bush and grass," or "open bush and thick scrub."[115] The fact that all the emergence sites were located within what is known as Asi territory and within the woodland ecologies of hunter-gatherers, reinforces other evidence that early agricultural settlers came into direct competition with hunters for the same land.[116]

The evidence also seems to indicate that as the farmers increasingly marginalized the Asi from the hill ecologies, Asi identity became a class designation for the poor, or those without livestock.[117] Older, regional patterns consistent with present attitudes seem to indicate that people who lost their livestock or crops in a time of famine would "become Asi," as they relied on hunting and forest products to survive, while Asi that gained livestock wealth could, over the generations, assimilate into farmer or herder communities.[118] One Ikoma elder said, "the Asi are only poor Maasai; when they get wealth they become Maasai."[119] Many western Serengeti elders spoke disparagingly of the Asi, saying they had no home or land of their own; they just wandered—and farmers never engaged them in friendship, interaction, or intermarriage.[120] Although many western Serengeti clan names and lineages indicate Asi ancestry, few will admit to the connection personally. As one elder reminded me, "people will get mad if you call them an Asi."[121] Because the farmers had to gain authority over the land through the original owners, the Asi are part of their origin traditions but are no longer acknowledged as direct kin. The Asi are now disenfranchised, as one elder put it, with "no land of their own."[122]

Yet western Serengeti peoples have lived closely with the Asi as long as they have been in the area, and elders recount personal stories of interactions with individual Asi. Contrary to the stereotypes, one elder told of a famous wealthy

Asi named Mbahi who lived near Burunga keeping many sheep and goats.[123] A number of Ikoma elders remembered close family relationships with individually named Asi men who would come to trade tusks, ostrich feathers, and other wilderness products at their homes. Asi taught the Ikoma Himurumbe clan to hunt elephants and had ongoing hunting relations with them that extended to hospitality in their homes.[124] One Ikoma elder said that Asi used to come to eat at their house, since his father was their friend. They brought lion skins and traded one for five goats.[125] A number of western Serengeti clan names refer to Asi ancestors, though elders dispute whether this was real or mythical. One of the first German sources distinguished between Ndorobo in the Serengeti area, who spoke a Maasai language, and the "pure Wandorobbo [sic] in the Zerengeti steppe on the Syonera [sic] living as nomads" who spoke a different language, for which the "Washashi in Ikoma" acted as interpreters.[126]

Oral narratives of the western Serengeti acknowledge their debt to the Asi in learning the lore of the woodlands and hunting.[127] Elders claim that the Nata people learned the secret of arrow poison (obosongo) from the Asi.[128] Inconclusive evidence also suggests that western Serengeti peoples borrowed their common style of bow from the Asi, though the arrows look somewhat different today.[129] Ritual feasting for taking eldership titles still requires a prescribed number of pieces of dried wild animal meat. Before the Germans came, western Serengeti peoples paid bridewealth and made clothing with wild animal skins[130] The Nata give each of the wild animals a separate name according to sex and age, as the chart for a few selected animals demonstrates. An awareness of the hunting season permeates the traditional agricultural calendar. The topi, one of the resident ungulate populations in the area, give birth to their calves in September and October. This is the sign to prepare the fields because the rains are coming. The calendar also shows that it is time to plant sorghum when the wildebeests return to the short grass plains.[131]

Although many people told me that the Asi "have no home," oral tradition and ritual practice recognizes them as the original owners and guardians of the land. Attesting to this fundamental connection between the land and the hunter-gatherers, the name Asi itself may be derived from the Bantu root for earth or soil (ase or ahaase in Nata).[132] Early farmers who inhabited these hill sites within Asi territory had to accommodate and assimilate the Asi if they wanted to take control of the land themselves. Elders remember the Asi hunters as those who performed rituals to propitiate the ancestors of particular lineages as "guardians of the land" in order to maintain the prosperity of the land. Many of those ancestors were from the Asi hunter clans of Gaikwe and Hemba, which became farmer clans. Either by intermarriage, oathing, or ritual adoption, western Serengeti farmers became "children" in the lineage of the Asi ancestors who were connected to the land.[133] Thus, Bantu-speaking farming communities gained access to the land and successfully diversified their own economic practices,

animal	Ekinata name	adult male	adult female	male young	female young
hartebeest	abanosi	atiribati	anyabori	ang'ong'ona	ang'ong'ona
impala	asuma	abarogwini	anyabori	egisaka	amwati
gazelle -- thompsons	ambarahe	aborogwini	anyabori	egisaka	amwati
topi	asubugu	atiribati	anyabori	atororo	atororo
wildebeest	asamakiri ekiweri ndgosana	abaha	anyabori	atwabana	atwabana

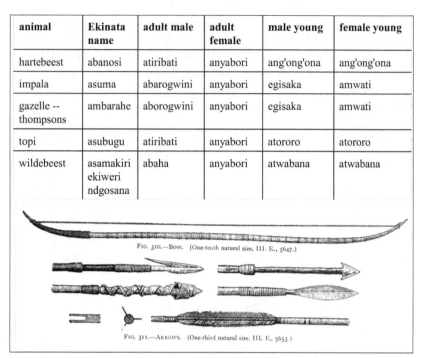

FIG. 310.—Bow. (One-tenth natural size, III. E., 5647.)

FIG. 311.—Arrows. (One-third natural size, III. E., 5653.)

Figure 1.6. Hunting vocabulary and tools. Names for wild animals are in Nata. Cultural vocabularies on wild animals and hunting provided by Nyamaganda Magoto and Tetere Tumbo, Mbiso, 23 November 1995. *Illustration from Kollmann,* Victoria Nyanza, *195*

leading to their dominance in the region, while the Asi lost their independence and rights to the land.

This long-term and intimate interaction of hunting and farming communities gradually came to an end beginning in the mid-nineteenth century when Maasai became dominant on the plains and the Asi or Ndorobo hunter-gatherers from the Serengeti became their dependents.[134] Colonial boundaries also increased these divisions. One Ikoma elder said, "The National Park cut us off from our Asi friends; they still walk through it but they don't come here anymore." The last time he saw an Asi was sometime around 1945 or 1946.[135] An early German report confirms that the Asi followed the Maasai, "during the great migration of the Masai, the Wandorobbo were either driven out or forced to submit."[136] Oral traditions both remember and deny this long-term history of settlement in which western Serengeti farmers established their authority over the hillside ecologies by marginalizing the original Asi owners of the land, who finally left altogether.

The landscapes envisioned in the asimoka stories also use gendered core spatial images in depicting interactions of interdependence that provide insight into the critical internal processes of frontier settlement involving gender relations. In the Nata story the woman welcomes the man into her home, while the man has no home, only a hunting camp in the wilderness. These images make use of the opposition between home and wilderness, inside and outside, female and male space. A new society emerges at the contact point between these dynamically opposed forces. Although the house belongs to the woman in the story, first man's presence from the wilderness domesticates the house and makes it a civilized place, as he brings fire to the house. Today the hearthstones are still considered the very symbol of home and family; no home exists without a fire burning in the hearth.[137] First man knew the secret of fire and, in many versions of the emergence story, told the woman that he excreted the fire to conceal how he really produced it. In other versions he taught the woman how to cook. Meat from the wilderness and grain from the homestead are both necessary for building this new community.

Although the story gives the man authority over the homestead, it also describes spaces of interdependent mutuality between genders. In many of these stories women are in control of an autonomous sphere of authority in the house, and men are dependents in women's houses. The emergence stories often use fire and water to symbolize the different kinds of authority men and women control. Rituals and narratives also use the symbols of water and fire as transformative substances of power.[138] In one Ikizu version of the origin story that comes after the Muriho story, first woman's secret was water, or rain, while man's was fire. The woman won in a contest between water and fire and achieved authority as rainmaker "chief" over the Ikizu. In the story Nyakinywa, the exiled daughter of a chief comes from Kanadi, in Sukuma, to the cave of Gaka after she has left her sisters in other places.

> When she entered the cave of Gaka and stepped on the rock it cried like a drum and when she hit the rock, it sounded like a true drum. That is how she knew that she had come to the end of her journey. She laid down her bundles in the new house that had been prophesied by her father. The next day she went outside to look at her surroundings when she saw some smoke far off in the distance. After careful investigation she found out that the smoke was coming from the hill of Sombayo. She went to find it over the Kibangi River, and inside the cave at Sombayo. Right away she met a man coming out named Samongo who lived in the cave and was of the clan of Muriho. Samongo asked his guest where she had come from. Nyakinywa replied that she came from her house to see where the smoke was coming from. He asked to see her house, so they went together to Gaka, to Nyakinywa's place, where he asked her to be his wife. She refused but they lived there together as lovers.

Those two people lived together, each had their own area of expertise. Samongo had the secret of making fire and Nyakinywa had the secret of water, that is, she could bring rain. Each asked the other to show them their expertise, but neither would agree. Things went on like this until one day Samongo went out to hunt. Nyakinywa brought a big rain that completely soaked Samongo out in the bush. Back at home she had put out the fire. When he returned, cold and wet, he found the fire out in the cave at Gaka. Samongo had to show Nyakinywa how to make fire. He took out a board and a stick that he twirled until he had kindled a fire. So you see, he had already shown her the secret of fire. He then asked Nyakinywa to show him the secret of rain. She did not refuse but first asked him to bring her sister Wang'ombe from Hunyari. Then he should go to the bush and kill a bushbuck, skin it and bring the skin alone to her. Samongo did all that she asked. The next day she asked him to make the pegs used to stretch out the hide on the ground to dry. When this was done the three of them left with the bushbuck skin and the pegs for stretching a drum and went to the pool at Nyambogo. When they arrived, Samongo was asked to peg out the skin on top of the pool. He tried and failed. Nyakinywa tried it and succeeded in laying it on top of the water. They asked Samongo to try it again but it still didn't work. Then Wang'ombe was asked to try and she also succeeded in pegging out the skin on top of the pool, like her sister Nyakinywa. So Samongo was told that he had failed the test and would not be shown the skill of bringing rain. They returned to their house at Gaka.[139]

In this narrative the interaction between first man and first woman explains the establishment of ritual authority in Ikizu over rainmaking and prophecy. Ikizu recognize Nyakinywa's line as the chiefly line of rainmakers, while Samongo's is the line of prophets. The female power of water and fertility triumphed in this story but only with concession and compromise. In other versions we hear that to become the *mtemi*, or "chief" of Ikizu, Nyakinywa had to agree to certain provisions laid down by Samongo, including that Ikizu would retain its basic social institutions of circumcised age-sets, generation-sets, and eldership ranks.[140]

Spatial organization of the homestead by first man—when he left his nomadic camp and came to live in the enclosed space of the cave, which remained the woman's house—represented separate, though interdependent, spheres of gendered authority. Within the "traditional" homestead, elders claim that male space is outside, in the courtyard, while female space is inside, in the house. In East Nyanza languages, "house" (*anyumba*) refers not only to the physical house but to a woman, her offspring, and the property and dependents attached to her.[141]

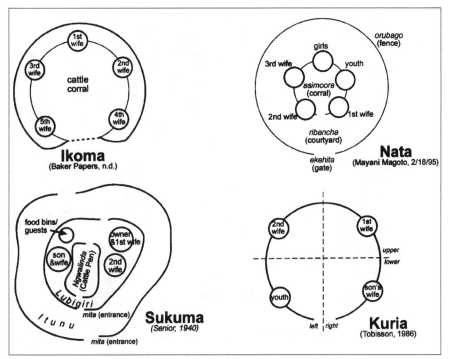

Figure 1.7. Homestead layouts. *Diagram by Peter Shetler, 1998*

Men have no houses of their own and sleep in the houses of their wives. Men's conversation takes place in the courtyard that surrounds the central cattle corral, as the symbol of male property.[142] A Nata elder told me that the man, Nyamunywa, was an *omotware*, a term for a man who goes to live in an independent woman's (*omosimbe*) homestead and is married by her (note the passive voice) — often called a male wife.[143] He could never be the head of the homestead and was beholden to her goodwill, as she could ask him to leave at anytime. In the Ikizu origin story Nyakinywa refuses to marry Samongo, but he stays in her house.[144]

Although these interdependent spheres of authority functioned to give women some autonomy, it is clear both in the oral traditions and in ethnographic accounts that the men still controlled production in the homestead and that women bore the burden of work.[145] The gateposts and the log crossbar used to shut the cattle corral symbolized male control over homestead wealth.[146] The gendered division of labor evident in the origin stories legitimizes a pattern where women farm and men hunt. Today, women of one homestead do most of the farm work; they control their own fields and jointly farm their husband's field. In rural Tanzania as a whole, 70 to 80 percent of the food for household consumption is produced by women.[147] Just as the farmers created interdependent and mutually

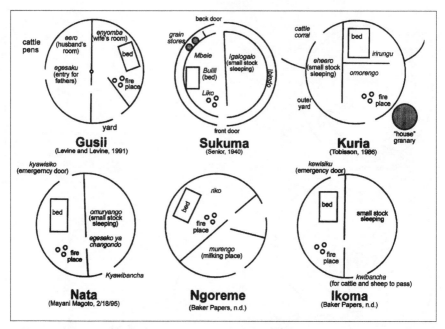

Figure 1.8. Interior house designs. *Diagram by Peter Shetler, 1998*

beneficial interactions with Asi hunters while at the same time marginalizing and dominating them, so it is with the interactions between men and women. In both cases these contradictory relations were legitimized in the asimoka traditions of ethnic origin. The parallels between these stories and what we know from archaeology, linguistics, ecology, and ethnography seem to place these patterns in the realm of long-term historical processes that cannot be dated exactly and yet are part of the ongoing substrata of the way that people continue to imagine the landscape today.

People, wild animals, and plants, developing in relation to one another over the millennia, have continually reshaped the Serengeti environment. The evidence for these processes comes from the grasses, trees, wild animals, human social habits, and the landscapes themselves. As people provided for their own subsistence and health needs by creating open landscapes, maintaining watering points and encouraging certain kinds of vegetation, they also benefited the wild animals. The people, as well as animals, have had to adapt to the hard realities of the Serengeti environment but they have also helped to create that environment. At the same time, people have not necessarily sought to conserve the environment, and with increased population and a different context today the old patterns do not necessarily function anymore. However, addressing the challenges of conservation must begin with the understanding that the Serengeti is a profoundly humanized landscape.

The core spatial images of interdependent interaction of ecological zones were recontextualized as different peoples moved into this socially created landscape over the last two millennia. Using the generative principles embedded in this landscape they learned from those that were already there how to prosper in this environment, developing economic strategies shaped by the ecological realities of the hills, woodlands, and grasslands. In doing so they created an interdependent regional system in which farmers, herders, and hunters each played their part. Although East Nyanza Bantu–speaking farmers dominate the region today their economy, society, and culture owes much to the region's Tatoga herding, Asi hunting, and Sonjo heritage. So, too, early settlers created interdependent spheres of authority in the homestead, with women and men exerting authority over different aspects of the domestic economy. At the same time farmers did dominate and marginalize hunters while women submitted to the overall homestead authority of men. However, addressing the challenges that Serengeti communities face today must begin with the understanding that the frontier settlement process was as much one of cooperation and integration as it was of competition and division. Seeing these ecological landscapes from the perspective of the asimoka story of Nyamunywa and Nyasigonko brings the hills, woodlands, and grasslands to life as places of both struggle and adaptation.

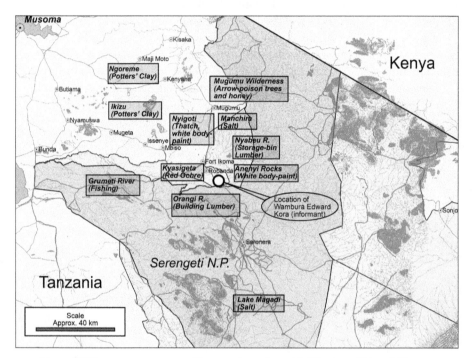

Map 4. Wilderness resource use. From an interview with Wambura Edward Kora, Ikoma. *Map by Peter Shetler, 2005. Underlying GIS data courtesy of Frankfurt Zoological Society*

2 ᔰ Social Landscapes

Forging Food Security Networks (Hamate), ca. 1000 CE to Present

THE NEXT HISTORICAL WAY of seeing the Serengeti landscape appears in clan traditions that name particular settlement areas and natural resources by their association with purported descent groups, linking diverse people together in extensive regional networks necessary for survival. When I traveled with the Ikoma elders in the game reserves, they identified each group of hills across the horizon and each abandoned settlement site that we visited according to the kinship group that had lived there. While the core spatial images embedded in the emergence stories depict an interdependent ecological landscape of farmers, hunters, and herders interacting from particular ecological niches, this is a social view of the landscape, creating networks through the core spatial images of diversification, inclusion, and distribution embedded in the generative principles of kinship association. Although more ideological than biological, descent was the mechanism through which people identified themselves, made connections to others, and gained access to resources. This diversified social landscape of kinship networks evident in the core spatial images of clan migration histories is also reflected in the ways that people talk about how they farm or herd, organize their homesteads, or find security in times of famine. These practices grow out of the generative principles that establish basic understandings of reciprocity in moral behavior. They are also congruent with evidence from the early frontier period when western Serengeti peoples settled in this challenging environment of periodic drought. Seeing the landscape in this way enabled them to maintain productive communities through social networks representing a moral economy that provided basic security and subsistence for everyone.[1]

The core spatial images of clan traditions are both those of diversification and inclusion, uniting diverse peoples from diverse places across the landscape as children of the original couple, first man and first woman. From those children, as representatives of the clans, all the descent groups were born and remain linked to each other as they spread out over the land. This version of the Nata asimoka story told by Megasa Mokiri ends with the creation of the four Nata clans, or *hamate*, by the children of Nyamunywa and Nyasigonko:

> The woman became pregnant and gave birth to a son. Then she gave birth to a daughter, and in total four boys and four girls. When they were grown, they were married to each other. This is the reason that Nata inherit through the woman's side. The children made the clans of Nata. The place where they lived is called Bwanda. When they got to be too many, they divided into the *saiga* [age-set cycles].[2]

In this version the boundaries of group inclusion are entirely self-sufficient, as its sons marry its daughters. Similarly, Ikoma elders said that the sons of Mwikoma founded the eight Ikoma clans.[3] Inclusion brings with it access to the resources of the group. Ethnic origin stories often mention the clans of first man and first woman, giving those clans' members a particular role in the community; the clan of the first child might have the highest political status, while another clan might serve a particular ritual function.[4] Not coincidentally, Megasa Mokiri includes the issue of inheritance in his narrative. Wealth, particularly cattle and moveable property, was inherited through descent, either on the mother's (matrilineal) or the father's (patrilineal) side. Clans controlled land by parceling out crop fields and maintaining common pasture and wilderness resource land. Yet within this image of the autonomous group, the generative principle of diversification is evident. Since each clan controls a particular territory, this way of seeing the landscape maps connections to a variety of resources and communities.

Other clan stories exist outside an ethnic narrative as the migration history of that clan alone and more clearly illustrate the core spatial image of distribution as the way that this view of the landscape promoted the formation of social networks. An excerpt from a Hemba clan history by Samweli Kirimanzera, which will be analyzed later, gives a strong sense of the spatial elements of these clan histories:

> Of the clans that were left, others divided and came to Ikizu. [Pause.] When they got to Ikizu others dispersed, small groups went to Sizaki, Changuge, the Hemba of Changuge, the people of Mbasha Megunga, they were — one. Then other Hemba dispersed and when they got to Bigegu, they went to the area of Mwibaghi, where there were lots of animals. After Mwibaghi they came to Rindara, in Majita. When they lived in Rindara, one elder named Guta built at the place now called Guta, he died there in Guta, this Hemba man. They went to Kirio, near Guta. Then when they were

finished, other Hemba of Rindara, after some years, went to Majita. That is why you hear then talking about Wiyemba, Wiyemba—they are in Majita and are the same as we Hemba. We came from one place and dispersed.[5]

This story and similar clan histories provide a view of the landscape that connects diverse kinds of people, unified by purported clan membership, across the region. The land itself came to embody these social networks as people used the generative principles inherent in the core spatial images of diversification, inclusion, and distribution in the ways they interacted with the environment.

Yet these conclusions are not self-evident in the oral traditions, which have lost connection with their original context. Although western Serengeti peoples now understand clans as subsets of ethnic groups, an analysis of these stories suggests that this was not always the case. In fact clans seem to have preceded ethnic groups; as suggested by the evidence for the formation of ethnic groups in the nineteenth century (see chapter 4) and clan stories that seem to have been the building blocks for the construction of ethnic stories. Furthermore, the same clan names and associations are found throughout the region within various ethnic groups of diverse origins and economies, indicating that they are not discrete subsets of a single ethnic group. Ethnic group stories of the division of clans, like the Nata version above, must have been constructed after the fact to explain the relationship among the diverse sections of the newly formed ethnic group. As we have seen in chapter 1, the oral traditions that are told as ethnic origin stories contain the core spatial images of groups defined by the economies of farmers, hunters, and herders rather than ethnic groups. All this evidence suggests that ethnic groups originated in the nineteenth century from the joining of already dispersed clan networks in particular areas rather than clans dispersing from an ethnic center, as stated in the clan stories.

The core spatial images of clan stories are particularly difficult to recontextualize in any specific chronology, since they fall into the time frame of stories from the middle period of social process, or cyclical time. While we don't know how far back the particular clan stories date, we do know that the earliest East Nyanza Bantu–speaking settlers brought with them the idea of descent group membership as a way to gain access to and divide resources. The core spatial images of clan stories are also congruent with long-standing patterns of resource use, agronomic practice, and settlement layout common throughout the region. Evidence for these patterns of diversification, inclusion, and distribution in reciprocal networks as strategies for survival in a marginal environment appear in archaeological, linguistic, ethnographic, and environmental sources. These generative principles, out of which the basic strategies for survival have been elaborated over the generations, cannot be exactly dated and are still embedded in social landscapes today.

However, logically this way of seeing the landscape dates to the period after 1000 CE, when East Nyanza Bantu–speaking peoples became dominant in the region. The first way of seeing the landscape as the interaction between farmers, hunters, and herders may have been more important between 200 and 1000 CE, when East Nyanza Bantu–speaking hill farmers were trying to establish their place in the wider regional system. Once they became dominant, their primary concern was not with the already marginalized hunters and herders but with other communities like themselves on whom they relied for marriage partners, trade goods, famine security, access to resources, and ritual knowledge. Naming and mentally mapping the landscape with reference to kinship allowed these early communities to make claims on each other and to remember the various links they shared as well as to provide strategies for farming and settlement organization. Much of the data for this chapter comes from conversations with elders today about what they remember of the old ways of making a living and establishing homesteads. Because these patterns are congruent with the core spatial images of oral traditions that seem to date back to at least 1000 CE, they provide evidence for a way of seeing the landscape from the past that is still alive in the memories of elders today. Because of the lack of data that allows us to see variations and change in these patterns over time, they are necessarily presented as a somewhat static model. Unless other kinds of evidence become available, the reader must keep in mind these limitations in imagining how these generative patterns might have reconfigured themselves over the past millennium.

The core spatial images of diversification, inclusion, and distribution are found in oral traditions, in other sources, and inscribed in the land through the ways that people organized their livelihoods and homes. One sees the core spatial image of diversification in the various subsistence strategies that western Serengeti peoples have used over the generations to spread risk rather than concentrate wealth. Descent systems, agricultural practices, settlement patterns, use of wilderness resources, and systems of land tenure illustrate the ways in which people used the land extensively rather than intensively. Western Serengeti peoples used the ideology of descent to define membership in local clan-based communities that controlled access to land and other resources. Yet, due to the need for more labor in a land-rich but people-poor environment, clan ideology was inherently inclusive. The core spatial images inherent in clan histories created an imaginative landscape of widespread regional networks of reciprocity that could be called on in times of hunger. Leadership developed in this moral economy of subsistence around people who gained authority by distributing wealth to followers. The challenges of the particular environment in which western Serengeti peoples settled shaped this view of the landscape, but the environment itself was also shaped by this imagined landscape as people applied its generative principles to extensive strategies of land use for maintaining food security.

The core spatial images of diversification in clan stories that depict various peoples and economies across the landscape united by clan membership are congruent with patterns identifiable in the archeological, linguistic, environmental, and ethnographic sources that are also pan-African and ancient. Historians have shown that people throughout Africa from the distant past came to depend on descent groups as ways to control access to resources in part because of the challenging environment that they confronted. Jan Vansina and David Schoenbrun have investigated the earliest Bantu words for kinship in central and East Africa and demonstrate how these words are associated with homestead building, alliance, and mutual aid.[6] Igor Kopytoff postulates the creation of "pan-African cultural principles" as a result of the conservative nature of the frontier process, including kinship as an idiom for political relations.[7] The archaeological evidence discussed in chapter 1 shows that early settlers on the frontier formed relationships of interdependence in order to minimize the risk of environmental disaster. As a result of both environmental challenges and demography the frontier culture emphasized extensive, rather than intensive, systems of production that valued maximum security over maximum efficiency, low risk over high production. Michael Watts, in his influential book *Silent Violence*, demonstrates the fundamental and long-standing strategy of diversification for risk aversion among African farmers.[8] Western Serengeti farmers too, seem to have coped with the problems of drought, soil infertility, disease, and pests by developing diverse farming and herding strategies that spread their risks over a large land area and among obligations to a variety of people. In this moral economy, the strong ties of family linked people together in order to provide basic subsistence for all. Although the particular ways that these generative principles were applied is specific to the western Serengeti, they are undoubtedly some of the oldest pan-African concepts.

Although savanna ecosystems such as the Serengeti are unusually productive and resilient, they present farmers with a number of challenges that are logically addressed by the generative principle inherent in the core spatial image of diversification.[9] Rainfall, the key determinant for Serengeti plant growth, is characteristically unpredictable due to the erratic movement of the intertropical convergence zone interacting with local factors.[10] The western Serengeti experiences a typical East African bimodal rainfall pattern, with the short rains starting in November and December and the long rains falling from March to May, producing an annual mean rainfall between 700 and 1,250 millimeters.[11] The land is prone to periodic local shortages, with rains falling evenly over the whole district only once every seven years and with a tendency toward more rainfall along the lake than in the interior. It is not unusual for the seasons either to fuse together into one long growing season or for the short rains to fail altogether.[12] Differences in rainfall patterns are highly localized, meaning that one hill settlement might experience drought while neighbors who live a day or two walk away might harvest in abundance.[13] In addition to erratic rainfall as well as the disease and pest environment discussed in the previous chapter, soil fertility and composition places additional limitations on people's ability to farm effectively. The fertility of most tropical soils is difficult to maintain, as they are prone to leaching and erosion.[14] However, the physical composition of the soil, which may be damaged by exposure to intense sun and rain or compacting, is more important and less easy to repair and has led to locally adapted farming techniques.[15] Western Serengeti farmers must deal with heavy and poorly drained clay soils in the bottomlands of alluvial and lacustrine origin, which are extremely fertile but difficult to work with a hand hoe.[16]

Marking the landscape by kinship networks was also a way to diversify in order to deal with the problem of the lack of enough people, or labor, to overcome environmental challenges. The continent of Africa has been underpopulated until the later twentieth century. Contrary to popular conception, the shortage of labor, rather than land, is still the major obstacle for agricultural development. The boundaries that elders described for the old Nata territory were enormous, encompassing approximately thirty-six hundred square kilometers. The colonial census in 1948 listed the population of Nata at only 1,519, less than half a person per square kilometer; adding in the populations of Ishenyi (2,428) and Ikoma (4,474) at the time still increases the density to only a little more than two people per square kilometer.[17] Even in 1993 the population density of Serengeti District was only 10.2 people per square kilometer, the lowest in the Mara region.[18] In western Serengeti history, environmental restrictions and low populations led neither to surplus production, nor to the development of political hierarchy and state formation for its extraction. There was plenty of land available and no way for those with more power to restrict and control access to that land. As in many other places in Africa, control over people, and therefore their labor, was more important than control over land or material wealth, other than cattle, which was used to procure productive and reproductive labor in the form of brides. Africans developed distinctively "land-rich cultural traditions" that emphasized fertility, diversification, mobility, and control over people or labor as the source of wealth.[19] Successful communities were those that attracted diverse new members.

Throughout Africa, kinship offered a method of regulating access to the land, labor, and capital, as well as making connections outside the local community, one that was suited to the overall strategy of diversification. Because food security always depends on much more than just the lack of rain or people, patterns of social organization and kinship have been the subject of intense scholarly interest since the colonial years.[20] In the classical anthropological literature on Africa, kinship in the form of genealogical descent is identified as the basic organizing principle of society, both widespread and of ancient origins. The segmentary lineage model proposed that ethnic groups, or "tribes,"[21] were made up of a number of clans unique to that ethnic group. A clan was defined as a group who claim descent from a common ancestor but cannot demonstrate the genealogical links. Each of those clans were made up of a number of maximal lineage groups, who were in turn made up of minimal lineages, down to the level of nuclear families, visualized as a neat set of nesting boxes or a genealogical tree. Lineages were descended from a common and demonstrable ancestor. This model assumes that ethnic groups, made up of discrete clans, formed isolated and autonomous units.[22] But closer analysis reveals that the ideal pattern seldom corresponds to practice. Descent is, therefore, more usefully understood not as a structure or a set of rules, but as a strategy in which people make choices in how to interpret and make use of the underlying and agreed-upon principles of the moral economy to gain access to the resources necessary for survival.[23]

Through linguistic evidence we know that the first East Nyanza Bantu–speaking settlers on the western Serengeti frontier also used purported descent as a critical ideological mechanism for controlling resources as well as for gaining labor by incorporating others into their communities. They inherited a bilateral (both matrilineal and patrilineal) descent system from their Great Lakes Bantu–speaking ancestors. In the western Serengeti strong matrilineal tendencies developed as people sought ways of learning from the previously existing hunting and pastoral inhabitants and incorporated them into their own communities.[24] In the origin stories discussed in the last chapter, the core spatial images of the male hunter moving into the female farmer's home represents the historical context in which stranger husbands moved into the matrifocal houses of their wives. Matrilineages enhanced their widespread networks of security by incorporating hunter men along with their kin connections into farming communities.[25] This strategy would help to explain the success of Bantu speakers to expand and eventually dominate in a region once shared by peoples of many different linguistic groups.

Matrilineality, or descent through the mother's side, was especially adapted for incorporating strangers and for expansion on this land-rich intercultural frontier. In a matrilineal system, production was controlled in the individual homestead, while distribution was communal. A man's sister's children—rather than the children of his wives, whose production he controlled, inherited his wealth—and his heirs most often lived in distant settlements. This disjuncture between the locality of production and distribution created widespread networks of security through the distribution of wealth, rather than through the accumulation of wealth within self-contained homestead units. By contrast, a situation of scarce resources concentrated on productive land that people must exploit intensively favors the patrilineage, since the goal in this situation is to keep close control over inheritance.[26] Although ancient western Serengeti societies had strong matrilineal tendencies, evidence of patrilineality also exists, indicating a nondifferentiated, or bilateral, descent system that suggests a flexible strategy for negotiating various kinds of relationships on the frontier. As we will see in chapter 5, patrilineage became increasingly important during the late precolonial and early colonial periods, with the increase in cattle wealth when men were concerned with concentrating and protecting their wealth as a source of local authority.[27] Inheritance through either side represented strategies for resource control and for establishing security networks rather than a formal set of rules.

The origin stories themselves also provide evidence for the historical depth of a bilateral descent system in the region. As in many places in Africa, the characters of these founding myths represent the genealogical prototypes of the kinship system.[28] Many of the stories indicate a patrilineal system by emphasizing the role of first woman as the wife of first man, yet in the Ikizu version Nyakinywa's sister, Wang'ombe, joined the struggle of first man and first woman. In other versions Nyakinywa left Kanadi with two sisters, all of whom were a chief's daughters. The presence of sisters as founders in the Ikizu story seems to indicate the strength of the matrilineage, while other emergence stories emphasize the brothers of a patrilineage. When a Nata elder taped some interviews for me while I was in America, following my instructions, he began by asking people about their descent. But unlike the interviews I had conducted where men told me that they were the son of X, who was the son of X, he asked them to name their mother's line. He said both sides are important and one can choose which side to go to depending on one's need.[29]

The same core spatial image of diversification found in kinship patterns and clan stories is also identifiable in agronomic practices for spreading out risks that are rooted in ancient generative principles. One example of this at the microlevel of the field is intercropping, used across Africa as a way to reduce the risks of losing the entire crop at once and to protect the physical and nutritional qualities of the soil.[30] Western Serengeti farmers used intercropping for eleusine (finger) millet and sorghum, the staple crops produced by first woman in the origin stories. One common farming pattern that elders remembered from their youth, was the *orotonga* field, in which farmers planted the taller sorghum plants around the periphery of the field and millet in the center, mixed with sesame and beans.[31] In 1947 the Agricultural Department statistics still showed over half the crops in Musoma District planted in this way; out of 84,990 total crop hectares, 8,180 hectares of legumes were interplanted with 40,573 hectares of sorghum and millet.[32] A very early British agricultural report commented favorably on intercropping as a "simultaneous rotation" of crops.[33]

In historic memory individual farmers diversified risk by using kinship or friendship to gain access to fields in many different places and microenvironments, often far from home. One Ikoma elder said that, depending on the rain, you might get a good crop in one field and get nothing four kilometers away.[34] People valued these far-flung fields even if they produced little because they kept social connections current and served as long-term insurance against crop loss.[35] This strategy also allowed people to grow crops in the appropriate soil type and climate. Careful research conducted in 1938 on agricultural practices in Sukuma, just to the south of Musoma District, described each of the many local designations and uses for soil types, concluding that people spread out their fields because they needed to take advantage of the various soils rather than because of localized rainfall patterns.[36]

Classification of soil types in the western Serengeti includes *ekebuse* (sandy upland soils) and *eseghero* (clay bottomland soils). The best soil is a mixture of both and is found on the low elevation rises. Because the heaviness of the clay eseghero soil made it difficult to dig, people had to wait for the short rains to prepare the soil. In the past farmers worked up the clay soil in the dry season, a practice called *kuharaga*, to get it ready before the rains. If the year was good, the dark, fertile eseghero soils were incredibly productive. Nevertheless, they either became too hard in a dry year or too swampy in a wet year to produce a reliable millet crop. They were good for sorghum, however, which tolerates waterlogging.[37] Various soil types in various mixtures were found nearby, usually depending on the elevation of the slope, so that farmers became adept at staggering the placement of their fields to find the right combinations.[38]

Western Serengeti elders remember that people also distributed their cattle among their friends and kin in distant places, using a widespread institution known as cattle trusteeship (*kuwekesha* in Swahili; *kusagari chatugo* in Nata) in order to find enough water and grazing for large herds and to reduce the risk of losing the whole herd to disease, drought, or raiding. The receiver of the cattle had rights to the milk and manure, while the giver retained ownership of the cattle and its offspring.[39] The wealthy Tatoga prophet Gishageta owned eight thousand cows and ten thousand goats and sheep in the early colonial years. His nephew said that the colonial veterinary officer identified his brand in Ikoma, Ikizu, and Ngoreme. He loaned his cattle to all kinds of people, not just those of his own ethnic group.[40] Another elder said that cattle trusteeship was practiced so that the poorest people could get milk.[41] In a similar manner, the Sukuma were reported to send cattle to partners in areas where the grazing was better or to leave their cattle behind when they moved to areas with tsetse infestation.[42] A North Mara officer noted that the common practice of cattle trusteeship in this area resulted in a situation in which "more than half the cattle in any one homestead will probably belong to the family and the remainder will be held on behalf of others." He approved of this practice because it encouraged "spreading of stock evenly over an area," facilitating "better farming and ranching methods."[43] While in many East African societies cattle trusteeship was an important form of patronage that was elaborated in complex forms and resulted in hierarchical kingdoms, here, where cattle wealth was not extensive, it remained more of a mutually beneficial partnership embedded in the moral economy.[44] People more commonly diversified their herds by mixing small and large stock to prevent famine. Agropastoralists could more easily consume or exchange sheep and goats for grain when the need arose, and small stock were better than cattle at browsing the scrub when the grass dried out. In case of extreme drought they could even graze in tsetse-infested bush, given their higher levels of natural immunity.[45]

Mobility was another important aspect of the generative principle of diversification in farming and herding practices that seems to originate long before the colonial years. Although western Serengeti peoples certainly must have known about more intensive farming techniques from their neighbors who practiced irrigation in Sonjo or manuring in Ukara, the system of shifting agriculture made the most sense in terms of productivity per unit of labor input and was best suited to soil requirements.[46] Elders often said they moved to find better farming and grazing lands, for example, when the fields were infested by a bad grass like the *oromboko* or the land was simply "tired" (*eketoha*). Within living memory farmers moved on to new fields in the area or moved their homesteads altogether every three to ten years.[47] Repeatedly elders told me, "We are a moving people, we never stay long in one spot."[48] In their life histories, many elders related all the moves they and their families had made. Oral narratives suggest that, ideally at least, moving a settlement was a decision made by the elders, or in consultation with a prophet. The decision was often precipitated by problems such as infertility, raids, human or livestock illness, deaths, famine, misfortune, or the promise of better conditions in another place.[49] In following a relative who found a fertile new area in which to settle, people used already established relations of reciprocity to gain a foothold in a new area.[50]

Over the long term, the need for new fields created a slow process of "natural drift," resulting in patterned mobility rather than random moving.[51] Elders said that new settlements were located fairly close to the old sites, usually within a day's walk. A settlement would be situated so that residents could move the outlying fields (*ahumbo*) for several seasons without moving the settlement. Older people described moves as festive times when people called all their relatives, friends, and neighbors to help carry everything on their heads or backs to the new site. They often sent the youth of the family ahead by a full season to plant and build temporary housing. Mobility was also possible here because of the relative ease of clearing the land for new fields and obtaining materials for building. Because of vegetation patterns in the area, people cleared new fields by burning brush and chopping down larger trees in the months before planting. This meant that farmers put little long-term labor investment into any one field.[52]

Oral narratives suggest that western Serengeti peoples held deeply personal attachments to the land but defined the land extensively and communally, over many square kilometers, rather than intensively and privately on a permanent farm plot. The resources that people needed to survive on a daily basis came from a very large land area in the wilderness that lay beyond the last fields and regular pastures. Descent groups claimed control over these areas as part of their ancestral lands, where they propitiated the graves of ancestors for fertility and prosperity. Men's seasonal hunting, raiding, and trading took them far from home and into the wilderness. These activities required an expert knowledge of landscape, terrain, and ecology. Men had to know where to find game at any time of the year and how to track game for days and still find their way home. They relied on generational knowledge for the location of the best hunting camps, water holes, shooting blinds, hunting pit sites, and arrow-poison trees. The retired generation was obligated to escort the new generation on their first hunting or arrow poison–gathering journeys. The youth carried the elders' packs and protected them at night as repayment for this service.[53] As they went, elders named each hill and rise and noted each river, seasonal water source, and pool. The names of these places often corresponded to people who had lived there or to incidents that had taken place there.[54]

Older women still remember how they collected wood for fuel, herbs for medicines, and wild greens for making relishes in the wilderness. When I walked with women along a path they would often point out medicinal or edible plants and stop to harvest some. Even in areas quite remote from villages, we often came across trees with some of the bark chopped off for medicinal purposes. Many women who were not professional healers knew the herbal medicines for particular illnesses and treated people in the community who came to them for a remedy.[55] Women also "read" the vegetation for what it told about the season or the forecast for rain. If a certain plant was blooming it might mean that a dry spell was coming or that it was time to plant the millet. While women or children were out watching the livestock they gathered wood or wild fruits to bring home. Although meat, often wild, and grain constituted the biggest part of the diet, women also regularly cooked greens with the meat sauces and offered wild fruits as a welcome snack. Women gathered small dead twigs from the trees nearest their homes for the cooking fire when possible but made a trip to gather larger quantities of wood at least once a week.[56]

Men also recount stories of how they undertook special trips to the wilderness in search of honey or arrow poison. Clans guarded the secrets of arrow poison (obosongo), prepared by boiling the woody portions of the obosongo tree (identified in 1947 as *Acocanthera fiersiorum*) and making a dark concentrate by evaporation over a fire. The active ingredient in the poison is the glycoside ouabain,[57] which western Serengeti peoples say "freezes" the blood of the animal. This arrow poison was extremely valuable; one small container sold for the equivalent of a goat. During the colonial years many people considered the arrow poison from western Serengeti of the best quality and they traded it as far as Shinyanga, Mbulu, and across the Kenya border.[58] There is a hill near Nyichoka that elders still remember as the place, protected by medicines, where they stored the obosongo when the men brought it home and performed ritual purification ceremonies so they would not become sick.[59]

Agricultural and pastoral peoples throughout Africa have also turned to the wilderness for food in times of drought and famine.[60] Of course, the first line of defense against the inevitable food shortfalls was storage; each woman was responsible to feed her own family with the food in her own grain storage bin. However, elders tell how everyone in the homestead also helped to farm the man's fields and how his grain storage bin was opened when the woman's was empty or when guests arrived. He could sell the grain as surplus only when it was apparent that the family would not need it that year. When women began to run out of stored grain they used alternative foods, principally wild plants, honey, or livestock blood. They gathered common wild greens like *chanderema* to cook and used a small succulent plant that spreads out close to the ground (*ekitando* in Ikoma) to dry and pound into flour for making the staple porridge (*ugali*). But most important, they hunted meat to dry and trade for grain.[61]

Two Ikoma elders independently mapped the various places that these resources were found. The map at the beginning of the chapter (map 4) illustrates the narrative of Wambura Edward Kora from when he lived around Robanda as a youth. Building materials, such as certain grasses good for thatch, were gathered far from home, where the cattle did not graze. Ebony trees, used for poles in houses that would not be eaten by termites, came from the Orangi River. At the Nyabehu River, southeast of Bangwesi Mountain, people cut trees for the construction of storage bins. The Ikoma found trees for making arrow poison and gathered honey around what is now Mugumu, in the woodland area at the Kyebosongo River where the Asi once lived and remained unsettled until the late 1940s, or up near Klein's camp in the Lobo area of the park.[62] They gathered salt either around the Moru kopjes at the alkali lake called Lake Magadi, now within the park boundaries, or at Manchira, on the Mokenyo River, near present day Bwitenge, where the salt lies on the ground and can be gathered and dried on rocks. White rocks, which could be ground up and mixed with water for ceremonial paint (*anenyi*), were found north of Robanda or near Nyigoti. Red ochre, also used ceremonially for painting bodies, was on Kyasigeta Hill, at present-day Fort Ikoma. People fished in the Orangi and Grumeti rivers and collected medicines throughout the wilderness.[63] This area for collecting wilderness resources described by Ikoma elders covers approximately ninety-six hundred square kilometers and demonstrates the reliance on extensive wilderness resources.

While lands outside the homestead may appear vacant or "natural" to an outsider, this area was a functional part of a very old local system of resource use. Wilderness resources were considered communal property and each clan territory (*ekyaro*) controlled the land within its extensive boundaries. One might make the argument that all these diversification strategies were adopted only in fairly recent times as a result of the nineteenth-century disasters or the stresses of colonial demands. The final three chapters of this book demonstrate how these core spatial images were adopted and elaborated in these contexts. But given their ubiquity throughout Africa and the evidence in other kinds of sources, we must assume them to be much older than that. The colonial government specifically worked against mobility and extensive resource use, assuming that everyone should be eventually settled on individual private plots of land (see chapter 5). The landscape appears "natural" today in part because of these long-term land use patterns that have shaped the environment over the last millennium.

Diversification strategies embedded in the core spatial images of clan traditions are also evident in stories about how people dealt with labor shortages. A homestead depended primarily on household labor, including wives, children, and other dependents, to accomplish its work. This is why families invested wealth in livestock in order to marry multiple wives, who would contribute to production by farming and gathering but also by producing more children. But homestead labor was often not enough, particularly in the labor bottlenecks during weeding and harvest, and the household had to reach out to extended family and neighbors for help. As in much of Africa, people gained access to others' labor within the community through the obligation within a moral economy of repaying social debts. Another way to understand this aspect of the moral economy is as a gift economy in which people, whether related by kinship or not, formed relationships by exchanging not only things like livestock or tools but also "courtesies, entertainments, rituals, military assistance, women, children, dances, and feasts." According to Marcel Mauss, an early scholar on the gift, to accept a gift was to accept something of that person's spiritual essence and to form a bond and a sense of obligation to return the gift without ever returning it exactly one for one.[64] Although these were serious obligations elders expressed them as voluntary acts of friendships. As one elder concluded, "Ikoma lived by helping each other. . . . Now people don't help each other anymore without being paid."[65]

This gift economy allowed people throughout Africa to organize work parties for the labor-intensive tasks of weeding or harvest.[66] In Ngoreme women remember reciprocal labor groups called *chesiri* that worked in each member's fields in rotation for a set number of days, helped members bring home building materials, like thatch, or came together to grind grain. Each group had a leader who organized meetings and kept order. The group amused itself while it worked by singing songs that exposed the misdeeds of community members.[67] Ikoma women's work groups (*kyama*) worked one day in each woman's field, moving together in a line and singing as someone led with the rhythm of a shaken gourd. A man might also call his neighbors to help him weed a field, cut poles, or gather thatch; in return he gave a beer party (*risaga*) to the fathers of the young men who worked, since in the past only the elders were allowed to drink beer.[68]

Women, especially, depended on the gift economy for maintaining a diverse set of reciprocal relations between the neighboring homesteads that gave them access to the food, household implements, and services necessary to accomplish their everyday work. When a woman ran out of millet flour for porridge, she went to borrow some from her neighbor, who in turn would ask her when she ran out of something else. Women understood these exchanges as gifts and did not repay them one for one. A woman did not return borrowed flour or a tool until the neighbor she borrowed it from ran out or needed it and came to ask. Ideally, a woman wanted to have both credits and debits outstanding at any one time. It was not to her advantage to balance all relationships. People who owed a debt represented potential sources for future needs; those who had credits would come to visit when they collected.[69] Women also practiced the gift economy when they took food gifts (*omotoro*) in their social visits to relatives in villages a couple of hours' walk from their homes on the occasion of marriage, death, illness, ritual, or celebration.[70] A generous woman called in her debts when she put on a large feast or a work party. Such events lasted for days and required a large amount of women's labor, both before and during the feast, in cooking, hauling water, gathering firewood, making beer, and grinding grain. This was the ultimate test of how skillfully a woman maintained her community networks. If a woman did not answer the call for help or distribute gifts to neighbors, no help would be given to her when she was in need.

The generative pattern of diversification in the core spatial images of clan traditions also appears in the evidence of archaeology, linguistics, ecology, and the more recent memories of elders about how their ancestors used the land. Although only directly applicable to the last hundred years, the memories of elders on agronomic practice, resource use, and labor recruitment provides a means of imagining how these generative principles might have been applied in the past. All this evidence tells us that this way of seeing the landscape dates back to at least 1000 CE, when East Nyanza Bantu speakers dominated the region, but it does not tell us exactly how these generative principles were practiced and recontextualized over the past thousand years. We only know that by practicing the strategies of diversification, ranging from kinship systems to farming techniques, western Serengeti peoples assured their livelihood on the land. The underlying principle of diversification inherent in this way of seeing the landscape allowed western Serengeti agropastoralists to minimize their risks and assure a basic level of prosperity. For western Serengeti farmers, diversification ultimately meant spreading energy among various enterprises and guarding subsistence resources from the unpredictable arrival of rain or disease. It meant investing in a wide range of people who could loan them land, take care of their cattle, or weed their fields when they needed aid. These patterns, in turn, are evident in the way the landscape appears today as the environment continues to be shaped by human hands.

Inclusion is another core spatial image of clan traditions that can be found in re-membered practice as well as in the historical data. This way of seeing the land-scape mapped access to diverse resources within networks accessible through in-clusive membership in kin groups. In interviews elders narrated clan histories to settle claims to land or other resources. The clan was the perfect vehicle for pro-viding entitlement to resources because it created a strong sense of obligation in the moral economy through kinship ideology while allowing for the inclusion of strangers. The images embedded in clan traditions indicate that people used the familiar understanding of mutual obligation within homestead relationships be-tween a man, his wives, children, and dependents to think about similarly struc-tured relationships in successively wider spheres. At the level of the clan, people were not necessarily related by blood but by a metaphorical sense of themselves as descendants of a far-removed ancestor or, for example, as people who honor a particular totem, like the bushbuck or the zebra.

A close analysis of clan histories demonstrates that the ideal model of segmen-tary lineage systems seldom fit the lived reality and that the images are rather those of the inclusive homestead. In the western Serengeti, as in many other places in Africa, the ways of calculating descent vary, with different terms used for different levels of segmentation in different ethnic groups and the same clan names found in different ethnic groups.[71] For example, the term for what might be understood as a clan in one ethnic group might refer to a lower level of seg-mentation, or the lineage or sublineage in another group. The flexible and adapt-able kinship definitions that people actually used varied dramatically from the fixed and exclusive structures that elders described in their narratives. Descent as ideology is one of the most useful and adaptable ways to understand and main-tain all kinds of relationships. Anthropologist A. Kuper suggests that the home-stead, rather than a set of nesting boxes or a genealogical tree, is an indigenous model for understanding how African societies have actually used kinship rela-tions. Although people living together in a homestead were usually biologically related, households also incorporated strangers and dependents, all of whom contributed to the domestic economy.[72]

The western Serengeti image of the inclusive homestead in clan histories as a way to think about kinship can also be found in linguistic and archeologi-cal evidence going back to the early frontier period and ethnographic evidence from recent times. The East Nyanza term for the homestead, *eka*, is an older term for residential groupings of people based on dispersed and exogamous lin-eages and derives from an even earlier Sudanic loanword for a head of cattle or a cattle camp. Western Serengeti peoples followed the distinctive house style of their Great Lakes Bantu–speaking predecessors: round with thatched roofs, sur-rounded by a tall fence with a main gate.[73] This long-term regional pattern is

Figure 2.1. House, Robanda, Ikoma. *Photo by Paul Shetler, 2003*

evident in archaeological sites in western Kenya, ethnographic reports through-out the region, and words for elements of the homestead.[74] The homestead was built around the male space of the cattle corral, signifying broader connections to kin with claims on that wealth and its use as bridewealth. Other words for par-ticular kinds of descent groups refer to elements of the homestead. Nata call a person's maternal line the *anyumba* (house) and Ikizu call it the *rigiha* (hearth-stones), while Nata and Ikoma call the paternal line the *ekehita* (lit., homestead gateway) and Ikizu call it the *ekeshoka* (gatepost).[75] The "house" included the bridewealth cows that the family received for a woman's daughters, over which she and her children had ultimate control, especially for milk and meat and pay-ing the bridewealth for her sons to marry.[76]

The core spatial image of inclusion through the metaphor of kinship is re-peated in what elders say about homestead and settlement patterns, which is still apparent in the physical appearance of the landscape. Elders remember that western Serengeti peoples built their homesteads and settlements on the hillsides to escape the swampy lowlands in the rainy season, which are considered bad for the health of people and livestock.[77] The lower hillside slopes also afforded a view out to the land around, which was useful for detecting danger from a long way off. This meant that women often had to walk a couple of kilometers to get water at the rivers or springs.[78] Elders described the *oruberi*, a small grouping of homesteads surrounded by a brush fence, as a homestead on a larger scale.[79] The oruberi fence, like the homestead fence (*orubago*), had one gate for livestock and a secret back door for emergency escape. By definition everyone in the oruberi was part of the same descent group, as they would be in the ideal homestead. Yet just like a homestead, many people who did not share a genealogical connection were incorporated. On a larger scale yet, the clan grouped their oruberi settlements within one area, or on one set of hillsides, which they called the *ekyaro*, or territory.

In the memory of elders distant grazing and farming lands were also considered part of the extensive resources available to a homestead through kinship. A common short-grass grazing area, called the *ekerisho*, lay just outside the homestead fence where livestock were allowed to graze when they were first let out of the corral in the morning and when they returned from the far pastures and watering holes in the evening. Heavy grazing of the cattle near the homestead trampled and mowed off the existing vegetation and encouraged a short sweet palatable grass that provided good visibility, prevented the regeneration of bush, and did not harbor ticks and other pests.[80] There were no fields near the house, except for a protected kitchen garden that the old or infirm tended.[81] Grazing lands were often out on the plains or near the river basins far from home, where farming was not possible. The herds from one community were pooled together for the day under one herder, taken from each family by turns.[82]

Elders also recollected that further out from the homestead, often on the gentle hill slopes between three and twenty kilometers and many hours' walk away, were the ahumbo fields, which descent groups farmed together in blocks surrounded by a brush fence, like the oruberi fence. Each family and each wife

Figure 2.2. Settlement with grazing livestock, Motokeri, Nata. *Photo by Paul Shetler,* 2003

Figure 2.3. Ahumbo fields with temporary shelter for the farming season, Mbiso, Nata. *Photo by Paul Shetler, 2003*

farmed their own fields but helped each other guard the fields from birds and pests. During the nights of harvest season, young people stayed at the field in temporary houses (*ekeburu*) to guard against wild animals. The old and the very young stayed back at the oruberi, along with enough youths to guard the cattle in case of a raid. Women came to the fields later in the morning, after taking care of the household chores and cooking lunch to bring to those in the fields.[83] The earliest German reports from the area state that, "the fields in some cases are several hours' journey from the houses, mostly lying in the low grounds amongst the rivers and brooks."[84] The homestead (eka) itself, the settlement (oruberi), with its related crop and grazing lands, and the larger territory (ekyaro) were positioned in the larger social landscape by kinship identification. The landscape was defined and mapped by these settlement territories of clan identity.

This pattern of settlement made a deep imprint on the appearance of the landscape, as fields and grazing areas were separated from houses and cattle corrals and each of those elements moved from year to year. There was no sense of a contiguous and self-contained piece of privately claimed land for a single homestead unit. The western Serengeti pattern of land tenure draws a striking contrast to the vision of the yeoman farmer brought by British colonial officers to their work in Tanganyika. They saw the way people used the land as inefficient. As one officer in 1930 put it, "The political condition of the people [of Musoma District] is very backward and will not improve much until they are taught to make better use of their land."[85] Even more pointedly, a 1949 broadcast extolled an imagined landscape modeled on Britain: "In place of the present untidy village lands we want to see the village lands so organized that they are used to the best advantage. . . . In short we need the development of a village community not markedly different from that which we knew in this country [Britain] thirty or forty years ago. But all this involves us in problems of land tenure, and land use, and the training of specialist workers and cannot be rushed."[86] Indeed the farms were "untidy," with farming and grazing lands removed from the homestead interspersed with uncultivated areas and land that had once been farmed returning to bush.

Because this social landscape of inclusion within the kinship settlement is derived from the core spatial images of clan histories, it is useful to distinguish between the different kinds of purportedly descent-based identities, all of which use the metaphor of the homestead yet represent very different kinds of identity. Clans, as opposed to lineage groups, were much more obviously based on ideology rather than descent and were therefore more useful for including strangers in a community. The lineage group, however, is derived from known ancestors. Western Serengeti elders can usually recite their exact lineage genealogies to four generations on both their mothers' and fathers' sides—naming up to their grandfathers' grandfathers.[87] Ethnic group or clan members, by contrast, cannot tell their exact genealogical relationship to each other nor to their founder, nor is this knowledge considered important. As is common in lineage reckoning throughout Africa, elders often ended their genealogical recitation with the subclan and clan founders' names as a matter of form in order to account for relationships to other groups within the framework of descent. Clan members relate to each other as equals within a common category, while lineage members must determine the exact nature of their relationship according to gender, generation, and distance. This is because lineage members have specific ritual and legal obligations to each other concerning, for example, land, livestock, inheritance, or funeral expenses. Knowing the exact genealogical relationship with other lineage members is necessary in order to determine the degree of personal obligation. Anthropologist E. Evans-Pritchard defined this difference between clan and lineage as one between relations of personal obligation versus structural relations between collectivities.[88]

The distinction between ideological and biological relations is also useful in the consideration of oral traditions about lineages and clans. Lineage traditions are frequently reminiscences about known ancestors. People still associate lineage ancestors furthest removed in time (five generations) with the particular places where they lived. They can communicate with these lineage ancestors through rituals performed at these sites, as will be discussed in the next chapter. On the other hand, people do not associate clan ancestors with exact places but with general directions or ecologies, nor do they propitiate the spirits of clan ancestors. The name of a clan ancestor often has symbolic meaning, such as Mwancha (lake; west) rather than the name of a remembered person. Although the localized clan could take political action based on the defense of corporate property, dispersed clan members had little corporate responsibility to each other. According to anthropologist David Lan, because clan members identified themselves as united by common totems, animals that they were prohibited to kill, clan was more accurately understood as a mechanism for "incorporation by common substance," rather than descent.[89]

Clan relationships developed out of the need to define obligations among people who shared common residence rather than common blood. The clan as a territorially based community was probably the highest level of social identity before the nineteenth century. Although clan organization and the words used to describe it varied across the western Serengeti, they shared the ideological view that a clan is a group of people who live together on clan-controlled land. In the ethnic groups farthest to the west (Ikizu and Zanaki), clan groups lived in one area and claimed clan lands. In the ethnic groups farthest to the east (Ikoma, Ishenyi, Nata, Ngoreme), clans dispersed among the three different age-set territories, and people of the same clan tended to settle as neighbors. In Ngoreme, traditions told explicitly of people moving out of clan lands during the disasters and gradually back again during the colonial period. Even in *ujamaa* villages organized after the Arusha Declaration in 1977, clan members tended to build their houses together.[90]

A linguistic analysis of *hamate*, one of the words for clan used throughout the region, demonstrates how western Serengeti peoples used clans as a generative principle of inclusion within a geographical territory and also to regulate access to resources. *Hamate* is a locative word that specifically refers to a politically independent territorial grouping based on descent ideology. Anthropologist Malcolm Ruel described nine Zanaki and fifteen Kuria neighboring clan territories, or hamate, as "provinces."[91] These territories are named by using the prefix for place designation, *bu-*, with the clan name, as in Busegwe (Zanaki), the place of the Segwe clan, or Bukiroba (Kuria), the place of the Kiroba clan.[92] There was little evidence of these autonomous clan territories within one ethnic group ever uniting for joint action.[93] The particle *ha* in *hamate* refers to an exact, particular place, as in *ahaase*, land. The clan praise names give a place of reference associated with each clan, not necessarily its origin place. The word *hamate* is also used generically throughout the region as "place." For example, in Nata one can use the term to describe a beautiful place. Because age-set cycles became territorially based in the east at the end of the nineteenth century, people in those areas often refer to both clans and age-set cycles as hamate. The word for clan, *hamate*, then literally means people who live on the same land.

The residential clan, or hamate, occupied, cared for, and ritually maintained the ekyaro, or clan territory, where they celebrated their own circumcision ceremonies and took public action when necessary. Eva Tobisson describes the essence of Kuria *ikiaro* (ekyaro) belonging as the "close link between agnatic descent (real or fictive) from an eponymous clan founder and affiliation to a particular territory associated with him as the 'first-clearer of the land.'"[94] There are certain corporate rights and obligations expected from those living in the residential hamate. They contribute to compensation in blood suits if another clan accuses a member of murder. At a Nata funeral the hamate of the deceased receive a cow out of the inheritance (*ang'ombe umwando*).[95] Thus the clan is functionally a group of people who live together on the same land and have obligations to one another defined by their membership, although ideologically they are a group of people related by blood. The generative principle inherent in descent ideology effectively uses the clan concept to incorporate everyone into the local community. Although clan traditions cannot tell us which particular clan groups existed in this region a thousand years ago, nor the location of their territories and how they changed over time, the generative principle of inclusion through kinship seems to have been an important way that western Serengeti peoples organized their homesteads and settlements, affiliated locally into clan territories.

The core spatial image of inclusion allowed western Serengeti peoples to organize their residential communities on the basis of the generative principle of clan identity within which everyone was supposedly related by blood, yet at the same time to employ mechanisms for the incorporation and adoption of outsiders. The ethnographic and linguistic evidence surrounding these practices suggests that there is continuity in these patterns since the East Nyanza Bantu–speaking farmers began interacting with others on the frontier. Because land was plentiful, and people scarce, communities gave inclusiveness high priority. The incorporation of strangers (*abasimano*) and even clans of strangers into the community rendered them natives (*abibororu*, those who were native born). Elders related that a young man looking for a new life or running from bad circumstances might enter a community through friendship with an agemate. If the stranger worked hard, a man who had no sons might be happy to adopt him into the family or graft the stranger's lineage onto his clan and pay his bridewealth. Men strangers were accepted as family when they underwent initiation into the local system of titles or age-sets, took a new name, and swore an oath not to leave the land or betray his adopted people. If a person was circumcised along with other age-mates, he was treated as one of them. Women strangers were adopted automatically at marriage. As an Ikoma elder said, "Once an *omosimano* [stranger] is adopted he has all the rights here, he is one of us and cannot go back to his original home."[96]

Elders claimed that the children of an omosimano are *omwibororu* (*abibororu*), or native born, and not differentiated from peers born of native parents.[97] At issue is not blood or biological inheritance but *where* a person was born. These devices quickly erased the origins of abasimano, and few signs of it remained for their descendants. The family did not discuss their stranger origins until a couple of generations had passed. While genealogies disguised this diversity, many elders identified some abasimano, even Asi or Maasai, ancestors.[98] Many people declared that they were "pure" Nata, but when I questioned them more closely they would tell me stories of a grandmother or a great-grandfather coming from another ethnic group. Because women's names often come from their natal homes, the evidence of stranger origins is evident in the local mix of names that originate in other ethnic groups. In fact, an Ikoma elder said that it was difficult to find a "pure" Ikoma required for certain rituals.[99] As one elder put it, "the Nata knew how to swallow a stranger," that is, to incorporate them without a trace of their origins in the genealogy.[100]

The life histories of elders today suggest that western Serengeti peoples not only tolerated but also valued the stranger.[101] One Ikizu elder said that he went to live as a stranger in Nata because he had already accumulated wealth, and jealous people at home were causing him trouble. The Nata community welcomed him after he was initiated into an eldership title as a wealthy patron who might assist them in time of need.[102] Although most men married women nearby, some took stranger wives as a result of friendships with men in other localities, whose daughters they married. Other women fled their homes, sometimes with young children, exiled because of pregnancy before circumcision or witchcraft accusations.[103] Where there was matrilineal inheritance, western Serengeti men sought stranger wives because their children would then inherit from their father rather than their maternal uncle.[104] The children of an omosimano wife inherited equally with their paternal uncles at the death of their grandfather. A stranger wife perpetuated the homestead of her husband after his death instead of his brothers, as was the case in a local marriage where the brothers also inherited his widow. Stranger wives also represented important in-law connections outside the community. These were useful in travel, in trade, and to gain support in political conflicts. Nata respected and feared an omosimano wife for her outside connections and strong internal power of inheritance. The liabilities of marrying an outsider were that she may be culturally and linguistically inept and cause embarrassment to the family; in addition, witchcraft accusations most often fell to the stranger wife.[105]

Despite the inclusiveness, the elders also confirm that a certain amount of stigma came with being an omosimano. One elder chided me that it was not polite to remind someone of stranger descent or of his home territory in another place.[106] In fact, in some cases strangers were synonymous with slaves. A common way in which the community incorporated abasimano, particularly during the period of disasters and famine, was as *abagore,* or people who were bought. Droughts were often local, and when a family ran out of food their only option might be to take a child to a neighboring group where they had connections and leave the child in exchange for food. If the child was a girl, the food would be considered as bridewealth; if a boy, as a sale. Elders claim that these children were not treated as slaves but as members of the family and were incorporated as other abasimano children.[107] During the colonial years, Chief Megasa bought his son Rotegenga from Simbete parents during a famine; yet when Rotegenga later succeeded Megasa as chief, few questioned his ability to represent Nata because of his origins.[108] One Ikoma elder told about his Maasai grandfather Matiya whose family brought him to Ikoma during the Great Hunger, when all the cattle died, and exchanged him for five goats.[109] The Ngoreme either did not use the term *abasimano* at all or used it to refer to slaves, as opposed to *kicheneni* (pure blood).[110] The Zanaki, Kuria, and Lakes peoples had a specific word for slave, *omuseese/abaseese*, whose origin was probably the Seese Islands in Lake Victoria, connecting it to the nineteenth-century coastal slave and ivory trade.[111]

Many clan histories based their narratives on the arrival of a stranger whose incorporation formed a territory. Western Serengeti peoples valued strangers in the homestead as wives and sons and also honored them as great and powerful ancestors. Among the important prophets of the past, Gitaraga of the Nata, who arrived as a child with the implements of rainmaking in his hands, was a stranger.[112] Elders said that the woman who made the medicine bundle of the bees at Riyara (Materera) had to remain an omosimbe (an independent woman) but take a stranger omotware (male wife) as her lover. The spirit propitiated at Nyichoka was a stranger wife. When the community needed to consult a prophet they often went far away to find one who was efficacious. The relationship of the Tatoga prophets to Ikoma is one in which strangers have become "parents" with ritual authority in some of the most important Ikoma ceremonies to maintain the health of the land. In each of these situations, the incorporation of the power of strangers was believed to promote the health of the local community.

The testimony of elders demonstrates that this inclusive ideology of descent, congruent with the core spatial images of clan traditions, was used as a way to incorporate diverse people in a landed community where they had access to resources controlled by the clan. Though the processes cannot be dated exactly the similarity of evidence across many sources indicates that this is a generative principle of considerable time depth in the region. One indication that this pattern of inclusive kinship was at least precolonial is amazement of colonial officers in discovering patterns that did not conform to their ideas of how "tribes" should behave. One British officer who observed these methods for including strangers stated, "Natives [who] drift into a new area and are absorbed into the local population under the local chiefs [are] treated as if they had always lived there [and] regarded on equal footing with [the] indigenous population."[113] His outrage at this situation was a result of a British view of the landscape that consisted of bounded and homogeneous "tribal" units (see chapter 5). The longstanding western Serengeti social landscape placed people and defined their position as members of local communities fulfilling rights and obligations within the moral economy.

A final core spatial image embedded in clan traditions and traced in a number of other sources is that of distribution. While descent ideology allowed people to form territorially based communities, understand their relationships within those communities, and include others in that community, it also allowed people to construct regional networks of security through the generative principle of distribution. Although these networks have largely disappeared today, histories of clan migrations provide evidence for long-term regional connections between people that gave them access to a variety of resources, including ritual power. When elders tell clan stories independently of ethnic origin stories, they do not invoke images of bounded territorial subunits related to each other by descent, but rather describe far-flung networks of affiliation. The story of the Hemba, told by Ikizu narrator Samweli Kirimanzera after we had been talking about matrilineal inheritance, is an example of this kind of clan narrative. He said that inheritance changed as people married into other ethnic groups, as it had, for example, when Zanaki people married Ikizu, leading to a history of the Hemba clan that makes connections to almost all ethnic groups in the region. A general idea of the spatial pattern of the Hemba clan described in his narrative is depicted in map 5.

> Samweli: The Hemba came—they were great hunters, fierce people who lived in the wilderness. They came from the east, around the area of Kilimanjaro. When they left Kilimanjaro they went to Tamoga, Singira, and Ragega.

> Jan: Aaah haa.

> Samweli: They came and passed Ndamio. Then they came to Ikoma, where they left behind some Hemba called Gaikwe people. Others divided off from Ikoma and went to Nata.

> Jan: So the Gaikwe are Hemba?

> Samweli: They are one thing. If you hear Gaikwe in Nata it is we Hemba, we are one. When they left there, they entered Tarime, while they were hunting. These people are called Nchage, and they are also Hemba, they are our people too, the Timbaru. When they—Have you written that?

> Jan: Uh huh.

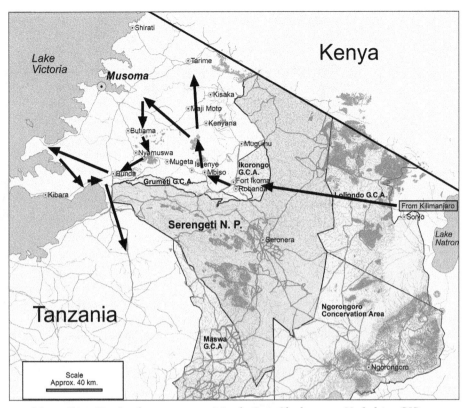

Map 5. Ikizu Hemba clan movement. *Map by Peter Shetler, 2005. Underlying GIS data courtesy of Frankfurt Zoological Society*

Samweli: When they left there, they passed here in the land of the Kenye, the Wahiri Hemba—they are our people too—in Bukenye. From there they went to Mobasi, the Asi of Kiagata. When they left there, they multiplied greatly and lived a long time and then separated again and went to Buhemba, Saani—the Hemba of Zanaki. While in the area of Saani others broke off and went to Butuguri of Zanaki. They are the Hemba of Zanaki. Again, Zanaki B. Zanaki A is here. They were hunters, going here and there to hunt. After that—clan—let's go slowly—

Jan: OK, I'm with you.

Samweli: Of the clans that were left, others divided and came to Ikizu. [Pause.] When they got to Ikizu, others dispersed, small groups went to Sizaki, Changuge, the Hemba of Changuge, the people of Mbasha Megunga, they were—one. Then other Hemba dispersed and when they got to Bigegu, they went to the area of Mwibaghi, where there were lots of animals. After Mwibaghi they came to Rindara, in Majita. When they lived in Rindara, one elder named Guta built at the place now called Guta, he died there in Guta, this Hemba man. They went to Kirio, near Guta. Then when they were finished, other Hemba of Rindara, after some years, went to Majita. That's why you hear then talking about Wiyemba, Wiyemba—they're in Majita and are the same as we Hemba. We came from one place and dispersed. Then other youth of Guta moved to Nyatwara and then to Nasa. When the Germans came, there was a youth named Kitubaha of Guta who was made Mwanangwa [headman] of Nasa. . . . All of them are Hemba, coming from up there. Those that stayed behind there and then dispersed in Tarime, they returned and are called those you hear of, the Ndorobo. They're still hiding out there, and even the foreigners can't capture them, they don't pay taxes. Even when Nyerere ruled they still didn't pay taxes. But they eat meat from the hunt, and honey.

Jan: So those that we call Asi—are they those? The same?

Samweli: The same ones.

Jan: They came from Tarime?[114]

Samweli: Ehhhee. They returned again, after coming from the east they came here, and then they went down again, they went again to the wilderness and stayed there. Don't you see? In a household, one child becomes a farmer, another a trader, another a hunter—so the household that turned back was of the hunters, they love meat, they don't want to herd or farm. Those who came down were the ones who wanted to farm and herd.

Kihenda: I have never seen a Ndorobo who was caught by the game scouts—

Jan: There where you came from in the east—where is it? Sonjo Loliondo or farther?

Samweli:	I told you—it's to the east, on the side of Kilimanjaro, there are a few hills . . . much farther east than Sonjo. Sonjo is our neighbor.
Jan:	So when the Nata tell the story of Nyamunywa and Nyasigonko, the first parents who gave birth to the Nata—they say that Nyamunywa was Asi.
Samweli:	Yes, he was of these people. This one, Asi—they used the name Asi, but when they began to disperse—you see the Hemba were the ones to discover fire. Fire was a problem then. They were hunters who ate raw meat until they discovered fire. So in some places they're called Bahemba morero, to light the fire. They're the ones to discover fire. Samongo of Ikizu was a Hemba too. . . . They shortened the name as language changed.
Jan:	So they came from over there and at each stop left people?
Samweli:	They left groups here and there and went on.
Jan:	Could you still go to those of the Ikoma and claim clanship with them?
Samweli:	The Nata and the Ikizu are one thing. Until now, even our eldership ranks are the same.
Kihenda:	Even if you go to my house you'll see that the brands for goats and cattle are the same as the Nata.
Samweli:	And he [Kihenda] is the same as me, he is a Hemba.[115]

The theme of this story, although referring to the relatedness of one family, is not the genealogical connections of a bounded residential unit but movement across the landscape among an extremely diverse group of people in many different places. As mentioned earlier, the Ikoma emergence story of Mwikoma moving from place to place before settling in Robanda presents similar images. One can identify clan stories, even those incorporated as origin stories as a migration route, by these core spatial images of dispersed pathways and networks. The sons and grandsons of the ethnic group founder move from place to place until they finally settle in their present home and become the local clans with connections to similar clans across the region. The narrators of these accounts project genealogical unity on the diverse origins of present-day ethnic groups. The listing of all these places and groups of people who live there unifies and makes connections between seemingly unrelated elements. Samweli's list cuts across ethnic lines, economic subsistence patterns, and geographical distance, reaching from Mount Kilimanjaro to the shores of Lake Victoria. In this account Asi hunters are the clan brothers of Bantu-speaking hill farmers and lake fishermen from all across the region. Mapping the points mentioned in the story produces a serpentine pathway, doubling back on itself across the region.

Clan histories were social maps that would enable the next generation to find the paths that led to people on whom they could depend.[116] There is little doubt that migration did play a central role in creating these dispersed clan associations.[117] The pathways described in the narratives were often literally the paths that people walked to return to known home communities or to those of distant relatives. Over time the connections between new, old, and still older settlements would diversify and extend across long distances. The particular connections of a few clansmen might become generalized to represent reciprocal connections between clan members in specific places, as they appear in Samweli's narration. Carole Buchanan's study of interethnic clan relations in Uganda demonstrates that people moved where they had secure clan relations, rather than randomly. She notes that clan affiliation "maximized the safe options for people in crisis and influenced groups to move less capriciously, particularly if kinsmen were known in comparable ecological zones." Clan identity provided a regional framework within which individuals could more easily adapt to new localities.[118] When western Serengeti elders talked about the moves their families had made, it was clear that they usually chose places based on the presence of close kin or friends rather than on economic resources only.[119]

Because clan affiliation was flexible and contextual, to a certain extent people could choose the associations from which they benefited. Igor Kopytoff, in his essay on the African frontier, argues that the "migratory flavor of African ethnic history" resulted from the impulse of newcomer groups to establish regional standing and legitimacy by claiming prestigious outside origins. At the same time, preexisting clans whose power was being usurped or co-opted by conquest often asserted their first-comer status as a way of maintaining some rights to the land over more powerful newcomers.[120] The chiefly Kwaya clan in Ikizu claims origins in Kanadi, Sukuma, while traditions from Kanadi testify that the connection was not a one-way, one-time event; clan pathways were established as the result of considerable traffic in both directions. Yet in oral traditions the chiefly clan in Ikizu continued to assert their outside origins, while the descendents of Muriho told stories that affirmed their ritual authority as first comers.[121] Ikoma, Ngoreme, and Ishenyi oral traditions claim outside origins in Sonjo. Within a regional context, Samweli's story connects these individually identified peoples, places, and resources that they represent, not by blood but by some symbolic identification as autochthonous Asi hunters, or first comers on the land.[122] Those who sought control over resources or knowledge in their home communities associated themselves with others in similar positions throughout the region through clan networks of first comers or powerful outsiders.

Clan praise names, designating avoidances, sometimes known by anthropologists as totems or taboos (*emigiro*, from the Bantu root *-gIdo*), use the same core spatial images to represent clans as symbolic associations of people who honor the same animal object with prohibitions concerning the totem's treatment.[123] In the past, young people sang or shouted the praise names at public dances. For example, within Nata's four clans (Gaikwe, Getiga, Moriho, Mwancha), the Gaikwe clan avoidance was the zebra, and praise names included the Asi hunter-gatherers coming from Rakana, Moturi, and Buhemba. The Mwancha (meaning "lake" or "west") clan avoidances were the fish and the leopard, and the clan named the place of Muganza, to the west. The Getiga clan avoidance was the *kunde* bean, and the clan named places in Sonjo. The Moriho avoidance was cattle; if a drop of milk spilled they touched their finger to it and put it on their forehead as a blessing. They named Bwiregi, a place in North Mara known for its love of cattle. One version of the Moriho clan praise names went like this:

> We are the Moriho, people who honor the cattle, the Iregi people of Mbisacha and Tunda: those who store freshly churned butter in the attic, together with the Iregi of cattle, coming from Itiyariro. Those who go to the fields are farmers; we took the branding iron and branded nine calves and the rams complained that they were not yet branded. We are the Iregi, who praise cattle and millet.[124]

The symbolic representation of animals, things, places, and economies allowed clans to associate themselves with other clans using similar praise names. The same clan names and names associated with clans, in conjunction with the same avoidances and places of reference, are common to many ethnic groups throughout the region. For example, references to the Iregi, named here in relation to the Moriho, are found all over the Mara region, far beyond the territorial boundaries of Nata. Ishenyi tradition says that the Iregi left Ishenyi during the disasters at Nyiberekira. The Kuria Iregi clan has a descent group called the Isenye with a tradition of migration from the Range hills near Ikorongo, in western Serengeti. Among the Southern Nilotic Kipsigis in western Kenya, a clan named Rangi takes its name from these same hills. One could extend these chains of association across vast geographic and cultural distances. As an example of the regional distribution of clan associations, avoidances, and place references, the four clans of Nata are compared to clans of other ethnic groups with which they had association around the region. As Kuria historian Paul Abuso explains, "these people who come to embrace that particular totem [avoidance] need not necessarily have the same historical origin. . . . [they only] agree to accept the myth of origin, as is indicated by the origin of the totem for themselves."[125]

Table 2.1. Regional distribution of four major clans

	Nata (Hamate)	Ikoma (Ekehita)	Ishenyi (Ekehita, Hamate)	Ngoreme (Hamate)	Ikuzu (Amagiha)	Sizaki (Ekeshoko)	Zanaki (Hamate)	Kuria (Hamate)	Other
GETIGA CLAN	Getiga (Sigera, Abatabori)	Getiga (Masgo) Hikumari also came from Sonjo, Regata, with millet prohibition, Kumari married Mbise of Serubati	—	Abatabori	Getiga	—	—	—	—
Avoidances	hende bean	—	—	hyena	kunde bean	—	—	—	—
Places	Sonjo	—	—	Sonjo, Regata, Ikoma	Urigata, Sonjo	—	—	—	—
GAIKWE CLAN	Gaikwe	Gaikwe. Gaikwe and Racha came from one mother. Himurumbe are also Asi in origin.	—	Baasi, Timbaru from Nyamongo	Hemba	Hemba	Hemba	Nchage Timbaru Kenye Abaasi	Sonjo Sagati hunter clan
Avoidances	zebra, millet, fire	zebra	—	zebra, sorghum	zebra, fire	—	—	zebra, fire	—
Places	Abaasi, Buhemba	Abaasi, Banagi, Siwi	—	—	Buhemba	—	—	—	—

MORIHO CLAN	Moriho	Serubati are Iregi of the cattle. Mbise married Kumari.	Sarega (Iregi)	Iregi	Muriho as founder, mountain of Chamuriho	—	—	—	—
Avoidances	cattle	—	—	hide rope, cattle, white-and-black-spotted cattle	—	—	—	—	—
Places	Iregi	—	—	Gwassi, Sabayaya	—	—	—	—	—
MWANCHA CLAN (meaning "west" or "lake, Lake Victoria")	Mwancha (Gosi)	Mwancha (Marakanyi) Rache also known as Gosi	Sageti, also known as Shora	Gosi, same as Gitare	Mwanza (Shora) or Mwanza Wakehwe and Mwanza Butiama	Mwanza	Mwanza	Mwanza	Shora in Mugango
Avoidances	fish and leopard	—	water, fish, mume	fish (mume or conger eel), bushbuck sisal, msuhe	pongo or bushbuck	—	—	—	—
Places	Muganza	—	—	Gwassi, Gosi, near Shirati	Watando, Mugango Wakehwe, Bumare, Ngoreme	—	Mwanza of Buturi from Simbiti	—	—

Table by author, 1998

These associations of people through a common set of symbols may have been formed by the desire to gain access to particular sets of resources, labor, and aid that they would not have otherwise shared. Historical examples show that groups who changed their clan name and/or avoidance because of misfortune, took instead the name of a prosperous clan.[126] Some peoples adopted new clan avoidances because of particular experiences. The Sweta of North Mara relate a story in which a group of baboons saved them, and so they adopted the baboon avoidance.[127] They made this choice with full knowledge of other baboon clans in the region with whom they would now be associated. Many Kuria wild animal avoidances were the same as those used in the Great Lakes kingdoms of Busoga, Bunyoro, and Buganda—states that came to influence what is now western Kenya and the Mara region in the nineteenth century.[128] The reputation of clan names and avoidances may have spread ahead of migrating people, a little like joining a prestigious club or lodge.

The Turi or blacksmith-potters of Ikizu and Ngoreme and other neighboring peoples, represent a critical example using these generative principles of access to resources through clan boundaries that are rigidly controlled yet ultimately surmountable. The elders present an idealized vision of the past in which blacksmith clans guarded access to the secrets of iron working and the iron trade, while nonblacksmith clans as patrons maintained ritual control over the land and farming. Yet in the western Serengeti, shared clans assured that blacksmiths and nonblacksmiths were never exclusive categories; endogamous blacksmith descent groups were incorporated into the clan structure of some western Serengeti ethnic groups. Turi say that they came from Geita, in what is now Sukuma, where they still go to get the raw iron, as there is no tradition of smelting in the western Serengeti.[129] In order to maintain clear social boundaries based on resource control Turi were kept ritually and socially separate from nonblacksmiths, or Bwiro.[130] If sexual relations occurred between them (or, in the past, if a Turi sat on the stool of a Bwiro or if a Bwiro picked up a Turi hammer), the offenders had to perform special rituals to protect the entire community from misfortune. In Ikizu, Turi could not accompany Bwiro to their sacred sites for propitiation of the ancestral spirits, even if they shared the same clan.[131] Although the Bwiro seemed to ostracize the Turi, elders compared the relationship between Bwiro and Turi to that between blood brothers (*aring'a* or *amuma*), among whom sexual relations and theft are prohibited. This relationship was a result of Bwiro respect for the secret power of those who work with iron. Through a ritual of oath taking, a Turi could become Bwiro and cross the ostensibly rigid boundary. Elders say that the Turi gave up political power to the Bwiro in return for the economic monopoly over iron working.[132]

Ultimately, the most important test of the ability to create distributive networks of security inherent in these core spatial images came in times of famine. When all other strategies failed, elders said that western Serengeti peoples relied on the practice locally called *kuhemea*, asking for food from a neighboring area that harvested well that year. As elders describe the practice today, people needed only go where they had a "friend"—that is, someone who would agree, or had obligation in the moral economy, to share food. People went in small groups, some finding "friends" when they arrived. They went to places where they had connections, especially through kinship, because kin might give you food without an exchange. One could also call on clan ties, even if rather far removed.[133] When a man went to trade for grain (millet, sorghum, or more lately corn and cassava to cook ugali), he took animal products, either domestic or wild, especially dried wild meat called *ebimoro*. One Nata elder quoted a proverb: "When the famine comes, each person digs a hunting pit."[134] This means that that people relied principally on hunting resources to trade for food in exchanges with neighboring peoples. The elder explained that "digging a hunting pit" can also metaphorically mean getting friends who will make these exchanges with you. Another elder said, "If hunger comes, you have to know how to hunt."[135] Although hunting products were most common, people also exchanged livestock and even children in times of extreme hunger.[136]

Since neither formal markets nor a class of professional traders existed in the western Serengeti, elders remembered that trade depended on similar personal links of friendship.[137] When a man arrived in a strange place with his trade goods, he needed a secure place to sleep, eat, and keep his wares. He made these arrangements with a previously established friend who would help him find others with whom he could trade his goods for the things he needed. His friend acted as an intermediary in a strange culture where people spoke a different language. In return, the trader left some of his goods with his friend and invited him to visit. The host visited his friend on another occasion and brought his own things to trade, expecting the same hospitality. The Nata words *kokeráni*, to exchange goods, and *kokérani*, to greet, are the same except for the accented syllable. This might indicate a play on words that shows the close connection between friendship and trade.[138]

Elders affirmed that men guarded their friendships more closely than kin relations because they were voluntary.[139] Because the trading system depended on these relationships, men often secured these friendships ritually through an oath. The most common way was through the ritual of blood brotherhood or by marrying their daughters to friends and forming a secure relationship between in-laws.[140] In this way kinship ties were again constructed to cement obligations within the moral economy. Whether the practices of kuhemea or kokeráni were based on preexisting kinship ties or new connections between people who became "friends" during the famine exchanges, they reinforced cross-ethnic and cross-territorial ties that were necessary for the ongoing food security of the region. People were willing to exchange their grain for dried meat because this created reciprocal obligations within the moral economy that could be called on when they faced hunger themselves or needed support for other reasons from their "brothers."

The generative principles inherent in this social landscape of regional clan networks and distribution that allowed people to find security was also used to gain authority in a system of unequal power relations. The linguistic data shows that western Serengeti peoples, like others throughout the Mara region, conceptualized prophecy and healing as the power of distribution and disbursement, rather than concentration and accumulation. The general word for prophecy, healing, divining, and rainmaking throughout the region is *obugabho*. This word is derived from the old Great Lakes Bantu root, *-gàbá, -gabira,* "to divide up, distribute," usually in the sense of one who gives big feasts or gives things away generously.[141] Mara languages use the verb *kugaba* only in reference to the division of inheritance and the noun form only in reference to ritual specialists such as prophets and rainmakers. In the Mara region, *obugabho* was not a gendered term, and some of the most famous rainmakers, like Muse (Zanaki) and Nyakinywa (Ikizu), were women, but all had access to their power as clan members. Other Great Lakes Bantu speakers use this root in reference to one of the oldest forms of authority in which "big men," and later chiefs, parceled out land among other resources to their clients in return for protection and ritual guardianship of the land.[142] The restriction of this word to prophecy and healing may indicate the value of prophets who shared power and blessings they gained through clan pathways and through their own idiosyncratic learning from multiple sources.

Prophets, healers, diviners, and rainmakers were some of the types of western Serengeti leaders who gained authority from specialized ritual knowledge within distributive clan networks. Although one could "buy" specific medicines, the power of prophecy was always a result of clan or descent connections. A particular ancestral spirit chose the person within the kin group to become a ritual specialist and actualized the individual gifts and inclination of the initiate by directing him or her through dreams. The knowledge base that this person tapped into was that of the dispersed and diversified clan. Using this kind of power to consolidate hierarchical authority was difficult, yet these specialists were the most influential in a society without chiefs or kings.[143] The glimpse we get of rainmakers through sources from the early colonial period demonstrates that their authority was regional and dispersed rather than localized and concentrated. Rainmakers could demand tribute in kind and in labor and had a powerful voice in the affairs of the people. Nevertheless, they were experts whom people called on in times of need rather than proactive agents of authority. When the rain came late or ended prematurely, the people called on rainmakers, often from other ethnic groups, for help.[144] In the emergence story, the Ikizu gave the head rainmaker authority only as she agreed to the ultimate authority of local institutions, demonstrating that the local community controlled the specialist as much as she or he controlled the community. Early colonial reports told of community sanctions taken against rainmakers who did not produce rain. In Ngoreme, when the rains were too late, the highest-ranking women would call together other women by beating a drum and would march off with them to hit and curse the rainmaker or take cattle until he or she promised to send rain.[145] In one colonial case the rainmaker was placed on a high rock in the sun until the drought ended or, as the officer added, until his "death from starvation and exposure proved him to have been an imposter."[146] Each community also had rainmakers in other areas to turn to, as well as their own sacred places where they propitiated the ancestral spirits for rain. No one person or office monopolized these powers.

Local languages used the term *omwame* (or *omunibi*, particularly for cattle wealth) to refer to another kind of leader as a patron, a wealthy man whose authority was based on his ability to distribute his wealth and to feed a large crowd of people at a feast, a community ritual, or a dance.[147] The expectations for what a patron would do for the people were rooted in understandings of a moral economy of reciprocity going back to the old generative principles of the social landscape. Elders tell stories of important past settlements named after wealthy men who represented a period of prosperity. Leaders who were selfish and did not feed the people brought destruction to the community. This is the lesson behind the story of the Sagarari clan that used to live on Kyamarishi Mountain, now within Serengeti National Park. Kyamarishi was the name of a very wealthy man who had so many cattle that he threw their milk into the river. But the rest of the Sagarari were also very greedy and did not want anyone to have their wealth. When people would come to visit and share a meal at Kyamarishi, the Sagarari would escort their guests to the river and then make them vomit before they left so they would not take the food away with them. As a result the Mbilikimo, autochthonous hunter-gatherers, defeated them in battle, and they had to leave and go live among other clans who soon assimilated them. There are no Sagarari clan members left today. As a result of their loss of prestige, the Sagarari may have been absorbed by the Serebati clan, who also lived on Kyamarishi Mountain and have gone back there to propitiate the spirits in recent years.[148] Hierarchical leadership never developed in the western Serengeti and leaders gained power based on their ability to distribute wealth, remaining accountable to a network of dispersed people on whose welfare their power ultimately depended.

By envisioning the landscape through the core spatial images of distribution, inclusion, and diversification since at least 1000 CE, western Serengeti peoples have learned how to live with unpredictable rainfall, varied soils, and low populations. The way they farm, build their homestead, organize their settlement, or assume leadership inscribes this way of seeing onto the land itself. They gained food security through strategies based the imaginative landscape that placed regional resources within kinship networks. In doing so, they spread out their risks by investing in a number of different economic activities and sources of labor, keeping mobile, and exploiting resources over a large area of land. They learned not only how to use descent ideology to bind people into relations of reciprocal obligation within a residential territory but also how to include newcomers. Regional networks assured security and leaders who would provide for their people in times of trouble. Although periodic times of hunger came, people developed ways of providing subsistence, and although they did not develop wealthy urban cultures, neither did they develop enormous class inequalities. However, the moral economy in which leaders must distribute wealth to their people developed out of unequal power struggles, whether in the form of strangers sold as slaves or the acceptance of other forms of dependence.

⇆

As Wilson Machota and Edward Kota, the elders with whom I visited old settlement sites that summer, surveyed the distant landscape across the hood of the Land Rover in the game reserve, they excitedly pointed out each past Ikoma settlement area in terms of its clan association. However, beyond the memories of elders like these, clan networks do not function in this way anymore, and new ways of imagining the landscape resulting in ujamaa villagization, game reserves, and village-controlled wildlife management areas (chapter 6) are obliterating the physical inscription of this way of seeing. Because these landscapes appear so "natural" today it becomes increasingly difficult to learn to imagine a different way of seeing and to find social meaning available only by learning the histories of clans, remembered subsistence strategies, and how the networks of obligation were mobilized in the moral economy. Yet our understanding of the natural environment is so much richer and deeper when we can also read its human history on the landscape.

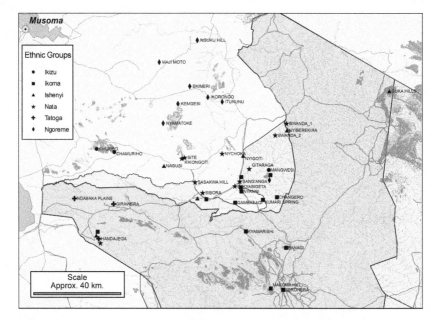

Map 6. Abandoned settlement and emisambwa sites. *Map by Peter Shetler, 2005.*
Underlying GIS data courtesy of Frankfurt Zoological Society

3 ᔓ Sacred Landscapes

Claiming Ritual Space of the Ancestral Land (Emisambwa),
ca. 1500 to Present

IN THE HISTORY of memory, a third historical way of imagining the Serengeti in chronological sequence is as a sacred landscape marked by specific places
in the wilderness where powerful spiritual forces remain accessible to those
seeking fertility, healing and protection. The core spatial images of this landscape—sacred places of power and encirclement of territory—are available in
the cryptic oral traditions that appear as simple lists of place-names, the rituals
performed at those sacred places and the walk performed by the generation-set
to heal the land.[1] The place-names can often be identified as the sacred places
where the ancestors of power, who have become synonymous with the land itself,
are buried. Descent groups responsible for the rituals that mediate between the
forces of the wilderness and home, propitiate these spirits for health and prosperity,
thereby establishing their claim to a land peopled by the ancestors. Generation
sets ritually repossess and define the territory of the community by walking the
boundaries of the land (ekyaro) and planting protection medicines. Western Serengeti peoples have used this way of seeing the landscape to unify people in a
common spiritual link to the protective forces of the land and also to legitimize
the elders' claim to unquestioned authority. This way of seeing the landscape
also changes the appearance of the land itself. Because the sacred places of
power must remain wild, people are prohibited from cutting trees or disturbing
the vegetation here, leaving sacred places, like small parks, in their "natural"
state.[2] By contrast, the ritual encirclement of territory in the generation-set walk
follows the indistinct boundary where wilderness and domestic space meet without leaving a visible mark.

After telling the stories of the emergence of ethnic groups and clans, elders
often immediately proceeded to recite a list of place-names, representing the time

period following first man and first woman after the clans moved out into different settlements. In Ikizu, where many in-migrating groups have come together, the list of place-names refers to the places where different immigrants settled. This is an example from the Nata: "We are the people of Gitaraga and Mochuri, Rakana and Moteri, Sang'anga and Kyasigeta, Torogoro and Site, Magita and Wamboye."[3]

The cryptic meaning of this list was not apparently discernable until I began to identify the social groups and ritual meaning associated with place-names by visiting the sites. As we walked out to those places elders spontaneously told other related stories and identified the socially significant elements in the landscape—each rock outcropping, hill, and stream with its own history. Places served as mnemonic devices to remind men of the stories behind them. The land is a text of history, and walking over it with the elders who tell their stories is an act of reading the past. The exercise of walking the places showed me that place-name lists referred to known and identifiable places where ancestors were buried with real gravesites, rather than to the mythical or forgotten places of the emergence stories of ethnic groups or the larger territories of clans. Men who had a thorough knowledge of local geography from hunting, trading, or raiding trips located these now uninhabited places in a wilderness without roads or maps. The spirits of those ancestors resided at these places and had to be propitiated by descent group members to bring security and prosperity to the living. As long as people remember and visit these places, mediating the wilderness forces, they will also remember the ancestors and their histories.[4] Without a physical connection to these landscapes, history is forgotten.

The land, however, is not only protected by ancestors in a landscape of isolated gravesites, the sacred places, dotted throughout the wilderness. The generation-set at their retirement ceremony performs a ceremony (*kukerera*) to encircle the boundaries of the ekyaro territory (understood as clan land) with a ring of protection. Although descent groups control access to land and natural resources, the land could be protected only by the much more inclusive generation-set group in power at the time. By telling stories about the performance of this ritual, and in the performance itself, people imagined a sacred landscape inhabited by the spirits of ancestors, protected by walking the boundaries and ritually planting the medicines of power. The protection of the ancestors created the ritual space of sacred places of power and the territorial identity of a people encircled within those wilderness boundaries.

Like the social landscapes of chapter 2, this landscape comes from the time frame of stories from the middle period of social process, or cyclical time, and is not precisely datable since it is rooted in older generative principles. These ritual patterns might go back to the time when the present form of generation-sets was established, around 1000 CE, or to the time when the Mara language groups were differentiating as these communities grew more isolated from one another, about 1500 CE. That date is a logical point for the beginning of the prominence of this landscape since the linguistic groupings had differentiated themselves as they identified themselves with and made claims on the land of their ancestors. Yet, unlike the clan traditions, we can follow specific references in the oral tradition of place-name lists to ancestors who lived in the early part of the nineteenth century, just before the period of disasters narrated in the next chapter. The location of the settlement sites from these traditions, both within and on the boundaries of the park, demonstrates the range and movement of western Serengeti peoples just before the disasters of the late nineteenth century and sets the immediate context for the historical traditions in the next section, even though this is a landscape of historical depth and is still in use today.

Oral traditions use the core spatial images of wilderness sites as sacred places of power inhabited by the ancestors who have become *emisambwa* spirits of the land. The rituals conducted at those sites that use the same core spatial images bring prosperity by mediating the dangerous forces of the wilderness with the civilized spaces of the domestic community and claim the land by peopling it with guardian ancestors. Descent groups are responsible for the ancestral sites, yet the more inclusive generation-set is responsible for the rituals of walking that heal the land and define its boundaries by employing the core spatial image of encirclement. Although various social groups have recontextualized generation-set rituals, the ritual itself is based on core spatial images consistent with oral traditions that have remained stable over time and over a space throughout the region. The walk of the generation-set unifies the community to bring health and protection under the orderly succession of the generations, establishing territorial identity through ritual space. While the territories themselves have changed drastically over time, the ritual process for defining territorial identity has not and must be recontextualized in each generation. Western Serengeti peoples use the sacred landscape to define and claim a territory and an identity in connection to the land. The symbolic meaning of the core spatial images in these rituals are unraveled in the generative principles of encircling, binding, and containment used to overcome internal divisions, bless, purify, and define the community, as well as to legitimize the authority of the elders. This view of the landscape makes a profound connection of people to the land, through their ancestors, that is necessary for the health of the land.

PEOPLING THE LAND WITH ANCESTORS: THE SACRED SPACE OF EMISAMBWA SITES

In trying to discern the meaning of the place-name lists in oral tradition, talk with the elders makes it clear that there is a particular descent group, or lineage, identified with each site and responsible for performing rituals of propitiation to the ancestors. As was discussed in the previous chapter, the lineage derives from a remembered ancestor at a known distance of remove. Elders recite their line-age as the son of X, who is the son of X, who is the son, and so on, often back to five or six generations. While elders identify some of these sites with the Aba-maina, Amatara, or Amasura generation-sets between 1850 and 1900, firm dates are difficult since they may have been abandoned and reinhabited over a long period. The memories of these sites may thus be telescoped into the memory of one generation, as is common with oral traditions of the so-called middle period, the time between the origin stories and the emergence of a clear historical nar-rative that can be accurately dated. Although many elders could not date the sites (they assured me it all had happened a very long time ago), these are among the earliest generation-set names that people remember in connection with spe-cific ancestors. Because the elders often use generation-set rather than the age-set names to refer to these sites, it seems more likely that the sites were in place before age-sets came to higher prominence, in the late nineteenth century.[5] These oral traditions of this imagined landscape then tell of the generation of ancestors in living memory who established recognized claims to the land. The list repre-sents the core spatial images of sacred places of power.

Identifying and mapping these places also gave further insight into their mean-ing. Consistent with landscapes discussed in chapter 1, these sites are most often located on hillsides and some farther to the west than the emergence sites. A se-lect few of the many sites that elders identify today are included on the map at the beginning of the chapter to demonstrate the extent of the sites of particular ethnic groups and the connection between them. Although many sites associated with the same ethnic identity today are grouped together, their placement does not seem to delineate ethnic territory as we now know it. If ethnic groups existed at all, they did not control a bounded and exclusive territory, nor were they iso-lated from each other. While the Ngoreme sites are more distinct, those of the Nata, Ikoma, and Ishenyi do not appear in distinct areas. Different ethnic groups claimed some of the same sites, such as Mangwesi (Bangwesi) Mountain, indi-cating the possibility of common roots and a later reformulation of group iden-tity. In addition, the sites claimed by different ethnic groups are mixed together across the landscape, with no bounded territory discernable. Although Nata elders often claimed that they were reciting the place-names in chronological sequence of the order of settlement, that ordering is also in geographical sequence, as one would experience it on a journey. What seems most likely is that the boundaries of different ekyaro territories changed a number of times as different groups came to control the land or as the groups themselves reconfigured. Mapping many of these remembered settlement sites gives some indication of the extent of west-ern Serengeti settlement before the disasters of the nineteenth century (chapter 4). It is clear from these maps that people used to live on land that is now part of Serengeti National Park and the Ikorongo-Grumeti Game Reserves. They lived in settlements as far south and east as Seronera, now the site of the park village and lodges and as far north and east as the Guka hills or Nyiberekira.

A few of the places in the list of place-names are simply old settlement sites or what the Nata call *ebimenyo* (lit., built places), where people remember specific ancestors and events of the past but do not propitiate the spirits there, nor do these spirits have power over the health of the land and its people. Elders remembered the sites from their particular descent group or lineage. Those remembered best seem to have been the most prosperous settlements of important or wealthy individuals. Oral traditions often name places after the well-known individuals who lived there, without formal title but with charismatic ability to attract people through their patronage. These leaders were often called speakers (*omukina/abakina, omwerechi/abawerechi, omugambi/abagambi*), men whom people respected for their ability to speak the mind of the community with wisdom and fluency.[6] The place-names Magita and Wamboye in the Nata list refer to such wealthy patrons and their settlements. The wealth and prosperity of these sites may also be associated with the generation-set that lived there. Elders remember Torogoro and Site because these settlements were so prosperous, with lots of food and leisure time to dance. The Abamaina generation danced so much that the youth pounded the dance field at Torogoro into a depression that one can still see today.[7]

Many more of the sites were connected to ancestors who possessed the power to provide protection and security for the living. The word for the ancestral spirits of a family is simply *omokoro/abakoro* (derived from the word for big, elder) or *ekehwe/ebehwe* (ghosts; shadows).[8] The dead were a part of the community and their descendants maintained relationships to them with as much care as they gave to relationships with living people and with the same possibilities for benefit and sacrifice. The word for the burial in Nata is *kutindeka* (to store), meaning that a deceased person has not gone away but has simply taken on a different form. Western Serengeti peoples buried their dead in the homestead and abandoned the graves when they left the settlement. Families or descent groups did not maintain common burial plots through the generations. For at least two to three generations, however, people were expected to remember the grave sites of their ancestors (*kusengera*, to beseech), return to the sites (which might now be in the wilderness), clean the graves, and offer gifts there at least once a year on the anniversary of their death. People also remembered their ancestors by naming children after a deceased grandparent or by communicating with them in dreams. If the family no longer used an ancestor's name or forgot his or her grave site, the spirit passed into the more dangerous realm of "loose" spirits, without community moorings.[9] When problems occurred in the homestead, such as illness or death, the head of the homestead consulted a diviner, who often diagnosed the misfortune as the result of forgetting the ancestors.

These sites became sacred places of power when spirits connected to re-membered settlement sites were different from the abakoro, or ancestors in general. They were the *erisambwa* (singular)/*emisambwa* (plural) spirits of important ancestors, often those who had been efficacious in life as rainmakers or prophets. These spirits were always connected to the land, and people propitiated them for rain, protection, and fertility. *Emisambwa*, in East Nyanza languages, derives from a Great Lakes Bantu root, *samb-(ua)*, territorial or nature spirit that protects first comers (often represented as an agnatic group). In other Great Lakes Bantu languages it means variously nature spirits of rivers, spirits attached to larger lineage groups and to areas associated with these groups, clan spirits and habitat for them (wild animals, rivers, etc.), and the protective spirit of a settlement (Rutara).[10] Both Michael Kirwen, who interviewed many diviners in the Luo and Zanaki areas of the Mara region, and Malcolm Ruel, who studied Kuria culture, interpreted emisambwa (*abasambwa* in Kuria) as forgotten, and potentially dangerous, malevolent spirits.[11] The western Serengeti elders with whom I spoke consistently referred to the emisambwa as powerful but nonetheless beneficent spirits.[12] The shift in meaning among peoples of the western Serengeti—from territorial, nature, or malevolent spirits to ancestral spirits that guard the land—suggests a particular kind of relationship to the land, one that may date to as early as five hundred years ago, when North and South Mara languages diverged.

Using the generative principles embedded in the core spatial images of the sacred places of power, western Serengeti peoples have combined the idea of spirits of particular important ancestors (abakoro) with the spirits of particular places (emisambwa) to the point where they are now indistinguishable as emisambwa. This explains why western Serengeti peoples perceive emisambwa as beneficent—because they are known ancestors, not forgotten and lost spirits. The elders themselves were not clear whether erisambwa meant the spirit of the ancestor or the spirit of the place, as the two meanings have become synonymous. The Ngoreme, who are geographically and culturally much closer to the Kuria than other western Serengeti peoples, joined these different conceptions of emisambwa. Many of their emisambwa did not demand sacrifices but represented places remembered for important events where people gathered the elements necessary for the ritual. Some of these sites, though unconnected to stories of ancestors, were inhabited by animals or monsters and marked the place where a particular descent group still brought offerings. The Ngoreme also had emisambwa that were both ancestors and spirits of the land. One of the most famous Ngoreme emisambwa sites was the hot springs at Maji Moto. An elder of the Kombo lineage, the group that was ritually responsible for this site, said that the people of a whole village lived under the water. Sometimes people heard a child cry at the site. The story goes that when the colonial government came to explore volcanic activity at Maji Moto, the earth swallowed up all their large machinery and those

Figure 3.1. Ngoreme emisambwa, Tatoga rocks, Sang'ang'a Buchanchari, Ngoreme. *Left to right:* David Maganya Masama, Wambura Nyikisokoro, Mayani Magoto. *Photo by author, 1995*

who did the work died once they got home. When the machines drilled into the rock, they brought up blood.[13]

Although the Southern Nilotic–speaking Tatoga herders denied that they had any emisambwa sites like their Bantu-speaking neighbors, many examples suggest shared regional understandings of ancestral spirits that cross cultural and linguistic lines. In Ngoreme a local farmer showed me a pair of large rocks worn smooth, presumably by the touch of hands, that he described as a Tatoga erisambwa to which Tatoga periodically returned to perform the sacrifices. Likewise, as Tatoga elders remembered the deeds of great prophets of the past, so the landscape was appropriated with their relics. For example, Tatoga elders say that the stone axe of the Relimajega prophet Gwataye is still embedded in a tree near the Mara River.[14] The hunter-fishermen Tatoga Isimajega have a profound attachment to the mountain that they call Somega, in the western corridor of Serengeti National Park, now known as Simiti Hill, across the Grumeti River from the Girawera game post. The Isimajega called the spring there Yiwanda, after the rainmaker prophet Ghamilay, who is buried there. People went there to ask for rain, fertility, health, or prosperity.[15]

Emisambwa belong to a polysemous category used in a variety of other circumstances, all of which connect people to ancestors who will help them.

Emisambwa spirits could also reside in particular objects or animals and might appear at the site as a snake or a hyena. Some elders called these animals the messengers of the erisambwa, others simply the erisambwa itself. One elder differentiated the "big" emisambwa as those propitiated at sacred places and the "little" as animals associated with descent groups or clans, the emigiro or avoidances that had particular prohibitions discussed in the last chapter.[16] The most common emisambwa animals were a particular kind of snake (often a python), a hyena, or a tortoise. Some Ikoma descent groups with the hyena erisambwa had a special gateway cut in the homestead fence for the hyena to enter. The erisambwa at Nyichoka was a snake, but also a barren woman.[17] Emisambwa clan totems were more than a symbolic representation of the spirit of the collective, they were the located spirit of an ancestor who provided for the welfare of that clan and the health of the land on which they lived. Ritual specialists such as healers, prophets, diviners, and rainmakers also had their own emisambwa, or ancestral spirits that directed their work through dreams, hence the word for prophet, omoroti/abaroti (one who dreams, from -rota, to dream).[18] Each erisambwa had its own rules and prohibitions, including carrying a special ornament or implement used by the ancestor, also called the erisambwa.

The mobile Ikoma erisambwa, a set of elephant tusks gained from consulting the Bachuta Tatota prophet Machaba, near the Ngorongoro Crater, further demonstrates the adaptability and flexibility of these generative principles, through which people maintained the fertility of the land. Although the Ikoma descent groups had their own emisambwa sites, at a certain point around the mid- to late nineteenth century, these sites became recontextualized as the collective Ikoma erisambwa, the Machaba.[19] The story of how the Ikoma got the Machaba takes place in a time of disease and famine, when the Ikoma and the Ishenyi were looking for a way to reverse the decline of their people.[20]

> They went to the Tatoga prophet, east in a crater but not Ngorongoro, another one near Mbulu called Mwigo wa Machaba. There was a lake in the crater. They went there because they had a problem with fertility. The Ishenyi, who were more numerous than they, came along too. The Ishenyi slept at the first place inside the prophet's gate, the Ikoma slept outside the gate. The prophet said they should grab a sheep as the sheep jumped over the gate. Ikoma got a skinny one, and Ishenyi a fat one; but when they butchered them, the sheep looked the same, and when they were cooked the Ikoma one was fatter. The prophet tried each of their bows and shot the Ikoma arrow far off and said they should follow it. He prepared the things that they should take along with them [mbanora] and showed them the path to take. When they saw vultures up ahead, the youth ran ahead to get the prize. The first to get there was Mayani

[a Gaikwe clan member of Ikoma], who took the top [right] tusk of the elephant, and second was a youth of the Ikoma Himurumbe clan who took the lower [left] tusk. The Ishenyi wanted to take it from them but the prophet had said not to fight. They tried to take it away but could not lift it or move it. [So the Ikoma took it home and kept the Machaba.] The Ikoma were at Tonyo at that time.[21]

A report by the Game Preservation Department in 1936 describes the story as taking place when the Ikoma, Ishenyi, and Nata went to Meatu, Sukuma (rather than to the Tatoga, near the crater), "to visit a famous witch-doctor, as their own tribes were dying from disease. They were told to return but on their way back they would find a token which they were to take."[22] At all important communal rituals the Ikoma brought out the Machaba tusks, and the people received a blessing by touching the tusks, which were smeared with butter. The Ikoma clans divided into two clan moieties, Rogoro (east) and Ng'orisa (west), with each moiety guarding one tusk.[23]

The Machaba story is part emergence story, part clan story, and part emisambwa story, recontextualized in the context of the mid-to-late-nineteenth-century disasters. The stories and ritual practice surrounding the Machaba define relationships between Ikoma clans as well as with the Ishenyi and Nata. In Ikoma the clans control the Machaba emisambwa. The Gaikwe (Ng'orisa, or west) have the right-hand, upper, or male, tusk, confirming this clan as first comers, similar to their status in the origin stories as the clan of first man, the hunter. The Himurumbe (Rogoro) control the left-hand, or female, tusk, often related to first woman, the farmer, from Sonjo (to the east). One elder noted that the Himurumbe clan was also Asi, or hunter-gatherer, in origin. These clans have special ritual functions when they bring out the Machaba; they also keep the tusks in separate places, one the east and one the west. The story also explains why the Ishenyi have remained such a small group while the Ikoma have become relatively larger. The Nata have a version of this story in which they, too, went along, and the prophet gave them a set of buffalo horns that were later lost when one group failed to pass them on at the proper time. The Machaba story illustrates the flexible and situational ways these stories have continued to define imagined landscapes in a variety of contexts.

Because the Machaba are ancestral spirits of Tatoga, rather than Ikoma, origin, the groups have a very special relationship. One Ikoma elder said that the Tatoga were "people of the oath [aring'a, our parents]."[24] The Ikoma commemorated the Tatoga role as spiritual parents in ritual. In one version of the origin story the first Ikoma man from Sonjo came because a Tatoga prophet told him to follow the animals until he found a place where lions lived, where he should then stay.[25] The Machaba as a Tatoga spirit is a mobile, rather than a located, erisambwa that dwells in the tusks, instead of a physical feature of the land. Yet the Ikoma domesticated this spirit, appropriated its power, and fixed it to the land by carrying the Machaba in the generation-set walk to seal the boundaries of the land.[26] In addition, they cannot take the Machaba across the Grumeti River, the "traditional" Ikoma territorial boundary. Among western Serengeti peoples, the Ikoma live farthest east and farthest out on the plains. The village of Robanda clusters around a hill that rises out of an otherwise flat and feature-less plain. The ecological setting suggests that the only way to prosper on this kind of land was to appeal to the spirits of those who own the grasslands, the Tatoga herders, but to transform that power using the core spatial images in the culturally recognizable form of the emisambwa.

The emisambwa ancestors are a real and personal presence in the commu-nity. For example, when I went with the elders to visit the grave of Gitaraga, a rainmaker, they did not address the spirit with formal ritualized speech but spoke to him as one would speak to a living person. We sat for a while at the grave site, talking, and then went to the place where the rain pots were buried in the ground under an overhanging rock. After discussing whether I should take pho-tos here and disallowing the use of pen and paper,[27] the lineage elder poured water out on the ground from a gourd he had brought along and spoke to Gi-taraga: "Mzee [elder] Gitaraga, we have come to greet you, we are your children, do not be angry with us but send us blessings, do not be astonished that some others of your children have come to greet you. They have not come for a long time, but they are nevertheless your children. They are from across the ocean."[28]

On the way home that day it poured down rain, indicating his blessing. On the other hand, another time when an elder that took me to the Nata emisambwa site at Geteku, he did not perform the prescribed rituals. The elders blamed drought in the following year on this incident.[29] A fine was paid and a ritual per-formed to appease Geteku.

The association of emisambwa core spatial images of sacred places of power with forces of the wilderness, as opposed to the civilized spaces of the homestead, provides further insight into the generative principles that identify the spirits of the land with the spirits of the ancestors. People could not cut the groves of trees that grew up around these sites, fostering untamed growth. Emisambwa sites were always located away from villages, in the bush, even those in the more densely populated areas of Zanaki. One elder described these as places inhabited by leop-ards and snakes and where lions give birth.[30] Traditions associate these places with the ritual symbols of water, fertility, women, and growth. Many other an-thropologists and historians of Africa have noted the recurrent ritual theme of mediation between the forces of the bush and the forces of the home. Steven Feierman and Randall Packard demonstrate the role of the king or chief in East African societies as an intermediary between wilderness and culture.[31] The peo-ples of the western Serengeti assigned this role to a variety of important ances-

Figure 3.2. Nata emisambwa site, Gitaraga. *Left to right:* Keneti Mahembora, Mokuru Nyang'aka, author. *Photo by Peter Shetler, 1996*

tors, located at specific places of power, to mediate between wilderness and culture. Because sacred places are often kept wild throughout Africa, archaeologists have identified these groves and found evidence of ancient burial grounds or abandoned settlement sites while environmental historians have looked at sacred places as an indigenous version of "parks" or protected places.[32]

Rituals, like oral tradition, use core spatial images as a way to communicate generative principles embedded in the landscape. The Nata ritual at the erisambwa site of Gitaraga is a clear example of the ongoing role of these spirits in mediating the dangerous but fertile boundary between wilderness and culture that makes habitation possible. When a problem arose, the community asked the lineage elders to do the sacrifices at Gitaraga so that they might have rain. Only men of the Abene Ogitaraga descent group performed the rituals. Taking along a black sheep and one young woman carrying a gourd of water and dressed in traditional skins and beads, they climbed the mountain to clean out and refill the rain pots. Elders sacrificed, roasted, and ate the black sheep, cutting it in half for the spirit Gitaraga and his wife, Nyaheri, at another nearby site. Then a young man climbed the tree (*omusangura*) near the grave of Gitaraga and poured water over the head of the young woman, bent over below the tree. While the others waved branches cut from the tree and sang songs, a descent group elder asked Gitaraga to send rain. Similar rituals happened at the other sites where people said that if the erisambwa was happy with the ritual they heard the beat of a drum (*ambere*).[33] These rituals reenacted the relationship of people and land. The young fertile woman standing beneath the tree represented the land receiving from above the male-spirit rain. The women danced the *eghise* to please the spirit who had blessed the land with rain.

The Ngoreme ritual at Kimeri and Nsoro, two springs up on the hill behind Maji Moto, also used core spatial images that marked the emisambwa as spirits associated with the wilderness, whose power is mediated by the ancestors. The Gitare descent group cleaned the springs periodically. They took tobacco, milk, and honey for the prayer, while the women took flour and the men a white-tasseled goat. They brought home with them the leaves from certain trees for rituals and water and white mud for the circumcision ceremonies.[34] The Gitare performed these rituals in times of trouble, such as lack of rain, ill health, infertility, or threat of enemies, rather than at regular intervals, and at the initiation of a new age- or generation-set. The new set went there to receive the blessing of the spirits for a prosperous period when their age-set was in power. The people brought the products of their labor on the land and offered them back to the spirit that made prosperity possible. In turn they took the powerful things of the wilderness (leaves, clay, water), now made safe for use in the civilized world by the spirit of the land, back to perform community rituals. People were prohibited from using anything but traditional implements at the emisambwa springs or pools.[35]

By collapsing the meaning of specific dead ancestors into the concept of a territorial spirit of a place at these sacred places of power, peoples of the western Serengeti made claims to the land they occupied and made a profound identification of themselves with the land, in the form of their ancestors. The descent group responsible for the emisambwa sites had a special connection to the land because their ancestors first claimed and gained ritual authority over the land by making a ritual accommodation with the land (perhaps through association with first comers like the Asi hunters).[36] The Ikizu described the emisambwa as both prophets and as spirits whom their founder, Muriho, "planted" there when he conquered the land. The Ikizu elders who wrote a book on their history listed twenty-one sacred places inhabited by the emisambwa. Out of that list more than half the ancestors came from outside Ikizu; for example, Nyambobe was a Luo woman who came in a boat with potatoes and bananas.[37] Because of the association of these places with symbols, like the python, that are ancient throughout the Bantu-speaking world and commonly used at sacred places like shrines, it may be that these are ancient sites of power that have changed hands many times as new groups took control.[38] However, they are now associated with fairly recent ancestors. One cannot become a people without a land, and the land must be ritually possessed rather than simply occupied. It becomes "our land" through the practice of naming it and burying ancestors there.

Some might wonder how people could claim land through grave sites when they moved every five to ten years and as many as ten times in a lifetime. And indeed, historians of the Kuria and Luo have confused mobility with a lack of territoriality or feeling for the land.[39] The sense of territory must be defined by the generation-set because it is ever shifting and contextual. Western Serengeti peoples might choose to claim as their territory any piece within the expanse of all land where the graves of their ancestors are located. People possessed the land, not in terms of ownership, but rather by identifying with it by ritual occupation, or by peopling it with the spirits of the ancestors who continue to respond to propitiation. That is why oral traditions represent the land as empty before first man and first woman came—because their own ancestors did not people it. The ancestors of others had to be either be expelled or co-opted in order to live peacefully there.[40] The land was empty not when it lacked people but when no emisambwa dwelled there.

The story of Muriho's taking possession of Ikizu is the most vivid example of the process of peopling the land with the ancestors among the western Serengeti narratives:

> Muriho himself was a healer and a prophet who had been promised authority in the area where there was a tall mountain. He built his homestead right there at Itongo Muriho. The goal of Muriho was to go onto Chamuriho Mountain, but he was not initially successful because there were people living around the mountain called Mbilikimo [dwarflike people], or in the Ikizu language, the Hengere or Nyawambonere. Muriho lived here with his followers while he was searching for a way to overcome the Mbilikimo. The Mbilikimo themselves also found out that someone else was living in the area of Rosambisambi. So they went to harass him, but when they got near to the homestead they could not see it nor any people. This is because Muriho had protected the homestead—he had encircled it with the medicine of the orokoba—which has the ability to make things invisible. Muriho then began his own plans for harassing his enemies. First, he went back to the mountain and passed the orokoba medicine around the base of the whole mountain and planted his spirits. The total number of spirits and their characteristics are described here. . . . Each of these five spirits he placed in his own stream and each stream is known by the name of the snake, which remains there even today. After placing the spirits in each spring, the water became bitter, so that the Mbilikimo could not drink it. The spirits were the weapon to defeat the Mbilikimo when they needed water and went to the streams to draw it. The Mbilikimo needed water but were afraid of being bit by the snakes, so they decided to move

away from Chamuriho Mountain. . . . Muriho and his followers, called the Kombogere, were following right behind the Mbilikimo and passed the orokoba protection medicines around the stream of Nyitonyi. . . . [The story then tells all the other places that he chased them out of, all the way to Lake Victoria.] When Muriho had made certain that his first enemies would not return he went back to his house and made a plan to firmly establish the authority that he had just won for himself. He went to Chamuriho Mountain to celebrate the ritual of purification, the ekimweso. The ekimweso is a celebration of protection and blessing for the courageous deeds he had done. . . . After this first battle Muriho had already attracted many new people, in addition to his followers and their leaders. He acquired eight wives after these events—people gave women to him in marriage as gifts of gratitude for his prophecy and medicines. . . . Finally, Muriho Nyikenge as the head leader of Ikizu and Sizaki, was forced to fight another battle with amazing beings, that is spirits which are called the amanani. . . . He used the same methods, that is to place medicines in water, to encircle each area where he had removed his enemies with the protection medicines of the orokoba so that they could not return again.[41]

Ikizu still make offerings on Chamuriho, the tallest mountain in the area, and know it as the origin spot and most powerful erisambwa of the Ikizu. Most of the other important Ikizu rituals either begin or end here. The Ikizu claim the land because Muriho planted their emisambwa at specific sacred places to guard the land.

The core spatial images of sacred places of power evoking the generative principles of the mediation between wilderness and culture thus occur through a broad range of oral traditions and the remembrances of elders, including lists of place-names, emisambwa rituals, words for leaders and ancestors, and oral traditions about ritual occupation of the land. Variations of this vision of a sacred landscape are widespread in the region, among those speaking other languages in North and South Mara. This indicates that the generative principles are quite old but that the particular western Serengeti forms now available date back to known ancestors in the nineteenth century. The sacred landscape has been re-contextualized with a number of social groups over time, the lineage or descent group in the place-name lists, the clan and ethnic group in the wider stories that use these sacred places of power, and the generation-set in dating and performing the rituals. In fact, although not directly connected to the named sites, the generation-set does seem to be the social group most often associated with the ritual space of these landscapes.

Although this sacred relationship with the land was claimed and maintained by each of the descent groups through their ancestors who acted as guardians of the land, people also saw the need to unify those separate descent groups to protect the boundaries of the land that they occupied together. This was accomplished through rituals then depended on age organization because this was the only collective identity that transcended even an inclusive kinship. Generation sets and later age-sets were mobilized to perform the orokoba rituals to encircle and protect the land that everyone occupied together. These rituals, with many variations, are widespread throughout the Mara region and seem to be quite old, although dating them is difficult. Because of their widespread provenance, with variations in each ethnic group, they seem to predate the split into ethnic groups in the nineteenth century, perhaps going back at least five hundred years, to the time that East Nyanza languages differentiated themselves. The variations in each ethnic group suggest that each group elaborated a given set of rituals from an older pattern.

Although the rituals themselves cannot be dated exactly, there is reason to believe that they also use symbols similar to the core spatial images that are passed on through time in relatively unchanging form even as their context and meaning change significantly. While social historians and historical anthropologists analyze rituals by looking at their changing meaning within particular historical contexts, sociologists and anthropologists have seen ritual as a means of communicating shared values and dealing with internal conflict.[42] All interpret rituals as a symbolic text whose present meaning they decode, beyond participants' claims to the commemoration of past events. Yet Connerton argues that "if there is such a thing as social memory, we are likely to find it in commemorative ceremonies," which he defines as rituals that ostensibly reenact the past.[43] Through an analysis of ritual language and gesture in Europe, he shows that the structure of ritual in "commemorative ceremonies" builds in a certain invariance because of the performative, formalized, and stylized language on which the reenactment depends. Those who perform rituals do so as members of a group that habituates them to certain bodily practices reserved for ritual. While anyone can narrate oral traditions, only members of the group can perform the rituals. The positions and gestures of the body in performance form the mnemonic system of core spatial images around which rituals are elaborated. Performers understand these actions as reenactments of past prototypical actions. Ritual suspends linear time and reconnects people with their past. The ritual performance of the "commemorative ceremony" conveys and sustains "an image of the past" through which a community understands its identity.[44] In western Serengeti ritual, the definition of the land encircled by the generation-set, the ekyaro, changed according to the historical context. However, the relationship between people and land embodied in ritual remained stable as the core spatial images, analogous to that of oral tradition, but inscribed in bodily practice.

Understanding the core spatial images of these rituals and how they have been used in different context depends on first understanding the western Serengeti age system. Elders in the western Serengeti today identify themselves not only by their ethnic, clan, and descent-group affiliations but also by their age-set (*saiga*) and generation-set (*rikora/amakora*) names.[45] According to anthropologists the generic terms *age-organization* or *age-system* describe social organization based on age, generation, or both. The terms *age-set* or *generation-set* are used where persons are grouped into hierarchically ordered sets with specific social responsibilities as a unit.[46] Age-set recruitment is based on age at initiation, or circumcision in this case, while generation-set recruitment is determined at birth by the father's set. Age-sets group together peers of the same age, making it possible for a boy and his uncle to be in the same set if they were the same age, while this is impossible in a generation-set system. Age- or generation-set cycles are systems in which a cycle of successive repeating names is assigned to each group as it is formed over time.[47] Western Serengeti peoples practiced both systems simultaneously, although historically the relative importance and function of each have varied.[48] Women had their own age-sets, but at marriage a woman became part of the age-set of her husband, while acquiring a generation-set name from her father at birth.[49]

All adult males were involved in the generation-set rituals, regardless of descent, making it an inclusive, rather than exclusive, institution in a society organized by kinship. These rituals were conducted in turn by either of the two generation-set moieties, Saai or Chuuma, which functioned as equal and complementary classifications and were used all over the Mara region. Every man was a member of the named generation-set in either the Saai or Chuuma cycle that followed the generation-set name of his father. The Saai cycling names were (1) Saai, (2) Nyambureti, (3) Gamunyere, and (4) Maina; the Chuuma cycling names were (1) Mairabe (Ngorongoro among the Kuria; Ghabasa among the Ikizu), (2) Gini, (3) Nyangi, and (4) Chuuma. For example, if a man was of the Saai cycle and his grandfather was of the generation also named Saai, his father would be of the Nyambureti generation, the man himself would be of the Gamunyere generation, his son of the Maina generation, and his grandson again of the Saai generation. Throughout the region, across all ethnic groups, a person's generation determined everything from daily greetings, marriage partners, eating arrangements, hospitality, respect, and responsibility. Relationships with people in the same generation or in that of their grandparents were informal and personal, while those with adjacent generations were formal and respectful. Men were supposed to marry women of an adjacent generation.

Based on historical linguistic evidence, generation-set cycles seem to be the older form of age organization among western Serengeti peoples. The cycling names Nyangi, Maina, Chuuma, Saai, and Ngorongoro are all loanwords from Mara Southern Nilotic speakers who had adopted them earlier, in the first millennium BCE, from Eastern Cushitic speakers on the Ethiopian borderlands.[50] The widespread provenance of a common generation-set system among all East Nyanza speakers, from Kuria to Suguti along the lakeshore, and the restriction of the age-set system to the eastern part of the Mara region is another indication of the precedence of the generation-set. Western Serengeti peoples would have had to adopt these generation-set words sometime between 400 and 1000 CE, since after that time Mara Southern Nilotic speakers no longer inhabited the region. These loanwords were part of the package that East Nyanza Bantu speakers adopted to gain access to livestock expertise.[51] When they met Mara Southern Nilotic speakers on the western Serengeti frontier, East Nyanza Bantu speakers would have already been practicing circumcision and initiation inherited from the earliest Great Lakes Bantu speakers (ca. 500 BCE).[52] The word used for the generation-set throughout the Mara region today is *rikora/amakora*, from Bantu *gokora*, to grow, implying that the generation-set system itself was a local innovation around the generational principles of growth, fertility, and successional development.[53]

While cycling generation-sets (rikora/amakora) seem to have developed before 1000 CE, western Serengeti peoples adopted the linear age-set system (saiga), at least in the cycling territorial form that exists today, sometime around 1850.[54] They adopted the word *saiga*, if not the concept, from Tatoga pastoralists who came into the region after 1000 CE. Tatoga do not use the standard cycling generation-set names but have a linear generation-set system called *saigeida*.[55] Although western Serengeti peoples may have adopted this word at any time up to the nineteenth century, it seems to date to an earlier period, since the Gusii, who were not in direct contact with the Tatoga in the nineteenth century, also use this word (*esaiga*) for the house of young unmarried men.[56] The Tatoga dropped the Southern Nilotic cycling age-set names sometime after 1000 CE, while their Bantu-speaking neighbors kept them. Throughout this long history western Serengeti peoples used various forms of the age system to reach across the boundaries of ethnicity and economy—to form interdependent relationships with pastoralists, to mobilize young men for collective tasks, and to unite people across boundaries of descent for rituals that would bring healing and fertility to the land. Overall, a regional pattern exists of two parallel institutions, age- and generation-sets, simultaneously in operation but with one more dominant.

Consistent with the core spatial images of the generation-set ritual of encircling, the generation-set served an integrative, unifying function on the community level as well as on the regional level. Ethnic groups united across the entire region by using the same generation-set names, while they based the particular ceremonies for circumcision or retirement on the individual ekyaro, or clan territory. People sharing the same generation names traveled and received hospitality over a wide geographical area based on commonly understood norms of generational relationship.[57] In Nata and Ikizu, the two generation-set cycles cross-cut clans and descent groups so that each cycle would be found in each clan. In Ngoreme and Ishenyi, each clan territory, or ekyaro, had only one rikora, and Ngoreme is divided territorially into Saai and Chuuma generation-set cycles. The age-cycle territory unifies a number of clans but corresponds to only one generation-set cycle. Unlike many other East African communities with age systems, the political and military role of generation-sets in the western Serengeti seems to have been minimal.[58] Although elders speak about the "rule of the generation-set," this seems to refer more specifically to its function of unifying and ritually protecting the community rather than to formal political rule.

Although neglected by scholars and kept secret by elders, generation-set rituals using the core spatial images of encirclement to protect the health of the land, rather than military or political power, seem to have been the most important function of the generation-set and one that gives the elders authority.[59] Elders claim that outsiders' knowledge of these rituals would endanger the health of the land and its people by opening it to malevolent intervention. The details about the generation-set rituals are second only in secrecy to the eldership title secrets about which initiates swear an oath of silence. Elders often described generation-set ritual in vague and generalized terms. Zanaki elders told anthropologist Otto Bischofberger that the rikora in power would "walk around" to keep in shape and strengthen the solidarity of the group.[60] Elders assert their power over generational succession by keeping these secrets and putting them into an unquestioned and unchanging category of "tradition," where they maintain authority over the definition of territory and a place-oriented identity. Connerton argues that in order for these "commemorative ceremonies" to reenact the past they must move into "ritual time," where an event is indefinitely repeatable.[61] Similar to the way oral traditions are assumed to be fixed, elders enhance the authority of the ritual by portraying their own contingent experience of the ritual as a timeless pattern. In fact these rituals have changed, in form if not in substance, over the last two hundred years as they have been recontextualized in various settings. They seem to have at first been generation-set rituals used to unify and protect descent group settlements within clan territories but during the radical social transformations of the nineteenth century were appropriated as age-set rituals or as the work of a generation-set to unify clan lands into one ethnic territory.

Among the seven ethnic groups from whom I have accounts, the practice of the generation-set ritual walk to heal the land varied extraordinarily around the consistent core spatial image of encirclement to create territorial identity and protection. Ideally, every eight years in Nata and Ikoma, the men who were in the "ruling generation" of the two cycles, Chuuma and Saai, alternately walked around the boundaries of the land, kukerera. The walk took place together with the retirement ceremonies for one generation as a way of preparing for the "rule" of the new set.[62] In Ishenyi and Ngoreme the walk took place in symbolic terms, around a tree or at a feast. Yet in each case, the bodily practice of the walk physically inscribed the boundaries of social identity in space. The appropriation of territory through ritual control over the land secured the health of the land and the people. The ritual creation of enclosed space by encirclement defined a territory and the identity of those who lived there as one people.[63] The walk in all its variation focused on passing the orokoba, or the medicines of protection, around the land. The leaders of the generation-set literally planted medicine bundles in the soil at intervals around the land or in the bark of certain trees that they slashed. The walk encircled the land with medicines to protect it from enemies and disease and ensure its fertility. The generation-set itself, sometimes with the instructions of a prophet from a distant land, kept and passed on this medicine. The walk also preserved the peace of the land; it "cooled the land," making it productive and healthy by using the ritual symbols of binding and enclosing. The word *kukerera* is likely a prepositional form of a proto–Great Lakes Bantu word (*kukila*) meaning to overcome, surpass, heal, unify. The prepositional form implies an object, in this case, the land and all that it symbolizes.[64] These rituals functioned to create territorial identity and group cohesion under the authority of the elders.

Using the core spatial image of encirclement, the ritual walk in Ikizu united the various clan lands that make up Ikizu by asserting the communal authority of the generation-set elders. Ikizu territorial unification in the nineteenth century transformed the ritual into a walk over ethnic territory. It began at a point in the westernmost part of the land and moved east. Each night the *abanyikora* (new generation-set members) feasted and slept at a different clan center, accepting patronage in wealthy man's homestead. On the third day they paused at a special tree and sent a small delegation to the mtemi, a descendant of first woman, to get the rain medicine (*omoshana*). The Ikizu prophet (*omunase*), as a descendant of first man, provided the medicines for protection and for the new fire. The whole group arrived for the ceremonial climax near Chamuriho Mountain in the easternmost part of Ikizu, where Muriho first claimed the land by planting the spirits of his ancestors (emisambwa). There they performed the *ekimweso*, the purification ceremony to pass authority from one generation to another. The new generation built a new fire, with the medicines of the prophet, after all Ikizu had extinguished their homestead fires. The elders sacrificed a white goat and cut its hide into strips that the new generation-set passed out and wore on their fingers.[65] When the ceremonies finished, the generation-set that had just taken power retraced their steps back to complete the circle, spreading the medicines for healing and rain and distributing the new fire to all the homesteads. By evoking the emergence story the elders used the walk to assert the unity of clan lands and to legitimize their own communal authority, as representatives of Muriho, over the rainmaker and the prophet. The elders spatially represented their interpretation of the political makeup of Ikizu through the walk as the generation-set paused in each of the clan lands of Ikizu.

In Ishenyi the walk of the rikora consisted only of a symbolic walk around a certain tree (*msingisi* or *msari*)[66] to prepare for the initiation of a new age-set into power and yet still worked to unify and protect the community. The clans (hamate) of Ishenyi were divided into two moieties, with each moiety representing one generation-set cycle, either Saai or Chuuma, who carried out the ritual separately. By contrast, in Nata, Ikoma, and Ikizu each clan claimed members of both rikora cycles. In Ishenyi the rikora did not crosscut and unify the clans. However, it did unify the age-set territories that developed by the end of the nineteenth century to replace clan territories.[67] The Ishenyi ceremony to install the new age-set took place at the homes of the newly chosen leaders, lasting for eight days of feasting, singing, and dancing and included the elements of strips of hide

Figure 3.3. Rikora leader Rugayonga Nyamohega, Issenye. *Photo by author, 1995*

Figure 3.4. Kang'ati of Nata Bongirate, Morigo Mang'oha, Kyandege. *Photo by Paul Shetler, 2003*

worn on fingers, blessing and purification ceremonies, and lighting a new fire. When there was need the generation-set encircled the land with the medicines of protection or rain, obtaining the medicines (*amusera*) from the rainmaker that they mixed with milk and flour. They spread the medicines with a cow's tail as one moved east and another west around the boundaries of the ekyaro.[68]

The Nata walk, occurring regularly every eight years, symbolized the generational principles of fertility, growth, and clan unity. When it was time for the ceremony, the elders sent the leaders (*abachama*) of the new age- or generation-set outside Nata—north, south, east, and west—to collect the ritual ingredients that the elders mixed together and sprinkled as a blessing on the new age-set, including water from Lake Victoria, honey from Riyara, excrement of an unweaned child, livestock manure, millet or other grains, and a young, unblemished black bull for sacrifice. The new age-set built a ritual fence around the homestead of their leader, whose father hosted the ceremony and fed the people. The spatial symbols of the ritual represented the mediation of external forces to protect the internal community: the youths left Nata territory to collect the symbols of prosperity outside the community and brought them inside the ritual homestead fence. During the ceremony the elder of the homestead took a black tail, eghise, as the symbol of his elderhood, and sprinkled a mixture of the ingredients brought from outside, along with the stomach contents of the slaughtered bull, on the new set as they stood in a circle with their wives, rubbing it on the breasts of the women. As the ritual father sprinkled he prayed that the youth might have children, abundant livestock, good harvests, and rain during the "rule" of their saiga. Elders then transferred the clan symbols of a bell for a young black bull and a heart-shaped trade hoe from the Sukuma Rongo clan, representing the two clan moieties. The eight *(aba)chama* leaders wore strips of hide from the sacrificed bull on the middle finger. As one leader from each clan moiety stood together on the hide and stretched a strip of the hide between them, an elder cut the strip in half and declared them blood brothers *(baragumu)*. After feasting and dancing all night, they moved on to feast at the homestead of the other clan moiety leader.[69] Even though this was an age-set ceremony for the ethnic group, its generative principles of unity were those of the generation-set: the leadership of fathers, the unity of clans requiring that everyone participate equally in order for the medicine to work, the emphasis on peace and prosperity, and the blessing for both women and men.[70]

The Ikoma also recontextualized the older generation-set rituals in the nineteenth century to unify the clans through an age-set ritual using the Machaba elephant tusks, their erisambwa, which they obtained from a Tatoga prophet. The Tatoga prophet of the lineage of Gambareku, of the Relimajega clan, carried the Machaba at the head of the walk, served as the age-set leader, and made the medicine, the orokoba, that the age-set planted around the land.[71] A Tatoga prophet of the rainmaking clan of Gaoga served as the top leader of the Ishenyi age-set and provided the medicines for rain and to bless the new year's seeds.[72] Both Ishenyi and Ikoma acknowledged the first-comer status of the Tatoga in their rituals to appease the land. The Ikoma ritual also spatially symbolized the unity of the clans into one land. The Ng'orisa clans, which controlled the male tusk, began the walk where the tusk was kept, in the west; the Rogoro clans began at the place where the female tusk was kept, in the east. They circled the land and met in the center, each going through the homesteads of the members of their clans. Each night a wealthy man in a different clan settlement hosted them overnight for feasting and dancing.[73]

Although no rikora walk existed in Ngoreme, generation-set rituals still emphasized the generational authority of elders and the protection of the land within the ekyaro. When the sons of most of the rikora matured (having their own wives, children, and homestead) it was time for the older men, who carried the black tails of eldership, to retire. The elders symbolically transferred authority by allowing the young men to eat the best cut of meat, the meat of the back, at communal gatherings. At any feast people sat in groups according to generation and gender and ate the appropriate kind of meat: the oldest retired men, sitting by the grain storage bins, ate the softest parts of the cow, like the lungs *(sarara)*; the ruling elders, sitting inside the cattle corral, ate the back meat *(omugongo)* and the head; and the young men, sitting outside the cattle corral, ate the chest meat, legs, or the hump *(sukubi)*.[74] Their position inside the cattle corral, eating the best cut of meat and the head, was proof of the elders' dominant authority, wisdom, and control over wealth. They were the "back" of society, leaving the youth to do the physical work. The young men sat expectantly outside the corral, eating the meat symbolic of their strength. The oldest men were "like women" because they sat at the granaries, where women underwent circumcision, and ate their portion of soft internal organs. Elders put prideful young men in their place by asking if they had tasted the back meat yet. At retirement the Ngoreme elders gave the *omotangi*, or first born, of each clan in the new generation, the horn that called the rikora together for a meeting or sounded the alarm in times of danger. They made this instrument from the long horn of the oryx or greater kudu[75] and wrapped it with special medicines embedded in the strips of buffalo and lion hide that were renewed at the time of transfer. The tight binding of medicines with hide strips employed the same core spatial images as the encircling of the land with medicines of protection.[76] The Saai generation in Bumare also passed on to the omotangi the generation-set medicines called the *omugongo wa mwensi*, or the protection medicine of the generation for healing the land.[77]

Generation-set rituals drew on common core spatial images recognized in a much wider area throughout the Mara region and beyond East Nyanza Bantu speakers. Kuria ceremonies used the recurrent symbols of regeneration and growth in the use of a tree, the symbolic homestead, the blessing of cows and wives, and the passing on of things from father to son. They closed the ekyaro boundaries to strangers and enclosed the symbolic homestead with a fence to define territorial identity.[78] The Zanaki generation-set retirement ceremony took place every twenty or more years in each ekyaro at the emisambwa sites of sacred groves or large rocks and also used the symbols of a new fire and passing the orokoba protection medicines around the land.[79] The Southern Nilotic–speaking Tatoga linear generation-sets, called saigeida, and the prophets also used the symbols of fire, strips of hide worn on the fingers, and medicine embedded in the trees for protection.[80] All these shared core spatial images of ritual between farmers and herders indicate a common view of the sacred landscape across linguistic and cultural boundaries.

FINDING TERRITORIAL IDENTITY IN THE
CORE SPATIAL IMAGES OF SACRED SPACE

The concept of territory inherent in the sacred landscapes of ritual boundary formation through the walk of the generation-set differs from the conventional academic understanding of territory as "the ecological locus of use or the political limits of domination and sovereignty."[81] Without chiefs or kings, western Serengeti peoples did not define their territory by centralized political authority, nor could the ecological landscapes of hill farmers or the social landscapes of clan networks make those claims. Territory and territorial identity were rather defined by ritual space. Robert Thornton's definition of territory—as "the symbolic differentiation of space and the appropriation of that space into a structure of meaning, so that it may be represented as a coherent and enduring image"—fits more closely with western Serengeti understandings of territory.[82] This definition does away with the false dichotomy so often posed by anthropologists and historians between "the social definition of territory" (assumed to be the precolonial African model) or the "territorial definition of society" (assumed to be the Western imperialist model).[83] Whether the society is organized on the basis of proximity or kinship, its people must first imagine and create the space of the territory, in this case through the performance of ritual and narrative.[84]

There were some aspects of western Serengeti territory that were defined ecologically as the space within wilderness boundaries where the generation-set walked. Ecology determined the boundaries of the larger unit of possibilities within which the specific ekyaro would be redefined ritually by each generation. This might correspond to all the land within which the group ever lived, harvested resources, or had ancestral sites, not necessarily the land that was immediately claimed. For example, Nata elders said that their "traditional tribal" boundaries were the Grumeti River with the Ikoma (east), first the Sanchate and then the Tirina River with the Ikizu (west), the Morega or the Somoche River with the Ngoreme (north), and the Mbalageti River with the Sukuma (south). Elders described the old Ikoma boundaries as the Orangi River with the Maasai (east), the Grumeti River with the Nata (west), Bangwesi Mountain with the Ngoreme (north), and the Mbalageti River with the Sukuma (south).[85] Both of these examples cover huge areas and follow a formula in which boundaries are defined by (1) the ethnic identity of who was on the other side of the boundary, (2) a cardinal direction, and (3) a physical feature that marked the boundary, such as a river or a mountain or long stretches of impenetrable bush.[86] Wilderness tracts of no-man's-land that marked the boundaries were often covered by tsetse fly bush.[87] The colonial files asserted that belts of tsetse fly bush fifty to fifty-five kilometers wide separated the Maasai seasonal grazing areas on the western edge of the Serengeti plains from the farming peoples and marked both the Ikoma boundary with Ngoreme to the north and the Sukuma across the Mbalageti River to the south.[88] Schoenbrun's work shows that from earliest times Great Lakes Bantu speakers carved out their farming communities between uninhabited buffer zones of tsetse bush.[89] Because rivers, mountains, and tsetse bush represented barriers to easy communication, people developed a sense of territorial identity within the boundaries of their daily interactions and routines. They perceived the peoples on the other side of these wilderness boundaries in stereotypical terms and in relation to idealized cardinal directions.

The people on the other side of the boundaries were often defined according to the ecological landscapes that differentiated herders, hunters, and farmers. They established boundaries with Maasai herders to the east and Tatoga herders to the south, where the grassland plains began and the hill/woodland country ended. Maasai did live in Ngoreme and even Ikoma at times, but this was never considered their territory. When the colonial government was trying to establish Maasai claims within what has become Serengeti National Park, the line they drew from Banagi Hill to the eastern end of the Nyarabovo mountain range, including the Moru kopjes, fell on the ecological boundary of the grasslands. One indication that the Maasai claimed no land west of this line was that there were no Maasai names for the area, while German maps show Maasai and Dorobo place-names for all areas to the east of the grasslands border and up to the Mara River where it crosses the Kenya border.[90] A 1974 landscape classification map from Serengeti Research Institute located most of the land that the Ikoma claim within these large ecological parameters, from Banagi Hill (now within the park) in the east to Bangwesi Mountain in the north (now within the game reserve), including the Sibora plains, Fort Ikoma and Robanda, within an internally coherent landscape category called Ikoma region.[91] Both Nata and Ikoma elders told me that there is no difference between the land in the park around Banagi and the land on which they now live, as it is "all one territory."[92] In the case of Asi territory in the "Oseru region," by the map's classification, or Ngoreme and Nata within the "Ikorongo region," concepts of the larger boundary limits of territory indicated by elders corresponded roughly with the map's classification of ecological landscape. In this classification scheme Sonjo is not separated from western Serengeti by the boundaries of the park but is the eastern extreme of a landscape that they share.

Oral testimonies also demonstrate that the concept of the "other" was integrally tied to these wilderness boundaries. People created boundaries in relation to what was outside those boundaries and formed their identity in relation to someone else.[93] Oral testimonies most often defined communal, now ethnic, identity in contrast to the "people of the wilderness" (Nyika) or the Maasai and Asi, who live outside civilized space. The plains and bush areas were outside possible homestead space and full of danger, from both wild animals and uncivilized people.[94] Civilization meant a ritual relationship with the land that transformed it from wilderness to home. Anyone living outside the boundaries was, by definition, the stereotypical "other." Oral testimony and ethnography reflect this understanding of boundaries as wilderness tracts and portray them as dangerous and liminal places where they disposed of polluted things. If an uncircumcised girl became pregnant she was forced to flee over the boundaries or she would pollute the whole land. In the past, the Ikizu generation-set leaders took breech babies or those whose top teeth came in first out to the wilderness, over the boundary with Nata or Zanaki, and left them there. In their job of caring for the boundaries, the Zanaki generation-set leaders disposed of polluted Kuria bodies deposited across their side of the boundary. Court cases in the colonial files describe dead bodies found on the boundaries, in fact on the colonial boundary stone, between two ethnic groups.[95] Today villages on the boundaries between two ethnic groups are often the frontier refuge for young men in trouble, who go there to gain a new identity. These places have bad reputations as the home of outlaws and people without respect for traditional authority. R. E. S. Tanner's colonial study of cattle theft in Musoma District showed that most theft cases occurred "along uninhabited district borders."[96]

For the western Serengeti, the only appropriation of space that can usefully be called a territory is ritually created through the core spatial image of encirclement, the bounded and enclosed space of the ekyaro within the larger ecological boundaries of the wilderness. Although the ekyaro was by definition a clan territory, the social unit used to create a territorial identity was the egalitarian generation-set rather than the descent group.[97] This allowed for a flexible definition of territory as the geographical area appropriated by a social group using the rituals of enclosure and boundary formation that are based on very old generative principles in relation to the land. The core spatial images of the rituals remain stable but the groups that perform them have changed radically, as they have been recontextualized in the last two hundred years.[98] At different times and in different places the territory has been defined as the land of a descent group, a clan, an age-set, a generation-set, or an ethnic group. The ekyaro was not a fixed unit. In ordinary speech the ekyaro referred to anything from the local community to the nation state. Whereas in Ikizu the space of the walk corresponded to the united clan territories, in Ishenyi elders defined the ekyaro as the age-set territory that developed in the second half of the nineteenth century and in Ngoreme, Kuria, and Zanaki it was the autonomous unit of clan territory.[99] The smaller ethnic groups of Nata and Ikoma defined ekyaro as the entire ethnic territory; however, they also called the smaller age-set territories *ebyaro* (plural). Western Serengeti peoples defined the ekyaro situationally, within the ritual boundaries of the imagined community described by the walk.

Western Serengeti peoples used the generation-set in the core spatial images of the walk to overcome the divisions inherent in descent-based settlement structures, to form territories, and to unify the ekyaro. Nata elders, while reticent about the details of the generation-set walk, emphasized it as a ritual for all of Nata—"it brings Nata together." Although Ikizu was also divided into smaller territorial ebyaro of the clans, the generation-set walk symbolically brought together all of Ikizu in the space of the walk. Nata elders said that the saiga, or age-set, divided, while the rikora unified.[100] Ruel states that Kuria generation-set designations in a communal setting emphasize solidarity and responsibility to the larger group rather than division and divergent interests.[101] The walk identified the rikora itself with the ekyaro, or the land, in some cases disallowing rikora leaders to move outside the ekyaro.[102] Whether the walk encircled the boundaries of the ekyaro or passed through the homesteads, it symbolically brought the parts into a whole. Ngoreme call the ceremony for the new Ngoreme age-set *kwitaberi asega. Kwitaberi* is derived from the verb *-itaberi,* to bless the land. This term explicitly connects the initiation of a new age-set (*asega*) with the prosperity of the land.[103] The walk of the generation-set formed and was formed by communal consensus that was necessary for the ritual health of the land. To carry out this large-scale ritual, the elders mobilized the whole community and each did their part—prophets and rainmakers provided medicines, wealthy men provided food, women prepared food and beer, elders and youth cooperated in the preparations and the huge investment of time and energy in the process. The walk itself was symptomatic of the state of relationships in the community and thus its health.

The meaning behind symbols used in the generation-set walk further illuminate the core spatial images in this view of a sacred landscape, where territory is defined by encirclement of ritual space. The central symbol of the walk, the medicine planted on the land in the walk called the orokoba, used the core spatial image of encirclement to create a sense of cohesiveness for a diverse and mobile population living together on the same land. The use of the term *orokoba* in other circumstances further illuminates its symbolic usefulness in the core spatial images of encircling to form protective boundaries and binding up exterior forces to control outside power. Descent groups sometimes possessed an orokoba in the form of an *ekitana,* a medicine bundle, that could be used by others in times of need to protect homesteads from theft, illness, and witchcraft, or to protect young men from harm when they went after cattle thieves. The word *orokoba* literally means a cowhide thong or rope, used for tying a cow's legs for milking. The hide used to make the rope is an exterior cover, the boundary or enclosure of the animal. Medicine bundles were often tied up with a thong. Elders also used *orokoba* to refer to the strips of hide from the sacrificed animal that generation-set members ritually wore on their fingers, a practice common in boundary ceremonies throughout East Africa, irrespective of language group or culture.[104] In the western Serengeti, men became blood brothers by holding the end of an orokoba, or hide thong, that was ritually cut by an elder, between them. In Nata *orokoba* also means the matrilineage, your mother's kin, symbolically compared with the unbroken line (rope) of inheritance through the mother's side. A Ngoreme dictionary defines *orokoba* as the umbilical cord.[105] People described the closest kind of relationship between people as that of "one womb."[106] In the rare and drastic case where brothers of one womb disagreed and could no longer be reconciled, lineage elders performed a ritual in which they passed an orokoba through the maternal house wall, each brother holding one end of it. An elder cut the thong in half so that the house (anyumba), divided rather than united them as "children of one womb." In Nata *orokoba* is also another word for clan lands (ekyaro), the land around which the orokoba was passed, the land over which the clan had ritual control.[107]

In all the rituals described above, the core spatial images of encircling, binding, containing, and interiority were repeatedly used as a way to bring healing and prosperity.[108] Schoenbrun describes old Great Lakes Bantu ideas about medicines that control malevolent power by binding it that are evident in the common practice of tightly binding various substances together in medicine bundles and charms. In many Bantu languages, heat is associated with witchcraft while coolness is associated with peace and prosperity. The Nata word used to describe people or land that is whole and healthy is *buhoro*, from Great Lakes Bantu *podo*, quietness, cold water, good health.[109] Yet the control of heat through binding is also necessary to activate medicines of healing. Elders said that the orokoba works to "cool the land" by tying up the powers of disorder. The spatial metaphor of outside/inside is operational in all these rituals by using the stomach contents of the slaughtered animal (inside), the womb (inside), the hide strips placed on everyone's fingers (outside), the building of ceremonial homesteads (inside), and fences (outside).[110] These rituals civilized the power of the wilderness, or outside forces, by bringing them inside through control and containment rather than isolation or extermination. However, an encirclement or enclosure without an opening is associated with death. The prosperity of the house depends on its gateway, which leads to the outside. The term for the extinction of a descent group in Kuria means to be stopped up or blocked off. Elders performed the ritual "piercing," or opening, of a cow's stomach and sprinkling the contents as the central act in any ceremonial sacrifice. Circumcision itself is an act of opening or cutting. The community must be enclosed for protection, but it dies without links to the outside, for wives, dependents, and security. A healthy community mediates the dangers of outside forces by controlling, not closing, its boundaries.[111]

The image of fire purifying by covering with smoke also has rich and multivalent meaning in ritual evocative of the same core spatial images of encirclement. Fire was the civilizing gift of first man, but was brought from outside domestic space. The hearth constituted the moral center of a house, and lighting a fire established a household in its own right.[112] People feared the purifying, transforming fire of the blacksmith and associated the smithy with the womb of the woman. Elders built a fire to see if it was an auspicious time to act, and if the smoke went straight up, success would follow. For example, when the new Nata saiga went to emisambwa site at Riyara to gather honey for their ceremony, they first built a fire and watched the smoke.[113] Ishenyi elders lit the ritual fire for purification and blessing (*ikoroso*) for a single clan (hamate) or a single homestead when confronted with the problems of death, sickness, or infertility. In Ngoreme the ikoroso fire was kindled from particular trees (*esebe, omoreto, omorama*) that grew at the erisambwa site.[114] After being instructed by a prophet, the elders made the fire in the wilderness and added medicines, producing a smoke that "covered" the whole land for its purification.[115] Rikora leaders performed the fire ceremony, *ekimweso* in Ishenyi and *shishiga* in Ikoma, to bless the land (ekyaro).[116]

Water from emisambwa springs was also used in the rituals to bless the land.[117] Water was the gift of first woman in the emergence stories. Both water and fire were transformative and transitional symbols used to mediate the boundaries, the inside/outside dichotomies of the orokoba, the womb, and the house. A fundamental ritual act in these ceremonies of purification, protection, and sanctification was the sprinkling of a mixture of contents (water, milk, honey, millet flour) on to the gathered people as a blessing and a prayer for fertility (komusa). Sprinkling covered the bodies outside, on their physical boundaries, with the symbolic things of sustenance and life from the inside. Elders said that the smoke, like water, covered and protected the participants in these rituals. Brad Weiss points out that fire and smoke are often used in binding rites, both to surround a house and to drive off malevolent forces.[118] These symbols operated on all levels, from the individual homestead to the territorial ekyaro. It is this contextual aspect that makes these symbols so powerful. Testimonies compare the health of the homestead to the health of the larger territory, symbolically equating the enclosure of the womb, the homestead fence, and the orokoba around the land. The core spatial images of these ritual practices created an identity embedded in a bounded territory controlled by elders. People use the same symbolic language regionwide to express these concepts of territorial identity.

The requirements for and character of generation-set leaders indicate that the authority of elders was tied to the generational principles of fertility, growth, and peace, as opposed to the warrior symbols most often associated with age-sets. The leaders of the generation- and age-set, called abachama, usually numbered eight members in all, with one top leader, whose job it was to "guard the land." An Ikoma elder said that these leaders were responsible for anything that concerned the land—rain, disease, peace, war, and hunger. The Ikizu called every member of the rikora in power an omochama while the Nata used this term only for the eight chosen leaders.[119] Chama is an old Bantu word with wide use, usually meaning a group that works together or a council. Outside the western Serengeti, the Kuria secret council of elders is called the injama, while the Kikuyu elders' lodge and the Maasai elders' meeting are both called the kiama.[120] The body of the generation-set leader, the kang'ati (leader, from (gu)kungata, to walk ahead), had to be unblemished, without scar, sore, or disability.[121] His parents, children, and wife had to be alive and healthy, while he had to be a man of good character and of peace. Once he was chosen, no one was allowed to see him naked, even to bathe, nor was he allowed to touch blood, handle raw meat, drink water other than spring water, or take lovers outside his homestead. His wife (wives) observed similar prohibitions. The age-set leader could never fight and on a raid took up the rear position; instead of carrying a weapon he carried a long stick (orutanya) as the sign of peace. He had only to lay the stick between two people to stop them from fighting.[122] The leader of the rikora commanded more respect than the leader of the age-set; far from the warrior hero, he was a man of peace, embodying the fertility and health of the land.[123]

Yet the power of the elders was not entirely benevolent, and the core spatial images could be used to promote their own interests rather than those of the whole community. By keeping the rituals in the realm of secret knowledge and maintaining ritual content that emphasizes generational power, the elders controlled access to and authority over processes fundamental to the health of the community. Youth could not question the decisions of elders or the wisdom of "tradition" because this might bring danger to everyone. Neither could people question the livestock fines elders meted out for infractions of the ritual prohibitions, for fear of bringing on hardship. In the scheme of generational succession, elders clearly had the upper hand. They decided when their sons would marry and establish their own homesteads through their control over cattle for bridewealth.[124] A father might decide to marry a third wife himself rather than providing the cows for his son's first marriage. Although youth eventually took their father's place, elders could also delay the time and rate of promotion to a new generation- or age-set.[125] In fact, although youth had important roles to play in generation- and age-set ceremonies, they were more often the "agents of force" rather than the "controllers of force"; that is, the elders sent the youth on errands but ultimately made the decisions themselves.[126] The Ngoreme also have abachama of leaders for the ekitana or orokoba, sometimes called the abachama of the land, or ekyaro, and the abachama of the rain, who carried out the orders of the elders, prophets, or rainmakers on how to perform the rituals. Elders said that the most important attribute for the job was the ability to keep the medicines secret, for which they swore an oath.[127] Elders also represented a gendered male authority, although the generation-set rituals promoted the feminine values of peace and fertility over the masculine age-set values of war and strength. Women's role in reproduction was valued, but women did not perform either the rituals at the emisambwa sites or the walk of the generation-set except in parts for which they were instructed. The generation-set legitimized and naturalized the authority of the elders over women and youth as a necessary part of protecting the land and the community from harm.

Western Serengeti peoples creatively used the process of ritually defining territory by the core spatial images of encirclement to imagine different kinds of territories with different boundaries at different times. Territories were flexible units that were defined anew each time the land was ritually walked over; both the boundaries and the social groups contained within the boundaries shifted over time. But saying that territorial boundaries were flexible, situational, multiple, and shifting does not imply they were not meaningful or enduring indications of identity.[128] The western Serengeti rituals of walking the land show that place did matter and that a people's identity was deeply tied to the ways in which they had appropriated and created the space around them, whether territorial or not. People understood the ekyaro as the land that they ritually controlled, intimately linked to their own health and well-being.

Mwalimu Julius K. Nyerere, the nation's first president, born in Zanaki, Mara region, used this sacred landscape as a way to define a national identity by extending the concept of the ekyaro as a ritually maintained territory of a people to all of Tanzania. He was the son of the second colonial chief of Butiama, Nyerere Burito (r. 1912–42), and the half-brother of Edward Wanzagi Nyerere, the last colonial chief of Butiama and of the reunified Zanaki Federation (r. 1952–61).[129] In spite of his training and later his baptism as a secondary school student in the Catholic Church, Mwalimu Nyerere grew up with the concepts of the ritual health of the land as a member of the generation-set. During my stay in the region, I heard many stories about the auspicious signs surrounding his birth, his inheritance of prophetic powers, and his use of these in the politics of state. Given his background in generational authority, it is no wonder that Julius Nyerere was one of the first and only African presidents to retire, which he did using the Zanaki word for the retirement of the generation-set, *kunyatuka*, and instituting the most important symbol of national unity, the annual ritual "walk" around the nation carrying the ritual torch, the *mwenge*, from the top of Mount Kilimanjaro. In the Mara region, people explicitly discuss the mwenge as an orokoba for the nation. Some claimed that the mwenge contained medicines that were spread across the land for protection and healing. The torch's fire symbolized the spread of the new fire to each homestead throughout the ekyaro.

Mwalimu Nyerere's appropriation of ritual space to define the national community is a good example of the flexible nature of these older landscapes. Over time, sacred landscapes have been recontextualized to help people protect and heal their communities, to define new communities, and to legitimize the authority of elders. People claimed rights to the land based on their spiritual connection to it through the ancestors.

~

This way of seeing the land became clear to me as I trooped around after elders Wilson Machota and Edward Kota, looking for the emisambwa site at the Kumari spring. Because the site was in the game reserve, no one from the Hikumari descent group had come to clean it out and do the rituals for propitiating the spirit for thirty years. One elder remembered that they used to go there to kill a cow and a goat, clean out the spring, and dance for the snake, Kumari, to make him happy. When they went home, it would begin to rain. When we went, the spring was dry and barely visible, a sign that Kumari was slipping from being a benevolent emisambwa spirit of the land to becoming a forgotten and "loose" spirit, dangerous to the community. The elders lamented that if these places are forgotten and no one goes there anymore their children and grandchildren will forget the ancestors and not be able to gain their protection, causing the land to lose its fertility.[130] The sacred landscapes of oral tradition and ritual provide the most compelling evidence that the collective historical memory of the western Serengeti is indeed written on the landscape.

PART II

Landscape Memory and
Historical Challenges

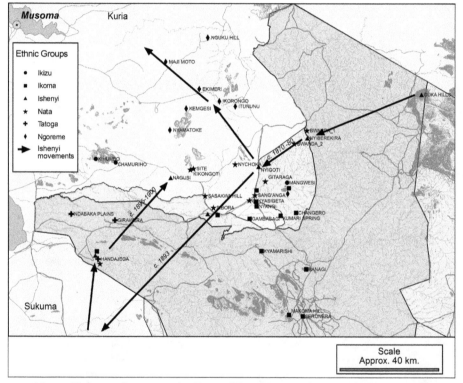

Map 7. Ishenyi settlements in the Time of Disasters. *Map by Peter Shetler, 2005. Underlying GIS data courtesy of Frankfurt Zoological Society*

4 ⤳ The Time of Disasters

Creating Wilderness, 1840–1920

IN THE LATE NINETEENTH CENTURY a series of disasters brought significant challenges to the imagined ecological, social, and sacred landscapes of the western Serengeti. As in other parts of East Africa, drought, epidemic disease, interethnic warfare, and ecological collapse marked western Serengeti incorporation into both a regional system controlled by the Maasai and a global Indian Ocean trading system. It is in the oral traditions about the time of the disasters that we can see how the generative principles embedded in older core spatial images get recontextualized in new kinds of oral traditions and how new core spatial images of loss and dispersal are generated. It is here where the static ways of seeing the landscape presented in earlier chapters take on historical motion. Western Serengeti peoples successfully responded to the disasters by integrating the generative principles of core spatial images from older imagined landscapes into new historical realities that radically transformed their societies. This time period also introduces a new way of seeing the landscape from the Maasai, which turned the older ecological landscapes of interdependence into a system dominated by pastoralist control. Maa-speaking people came to eastern Africa from southern Sudan sometime during the first millennium CE, gradually specializing in pastoralism as the defining mark of their ethnic identity—in contrast to Bantu-speaking agriculturalists and Okiek-speaking hunter-gatherers. Maasai defined a landscape centered in the plains but reaching out to incorporate other ecologies into its system through raiding.[1] Western Serengeti peoples both accepted and resisted this landscape by integrating it into their own understandings and strategies of survival as well as generating new core spatial images. The time of the disasters also left a significant impact on the appearance of the Serengeti: the dispersed pattern of homesteads was consolidated into fortified settlements, large areas were

135

depopulated and recolonized by the bush, epidemic disease killed wild animals as well as livestock, and an empty wilderness was created where people had once lived.

Dateable history from the perspective of western Serengeti oral traditions begins in the second half of the nineteenth century. While the oral traditions discussed previously provide no dateable account of events or personalities, elders narrate the stories of this period, including battles with Maasai, suffering during famines and migration to Sukuma, with elaborately precise details that written sources corroborate. This historical moment, when western Serengeti traditions became historically grounded in verifiable events, represented a time of transition in social identity. Judging from observation of other societies, because specific social groups transmit oral traditions to promote their own interests, then groups developing new forms of social identity will transmit new kinds of oral traditions generated from recontextualized earlier material. Among societies with centralized states, the traditions of historical time often begin with the consolidation of the kingdom under a known king, even if recontextualized genealogies extend the antiquity, and thus the legitimacy, of the kingdom in oral tradition.[2] In the oral traditions of the Maasai, historical time begins with the leadership of prophets in the late eighteenth century.[3] Historical time in the western Serengeti begins with the events of the disasters, representing a rupture in previous ways of understanding social time and in social identity.[4]

In addition to recontextualizing older core spatial images, the oral traditions describing the disasters incorporated new core spatial images of loss and dispersal. An Ishenyi example of a disaster narrative provides a view of what this rupture in time meant for historical consciousness and local interpretations of the events of the late nineteenth century in terms of past landscapes. The story begins with familiar core spatial images of the interactions of herders and farmers and the sacred places where spirits are propitiated. It takes place at Nyiberekira, a settlement marked by Bwinamoki, a tall rock outcropping that served as an outlook for Maasai raiders and associated with a pool on the Grumeti River, an erisambwa site of ancestral spirits that today lies within the western boundaries of Serengeti National Park.[5] It ends with the necessity to leave that fertile place and the dispersal of Ishenyi people. Mikael Magessa Sarota, the son of one colonial Ishenyi chief, told this version of the story that is illustrated in map 7:

> Long ago the Ishenyi lived at Nyiberekira, over to the east of
> Mugumu, inside the park. There are hills there, a fertile land that
> cries buubuubuu . . . when you walk on it. The land was called
> Nyiberekira. This is where we came from. When we left there, we
> came to Nyigoti. The Maasai drove us out in the time before my
> grandfather. The Maasai raided us. We were farmers, and they were
> herders. The Maasai came to steal the few cattle that the Ishenyi

kept. The Ishenyi had a dream prophet at that time named Shang'angi. When the Maasai would come, just enter their land, Shang'angi would make biting ants appear, which would fill the river and prevent them from crossing. Then he would make bees that would swarm all over and drive them back. They would be unable to raid the Ishenyi. Then one day the Maasai sat together to decide how to defeat this Ishenyi prophet who sent ants and bees against them. They went to their own prophet, who could stop the rain from coming. This prophet stopped the rain from falling on Nyiberekira. At this time the Ishenyi were farming with wooden hoes because the soil was so fertile and loose. They had ample food, and there was no hunger. This Maasai prophet stopped the rain from falling for eight years. All the food stores were exhausted. When this happened, the Ishenyi went to their prophet, Shang'angi, and asked him to send rain. He said that he was not a rainmaker and only knew the medicine of war against the Maasai. Nevertheless, they would not listen and sat in their meetings over and over again, asking him to make rain. Finally they decided that Shang'angi must be lying to them—how could he be such a powerful prophet and not know how to make rain? Surely it was not true! When the drought continued, they decided to kill Shang'angi. They tied him up on a tree and chopped down the tree, which fell and killed him. Yet of course once he died there was no one to protect them, and they were driven out, dispersed here and there by the Maasai raiders. Some came to Nyigoti, near where the Nata were living. Others refused to come to Nata because the Nata were sick with *kaswende* [syphilis]. Many Ishenyi warriors impaled themselves on their own spears rather than go to Nata and suffer the slow death of kaswende. Others went to Kuria, where they became the Iregi of today, those of the clan of Seriga. The Iregi are really Ishenyi people. Those who moved near the Nata settled and began living there. Then another famine came and they were forced to go to Sukuma to beg for food. After the drought lifted, they returned from Sukuma and came to live where they are today, at Nagusi. They did not return to Nyigoti. This terrible famine was called the Hunger of the Feet. It was called that because of the sores people got in their feet as they walked to Sukuma in the dust.[6]

This story both reaffirms the importance of older ways of imagining the landscape as well as introduces radical reinterpretations of those landscapes within a new context. The ecological, social, and sacred landscapes described in the first three chapters are all obvious in this passage: the distinction between the space of farmers and herders, the use of wild creatures as weapons, the distributive power of prophetic leadership, the mobilization of clan networks in times of trouble, and the sacred places of the wilderness. Yet the new context also generates new core spatial images of loss of dispersal, including killing a prophet who cannot meet the new challenges from the Maasai, migrating to new areas rather than just going to look for food, and considerable social stress. New forms of leadership, new kinds of settlements and confronting an upset in the interdependent balance between farmers and herders are the result of using older principles to generate new social forms. This reinterpretation was precipitated by the unprecedented disasters mentioned in the Ishenyi text—cattle raiding, drought, famine, and epidemic disease, in the context of the Maasai's emerging regional role. Many western Serengeti peoples, like the Ishenyi, moved some distance west of their nineteenth-century settlement sites to escape Maasai raids, dispersing and regrouping in a new place with a new identity. Shang'angi's murder was the Ishenyi's final break with Nyiberekira and with their identity as a people under his leadership. It is only in recent years that some Ishenyi returned to Nyiberekira to propitiate the spirit of Shang'angi, the murdered prophet. The Ikoma also have a story in which a prophet forbade them to return to Sonjo, forcing them out to new lands and the formation of new communities.[7]

In spite of the lack of chronological dates in historical traditions, many elders agreed that the prophet Shang'angi belonged to the generation of Maina (ca. 1840) and that the Ishenyi left Nyiberekira during the generations of the Saai and Chuma (ca. 1860–80). This matches Mikael's assertion that the dispersal from Nyiberekira happened in the time before his grandfather. The Ishenyi lived at Nyigoti during the generation of the Saai and the Nyambureti (ca. 1870–90). Ishenyi elders further agreed that the Hunger of the Feet and the eventual move to Nagusi took place during the Kihocha age-set, just before the turn of the century.[8] Traditions of other ethnic groups in this area relate similar tales of disaster during these age-sets. It is also during this period that the elders begin to identify events with age-set rather than generation-set names, indicating a fundamental shift in the way people perceived the passage of time. One Nata elder said that they formed the first saiga, or age-sets, during the generation of the Maina (ca. 1840), living at Site, where they divided that generation into the three age-set cycles of Bongirate, Busaai, and Borumarancha.[9] An Ishenyi elder confirmed that they divided the cycling age-sets when they left Nyiberekira, during the generation of the Maina.[10]

Western Serengeti peoples faced the challenges of drought, disease, famine, the caravan trade, and Maasai raids during the nineteenth-century disasters. The old ecological landscapes began to break down as people lost control of the environment, allowing the encroachment of tsetse fly bush that forced them to move west out of ancestral lands. The interdependent ecologies of Bantu-speaking farmers, Tatoga herders, and Asi hunters also collapsed as the Maasai forged a militarily and culturally dominant position based on a new view of a regional landscape. The marginalization of western Serengeti peoples in the new landscape generated core spatial images of loss and dispersal. Though they still envisioned ecological landscapes in their attempts to cope with environmental collapse and to order their relationships of both accommodation and resistance to Maasai power, western Serengeti peoples were less successful than in the past. The social landscapes of regional networks, embedded in clan traditions, allowed western Serengeti peoples to cope with famine and to form new kinds of identities through regional connections. By recontextualizing clan traditions as emergence traditions, western Serengeti elders legitimized new kinds of authority that developed as a result of the disasters. Western Serengeti peoples used the symbols of older sacred landscapes to encircle and enclose new kinds of concentrated and fortified settlements as protection against Maasai raids and to enhance the power of the elders over youth, who had found some independence in their warrior status. A ritual definition of territory allowed them to create new kinds of territorial age-set cycles that both addressed Maasai dominance and provided continuity with the past. When Europeans first entered this area around the turn of the century, they formed their own view of the Serengeti landscape when they "discovered" a vast empty wilderness in the Serengeti, only recently vacated as people moved to escape raiding and the encroaching tsetse fly bush, and surmised that permanent human habitation had never been possible in this inhospitable land.

THE DISASTERS:
DROUGHT, DISEASE, CARAVAN TRADING, AND MAASAI RAIDING

The new context of the late nineteenth century was, by the accounts of people who lived through "the generation of disasters," a most challenging time when a series of devastating droughts with subsequent famine and the spread of epidemic disease eventually led to ecological collapse across the western Serengeti.[11] One Ishenyi elder said that they called the famine at Nyiberekira the Hunger That Finished the Cattle (*rimara ng'ombe*), in which the Sonjo or Regata peoples also left Nyiberekira and went to their present homes in Sonjo.[12] Without doubt there were climatic factors involved in these disasters. Precipitated by El Niño drought patterns and imperial policy, much of Africa, Asia, and northeastern South America experienced drought from 1876 to 1879, 1889 to 1891, and 1896 to 1902, resulting in twenty to thirty million deaths worldwide.[13] Although rainfall data is not available for the western Serengeti, we know that 1895 to 1900 and 1920 were particularly dry years in East Africa, following an unusually wet period.[14] The worst famine in western Serengeti is called the Hunger of the Feet (*agecha ya maghoro*), distinguished as a new kind of famine because of its widespread impact. The White Fathers, missionaries who had established themselves on Kerewe Island in Lake Victoria by 1893, dated this Great Famine on the mainland to 1894, when many people from the mainland came to the mission in search of food during the drought.[15] They reported that a small village of Christians had grown up around the mission station, most of whom were former slaves and famine victims from Maasai raiding in the interior, particularly Ngoreme.[16] As the famine abated, these converts returned home and the White Fathers made journeys in 1902 and 1904 to maintain contact with them, establishing the mainland station of Nyegina with the purpose of reaching Christians in Ikizu, Ngoreme, Zanaki, Majita, and Ruri.[17] By 1919 most of the famine victims had gone home, leaving the mission practically deserted.[18]

But drought does not necessarily cause famine unless there are other conditions that increase a community's vulnerability. At this time drought conditions became famine conditions when new diseases, to which western Serengeti peoples had no immunity, hit the already weakened people and livestock. The outbreak of rinderpest, introduced to Ethiopia by the Italians between 1880 and 1890, killed up to 90 percent of the cattle in East Africa. After this, pastoralist peoples such as the Maasai were reduced to "walking skeletons," as the German traveler Oskar Baumann described the starving Serenget Maasai who had taken refuge in Ngorongoro Crater highlands in 1892. He also found Maasai sick and dying on the Lake Victoria coast.[19] Oral traditions about the Hunger That Finished the Cattle refer to rinderpest as well as other cattle diseases, such as East Coast fever and cattle lung disease, that swept through both before and after the drought.[20] The Ishenyi tradition told above also mentions kaswende (syphilis) in Nata, where young men would consider suicide preferable to exposure. Scholars have assumed that the nineteenth-century caravan trade introduced sexually transmitted diseases (STDs).[21] Oral traditions, such as the Ikoma Machaba story (see chapter 3) about going to consult the prophet about decreasing populations, are another indication of the prevalence of STDs, which often resulted in infertility.[22] Many elders said that because of increased infertility, intercourse with a barren woman, assumed to have an STD, became taboo.[23] Other "foreign" diseases mentioned in narratives about this period may have been smallpox, cholera, and measles, collectively associated with dysentery and dehydration.[24] One Ikoma elder described going to consult Tatoga prophet Gamurayi during the first age-set of the Rumarancha (ca. 1890) about the outbreak of a disease (nyekekundi) that caused everyone in a homestead to die suddenly. The prophet, afraid of exposure, would not come out to meet the delegation until gifts of cattle finally induced his wife to open the door.[25] The Ishenyi story mentions the foot sores, most likely chiggers from the dusty path to Sukuma, that gave the Famine of the Feet its name.

Another cause of famine was the disruptions, even if indirect, caused by integration into the global economy that exacerbated vulnerability to famine by introducing new diseases and crops, taking slaves, creating new alliances and conflicts, and introducing a market for food in exchange for desirable trade goods. The first contact with the coast came from trading routes pioneered by Nyamwezi porters in the second half of the nineteenth century. Swahili traders on the coast also pushed further and further inland, trading beads, wire, cloth, and guns for slaves and ivory.[26] Few western Serengeti elders knew stories about caravans in the region except in Ngoreme, where some remember tales of Arab slave traders. T. Wakefield's publication of "routes of native caravans from the coast to the interior of Eastern Africa," based on Arab testimony, attests to a route from Sonjo, through Ngoreme, to the coast of Ukara, north of what is now Musoma. Otherwise, the western Serengeti remained a blank space on the trader's map until almost the turn of the century. The "native" routes across the plains from Maasailand ended to the north, in Kavirondo among the Luo, in what is now western Kenya, near Kisumu.[27] However, the German explorer Baumann, when coming across the Serengeti into Ikoma in 1892, noted that local people immediately recognized his party as a coastal caravan and greeted him in a "Kinyamwezi dialect." Some of Baumann's porters deserted in Ikoma, hoping to stay "as slaves to the natives until another caravan passes." Baumann thought this foolish since many years could pass between caravans in Ikoma.[28]

The lack of caravan routes in this area was a result of geography, rumor, and the lack of an established market. Swahili caravans were afraid of crossing Maasailand to get from Kilimanjaro to Lake Victoria, and Europeans did not find a way through until Joseph Thomson's expedition in 1883–84, which went considerably north of the Mara region.[29] Early German reports warned that the Shashi (the people of the Mara region) were "predatory and warlike people" who "kept aloof from European influence, and often fell upon and massacred whole caravans that were merely crossing the country to purchase ivory in Ugaya [Luo, along Lake Victoria]."[30] Some historians surmise the Arab and Swahili traders exaggerated the dangers of entering Maasailand to keep the Europeans out of their trade.[31] The main caravan route went to the south in Sukuma and then across the Lake to Buganda, but its influence extended north as Zanzibari Saif bin Said established a trading station at Magu, just south of the Mara region, for regional slave purchases.[32]

Slave raiding and ivory hunting brought insecurity to the region, particularly as more slaves became available due to famine.[33] Locally, western Serengeti peoples used ivory only for bracelets as an emblem in the eldership title system and had a traditional prohibition against elephant hunting.[34] Yet during this period elephant-hunting associations brought new sources of wealth and authority. No local institution of slavery, except that of the dependency of strangers, existed in the western Serengeti, nor did western Serengeti peoples engage in slave raiding to supply the caravans or use slaves themselves.[35] Yet they did end up as victims of the slave trade. Colonial reports claim that during the famine of the 1890s the Kerewe Kingdom sent canoes to exchange food for children along the Mara

River.[36] Chief (*Omukama*) Mihigo II (1780–1840) began Kerewe involvement in long-distance trade by obtaining ivory from elephant-hunting associations in Kanadi, Sukuma. In the next decades Kerewe began to serve as the coastal intermediary for the powerful Ganda Kingdom across the lake. Kerewe Island was a source of famine food for western Serengeti peoples because of its higher and more reliable rainfall and the people's willingness to trade food for children.[37] Europeans first came to the Great Lakes region following these caravan routes beginning in 1861, with Emin Pasha, representing the German East Africa Company, bringing an end to the Arab trade on the lake and the slave market at Magu in 1890.[38] Evidence that the fear of ending up in the European slave trade reached into this isolated region comes soon after the turn of the century from a missionary who reported that his student told him that the porters "whispered around the campfire in the evening" that "there is a famine in their country and they are going to take our cattle and children to salt them down for shipment to Europe."[39]

Yet even drought, new disease, and the demands of the global economy may not have been enough to reduce the region to famine had they not come in tandem with the increasing regional hegemony of the Maasai. Oral testimonies, like those of the Ishenyi, most often blame Maasai raids for the disasters. Although the Maasai had been present seasonally in the region since the eighteenth century, territorial expansion based on a specialized form of pastoralism and new forms of prophetic leadership did not develop until the first half of the nineteenth century, and probably not before mid-century.[40] Maasai put themselves at the center of a new regional system composed of many different ethnic groups, and, through increased raiding and the incitement of fear by Maasai age-set warriors, they forced their victims to either move away or abandon pastoralism.[41] Maasai occupied the Serengeti plains and Ngorongoro highlands by 1850, having driven out the Tatoga from the Ngorongoro Crater during the Dwati age-set of 1836–51.[42] When the Purko-Kisongo Maasai, under prophetic leadership, completed their expansion into what is now south-central Tanzania between 1850 and 1890, those sections who had lost the wars began raiding to recoup their cattle losses and gain control over pastoral resources among their neighbors in the western Serengeti and Sonjo, who had few livestock. Sonjo traditions recall raids by the pastoral Lumbwa, among the earliest victims of Maasai expansion. The Ikoma and Nata came into increasing contact with the Maasai Serenget section, who by 1850 had begun using land for dry-season grazing in the western part of what is now Serengeti National Park.[43] The Serenget section was strong until 1890, when the Loitai Maasai absorbed the Serenget after winning the civil wars precipitated by competing prophets.[44] The Siria Maasai of the present Narok area raided and competed for pastoral resources in Kuria, and perhaps Ngoreme, as they were pushed into this area by Loitai Maasai expansion from the Rift Valley.[45]

We cannot understand western Serengeti creativity during this period outside their subordinate and peripheral position within this hegemonic Maasai system, both in terms of their acquiescence and resistance.[46] The Maasai pastoral lifestyle depended on interaction with farmers and hunter-gatherers within a regional economic system in which they controlled access to cattle as the main source of wealth and imposed economic specialization on everyone else in the region. At first, western Serengeti peoples seem to have accommodated and admired the Maasai, taking on many Maasai cultural innovations as their own.[47] But as the Maasai developed a highly specialized form of pastoralism, the older regional system — in which farmers also hunted and herded and the Tatoga pastoralists also farmed — broke down. The pre-Ishenyi community at Nyiberekira was one victim of this increasing competition for dry-season grazing grounds and water points, located as it was on the edge of the Serengeti plains. The Maasai forced the agropastoral community that once straddled the Serengeti plains back into the hills and woodlands on its margins, both east (Sonjo) and west (Ikoma). As is clear from the experience of the western Serengeti, Maasai power lay in both military and cultural domination.

Stories employing the core spatial images of loss and dispersal include Maasai raids as the most obvious and resented symbol of Maasai domination that affected all aspects of life. Raids took place mainly during the dry season, in the predawn hours, when the raiders surprised the village by stampeding the cattle, burning the houses, and sometimes killing men or taking the women and children captive. The Maasai fought with spears and shields, while the western Serengeti peoples mainly used bows and arrows, giving the Maasai an advantage in the close combat conditions of a surprise attack. J. E. G. Sutton attributed Maasai military superiority to the use of larger, socketed spear blades along with novel forms of military organization and tactics.[48] When the White Fathers took a trip to the interior of the country, in 1904, they reported that all along the lake people lived in fear of Maasai raids which resulted not only in loss of cattle but in burned houses and fields and general devastation.[49] In 1902 the Germans built Fort Ikoma in the western Serengeti to bring an end to Maasai raiding and provide a way station along the route to the plantation areas in Arusha.[50]

The core spatial images of loss and dispersal inherent in Maasai raiding stories from a number of different periods tend to become homogenized into one generalized narrative, even through the context changed dramatically over this period. Maasai raids to establish hegemony were altered when they, too, fell prey to the disasters between 1883 and 1894, with the epidemics of pleuropneumonia, rinderpest, and smallpox that brought severe livestock and human loss. The Maasai responded with increased cattle raiding to restock their herds, but from a weakened position.[51] By the beginning of the twentieth century, missionaries and German colonial officers just entering the region identified Maasai raiding as the most significant obstacle to the establishment of a productive colony. The White Fathers reported in 1902 that all along the lake people lived in constant fear of Maasai raids from the plains. As late as 1911 the White Fathers were still recruiting most of their believers from among the victims of Maasai raids.[52] A period of increased raiding was partially brought about by the improving status of Mara peoples. The next chapter describes the new relationship that was developed with the Maasai as western Serengeti peoples began to recover from the disasters in the beginning of the twentieth century. Western Serengeti agriculturalists began to gain more livestock after the famines and move out onto the plains, therefore coming into direct competition with the Maasai for the same pastoral resources and more open to raids. The events of World War I exacerbated this period of increased raiding when Kenya Maasai took advantage of chaos on the border to make raids into the German colony.[53] Narrators describing the earlier period of the disasters may have been influenced by their more immediate experience in this later period of more intense Maasai raiding.[54] And because Maasai raids continue today, it is difficult to know how much raiding actually influenced famine and disaster in the late nineteenth century.

The crisis of the late nineteenth century was thus a combination of many factors: drought, new diseases, the introduction of a market demand for slaves and ivory from the caravan trade, and Maasai raiding—the combination of which represented an enormous challenge to the survival systems that western Serengeti peoples had developed over the millennia. They narrated the story of this crisis by employing new core spatial images of loss and dispersal. Western Serengeti peoples responded in creative ways to the disasters by adapting the generative principles embedded in the core spatial images of ecological, social, and sacred landscapes to this new historical context.

THE BREAKDOWN OF ECOLOGICAL LANDSCAPES
AND THE FORMATION OF MAASAI RELATIONSHIPS

This sequence of disasters resulted in ecological collapse when western Serengeti peoples were no longer able to keep the balance between the forces of the wilderness and the domesticated homestead, so evident in the ecological landscapes of oral tradition. A view of the landscape in which peoples living in different ecological zones and practicing different subsistence strategies maintained interdependent relations was severely challenged and reconfigured as a result of the disasters. The system of sharing food when drought and disease devastated people in other ecological zones certainly continued to exist but came under increasing strain as Tatoga and Asi moved away and Maasai were increasingly seen as enemies.[55] As a result, western Serengeti peoples had to find a way to reestablish the links between ecological zones and economies within their own communities.

By the time the sleeping sickness epidemic broke out in the Mara region early in the century, ecological collapse was well underway. Throughout Tanzania, raiding and increased uncertainty resulted in the concentration of settlements, competition for accessible farmland near the settlements, and overworked soils and pasturelands in those areas. As depopulation reduced the number of settlements and farms, as well as the ability to burn the old grass and clear new land for farming, tsetse bush encroached into once clear areas and sleeping sickness resulted.[56] In 1902 the Germans near their base at Schirati on Lake Victoria identified the first clinical case of sleeping sickness in East Africa, after which time they gave the disease top priority in research and public health.[57] In 1913 early European hunter Stewart Edward White found that all of North Mara was under sleeping sickness quarantine. He reported that the sleeping sickness epidemic at Ikoma was of recent origin and found abandoned villages in Ngoreme and along the Ruwana River. There, according to White, "the deadly fly had of late years extended its boundaries" and was gradually "crowding the cattle-raising savages inward."[58] There were outbreaks of sleeping sickness in Musoma District throughout the colonial period, indicating that tsetse fly bush continued to expand. Early travelers' accounts indicate that the "sleeping sickness belt at Ikoma" was seen as an effective barrier to trading caravans trying to reach Lake Victoria from the Kilimanjaro Region.[59] Early white hunters on the Serengeti plains were afraid to go west into the tsetse-infested areas of the western Serengeti. The German gold mine at Kilimafeza ("mountain of gold") located, as the hunter said, "on the edge of nothing," in what is now the park, was known by Europeans as "Kill-a-man-easy" because so many Europeans had died of sleeping sickness or malaria there. Even Ikoma hesitated to make unnecessary trips to the Serengeti plains through the tsetse bush.[60]

The visible Serengeti landscape of woodland and grassland changed significantly during this period, rendering many areas that were once open plains into unhealthy bush and resulting in the loss of control over large areas of land once considered integral to the western Serengeti economy. Yet in the areas where people still lived, the land retained its orchardlike appearance. In 1913 European hunter White described the country around Ngoreme as "a perfectly flat green lawn of indefinite extent," with the "grass short as though mown," planted "sparingly with small trees with white trunks" that were "far enough apart not to spoil the open appearance, but thick enough to close in the view at a quarter of a mile" and the lawn scattered all over with flowers.[61] Although this evidence provides some indication that the ecological landscapes of the past were and perhaps still are intact, much had been lost, and this view of the landscape came to have less relevance in the years to come. As both the colonial and independent state claimed authority over the Serengeti, it increasingly stripped local people of their ability to exert control over the environment and eroded the leadership necessary for the task.

Ecological collapse also brought an end to the interdependent regional economy of Asi hunters, Tatoga herders, and western Serengeti farmers who had relied on each other as an insurance against bad times before the disasters. As the Maasai gained dominance in the region, the Asi increasingly entered into client relations with the Maasai and moved further east, toward what is now Loliondo, abandoning their interactions with western Serengeti farmers in the woodlands.[62] The Tatoga herders came under more pressure from Maasai raiders than the farmers since they were in direct competition with the Maasai for pastoral resources. The disasters also significantly reshaped Tatoga historical consciousness. One Tatoga elder divided his list of generations into those of the founding prophets and those beginning with the Great Hunger. In the Ngorongoro Crater, the Maasai defeated the Bajuta Tatoga and took possession of the crater around mid-century. Tatoga elders tell stories of how the great prophet Saigilo warned them three times to leave the region until the Maasai, famine, and disease soundly defeated them and they finally followed the prophet south to Nyamwezi. Saigilo's son Mahusa was killed by the Germans in Hanang', among the Barabaig Tatoga.[63] Bush encroachment also squeezed them out, as it rendered formerly productive pastures unusable and dangerous for cattle.[64] While some, but not all, Tatoga herders moved south, Asi hunters increasingly moved east as they accepted the patronage of the ascendant Maasai, and Bantu-speaking farmers moved west, further into the hills, to avoid raids. Oral accounts, however, make it clear that western Serengeti peoples continued to share their food when herding people were in trouble. Maasai and Tatoga herders settled near the western Serengeti farmers after the rinderpest outbreak and traded their children or worked as herders for food. Baumann went through Ikoma in 1892, just before the Hunger of the Feet, and described the surplus of grain brought for trade by the "peaceful inhabitants," enough "to pass through Masailand again if we wished." Kollmann, some years later, also reported full granaries and fat cattle.[65] But by the mid-to-late 1890s both resources and relationships were stretched thin.

Because the ecological landscapes of oral traditions about first man the hunter and first woman the farmer had influenced how western Serengeti peoples approached their relationships with people practicing other subsistence strategies in the past, they tried, with some success, to apply this to their relationship to the increasingly dominant Maasai. Yet this view was not entirely helpful since they now found themselves in a subordinate position. As a result they began to differentiate Maasai pastoralists from Tatoga pastoralists, naming people who lived in the wilderness *Nyika*, including the Maasai and some Asi hunters as enemies, *bisa*, as opposed to *rema*, or farmers.[66] The Tatoga were not bisa, as they and western Serengeti peoples cooperated together during Maasai raids. If a Tatoga killed a Nata, Ikoma, or Ishenyi it was just like killing another Tatoga, since Tatoga performed rituals of purification and paid a fine to the family.[67] The Tatoga were still spiritual "fathers" to some western Serengeti groups and participated in the rituals to "cool the land."[68] Both western Serengeti and Sonjo "farmers," as well as their Tatoga pastoralist allies, represented their animosity toward Maasai enemies in the *aghaso*, a ritual to purify and reward young men who killed a lion, leopard, or Maasai.[69] While still at the lion kill they cut out the heart of the beast, ingesting the tip, and took back the skin and claws as trophies. Maasai weapons and other things were also taken as trophies. Back in the village the community celebrated with eight days of feasting and gifting, after which the killers underwent a ritual of purification.[70] The aghaso explicitly categorized Maasai with the beasts of the wilderness, who are dangerous but also respected and admired. One elder said that the heart is the place of courage, the essence of the beast; by eating part of it one gains that courage. As in other rituals, western Serengeti peoples domesticated and managed the things of the wilderness by bringing them within.[71] Yet this was also a direct act of resistance to Maasai dominance. As one informant said, "the Maasai was boss then and the aghaso proved our triumph over them."[72]

In spite of these expressions of animosity toward the Maasai, western Serengeti peoples also interacted and cooperated with Maasai according to the older interdependent system of pastoralists and farmers that was held together not only by force, but also by the glue of common cultural understandings and social interaction with frequent boundary crossing.[73] Maasai ancestors often appear in Ngoreme and Ikoma genealogies, still propitiated using prescribed Maasai implements and cattle sacrifice.[74] One story of local origins in Ngoreme tells about the Maasai named Saroti, who was left behind by his family during the time of hunger. He became friends with an Ngoreme man, Matiti, who taught him how to farm and offered his daughter Nyaboge in marriage for the brideprice of one storage bin of grain. Nyaboge cut Saroti's hair to make him acceptable for marriage, and together they gave birth to the clan that now populates this area.[75] According to western Serengeti patterns of inclusion, once Maasai were adopted into the community, they were no longer named as Maasai and have become invisible in the genealogies. These interactions continued into the colonial years as Serengeti Maasai experienced famines in 1943, 1946, 1949, and 1952–53 that brought them further west for dry-season grazing near the Moru kopjes and to trade livestock for food with the Ikoma.[76]

These interactions created a new regional culture in which western Serengeti admiration and accommodation of the Maasai led them to adopt many Maasai cultural traits, to the point where many European observers of the time concluded that they must be "of Maasai blood."[77] Elders say the Ikoma, Nata, and Ishenyi practice resembled the Maasai in dance, ornamentation, and songs.[78] Although western Serengeti youth did not use the red ochre characteristic of the Maasai warrior in everyday ornamentation, young men would put it in their hair to dance, after circumcision, or for cattle raids. When the White Fathers visited the Ngoreme in 1904, they reported that the young men wore their hair in butter-and-ochre-smeared plaits, as did the Maasai.[79] At the dances held at the full moon, young people wore rows of brass or wire-wrapped anklets and bracelets, beaded headgear, and ear ornaments, all products of long-distance trade. It is not clear when Mara peoples began to pierce and elongate their earlobes, as the Maasai did, but people now think of it as traditional.[80] Present-day Kuria peoples shown the photos of Kuria taken by Max Weiss in 1910 could not believe that these were Kuria ancestors and not Maasai.[81] Early German notes on the "tribes" of the western Serengeti categorize Ikoma and Nata as "lands of the Maasai."[82] An early map of native caravan routes from Wakefield shows the whole western Serengeti region inhabited by the Maasai "Lumbwa."[83] Indeed, western Serengeti peoples must have been in appearance indistinguishable from Maasai.

The Ikoma in particular, who lived closest to the Maasai, seemed to continue to relate to the Maasai according to the older vision of interdependent ecologies. During the rihaha famine, or rinderpest of 1890, the Maasai came to "sell" their children in Ikoma for food. Many stayed and settled near Banagi Hill, well into Ikoma territory and now part of Serengeti National Park. Ikoma clans adopted young Maasai men and married young Maasai women, establishing in-law relationships of long duration. When the Maasai began to recover and the Hunger of the Feet hit the farming peoples in 1894, some Ikoma went to the Maasai for help. The Maasai were raiding on the lake during this time and used their Ikoma friends as scouts because they knew the land better. Even today western Serengeti peoples know some Ikoma as Maasai collaborators.[84] A 1933 report from Musoma District stated that the Ikoma were "on very friendly terms with the Serengeti Maasai," who brought wild animal tails for exchange. Furthermore, since the Maasai could "always rely on a bed and a meal" in Ikoma, one could not expect them to "abandon friendly relations which have survived the raids."[85] Collectively the Maasai were enemies, but individually they were friends and allies.

One of the most often told tales of Maasai alliance and enmity, the story of the Battle of Ndabaka, illustrates the ambivalence of western Serengeti peoples toward the Maasai as enemies, but also as a part of an interdependent system united by common age-sets and alliances. The battle took place in what is now the western corridor of Serengeti National Park, near the Ndabaka gate outside Bunda, where Nata, Ikoma, Ishenyi, Sukuma, and Tatoga people had fled during the disasters and built their homesteads together. Mariko Romara Kisigiro told this version from a Nata perspective:

> Nata, Ikoma, and Sukuma built together at Handajega. So many
> were killed there—the War of Handajega. The Maasai would pass
> there when they went to raid for cattle along the lake, in Buringa.
> Yet when the Maasai got to Handajega they saw that these people
> had built right in their path. So they went to the village of Handa-
> jega. At that time it was mostly Ikoma, of the Saiga of the Kubhura,
> they were called the Romore. The Nata living there were of the
> same saiga from the Borumarancha age-set cycle. The Maasai were

of the same age-set, with the same name of Romore. So they talked together. The Maasai said, "We want to go west, to the lake, to raid cattle. We are the Romore, let us pass and do not get in our way." The Ikoma agreed and swore an oath [*kula ring'a*] with them and the Maasai went on their way to the lake.

Nevertheless, after they left there was much talk about it, and others disagreed with the oath. The Nata and Sukuma who were there said that they should fight the Maasai instead of letting them pass to gain glory. They suggested that the age-set should go to see a prophet to find out if they could defeat the Maasai. So the Romore youth went to see the Tatoga prophet Gorigo. The elders told the youth that they should not fight because they had sworn an oath. The prophet slaughtered a lamb and cooked it for them, then he cooked a small bit of porridge. He also put a small bit of milk in a horn. He told the gathered youth that they should eat the food and finish it all. They ate but they could not finish the food. The prophet told them to eat, but they could not finish it. So the prophet told them that they should not fight the Maasai, because if they could not finish the food they could not finish the battle. So the youth went home. While they were still on the road they began to talk again: "Is it really true that we are not able to win the war?" Others said, "No, we could win, did not the prophet say we should fight?" So when they got back to the village, they told everyone that the prophet had told them to fight.

From that day on they prepared for the battle by sounding the alarm call to all corners, from Sukuma all the way up here to Nata. They waited two days, and still the Maasai did not return. On the third day they said, "If the Maasai do not come today we will go home." On that very day they looked to the west and saw many forms on the horizon. When the Maasai looked to the east, they also saw many forms on the horizon. So the Maasai sent a young man ahead to find out what was going on. The age-set at Handajega told the Maasai messenger, "We do not have any more words, we are ready for war." They sent the message to tell the Maasai that they wanted war. The Maasai discussed it among themselves. They offered to leave some cattle, but the Romore age-set refused. So the Maasai made a fire and the smoke went straight up. They said, "Well, the medicine says that we will win; let us fight then." The farmers all made a line that reached from the Mbalageti River to the plains. The Maasai made a line that reached across the Mbalageti River. On that plain they began a fierce battle. The Maasai pushed back the line in the middle. So many were killed. They went on fighting until everyone was mixed up on the battlefield. The Maasai fought with spears, and the Ikoma with bows and arrows. The farmers were defeated. So many were killed that day.[86]

Both sides lost so many men in the resulting battle that the Tatoga named the field Ndabaka, the plain of tears.[87]

Ultimately, however, western Serengeti peoples defined their newly forming identities during the disaster years in opposition to, rather than in interdependence with, the Maasai. The location of Ngoreme made them particularly vulnerable to raids from the Siria Maasai across the Kenya border, resulting in the conflict known as the wars of the Mairabe generation (ca. 1870–95).[88] Ngoreme clans reorganized their age-sets into five relatively autonomous age-set cycles during the disasters, between which animosity, rather than cooperation, was more often the rule. One elder described conflicts in which Ngoreme youth got help from the Maasai to fight other Ngoreme territories or fought on either side of Maasai conflicts: "They [Ngoreme] fought each other, clan against clan, and each clan had its own village. . . . At that time the Ngoreme did not assist each other in battle, one age-set fought and the others would watch without helping. That is why the Maasai won so often, because their age-sets worked together."[89]

In order to find sanctuary at Ikorongo, one Maasai clan took an oath of friendship with the Ngoreme, using the symbols of a hoe handle and a cow and learning their language and customs. Yet when other Maasai came to take revenge, some Ngoreme betrayed their oath and many were killed on both sides. The Ngoreme eventually overcame their divisions and united to drive out the Maasai and bring an end to the raiding. This story indicates that they "became Ngoreme" when they united to defeat the Maasai, rather than that they fought the Maasai because they already identified themselves as a united Ngoreme. The Maasai fell into the hole at Kimeri, an erisambwa site of the Ngoreme Tabori lineage in the bush, as they ran from the Ngoreme army. The ancestral spirit of Kimeri cried out when the Maasai fell in, "Uuuuuwiiiii, what can remove these corpses so that my children might sleep?" Then the corpses were all flung into the air and out of the hole.[90] The ancestral spirits, as "guardians of the land," appeared to bless this decisive moment of identity formation. The conflict itself led to the distinction of Ngoreme as separate from the Maasai, while previously a number of different agricultural clan territories coexisted in interdependent ecological niches with different pastoral sections.

The nineteenth-century disasters severely tested and in some ways invalidated the western Serengeti view of a humanized landscape, ecologically differentiated among peoples practicing different economic subsistence patterns. Western Serengeti peoples did not fully regain their ability to control the environment, nor did they find entirely beneficial ways of interacting with the Maasai as they had historically with other pastoralists. Just as Bantu-speaking farmers marginalized and assimilated hunters in the distant past, so western Serengeti peoples now found themselves incorporated as peripheral players in a larger regional system dominated by the Maasai. They formed new identities in opposition to Maasai as the enemy, even as the older patterns of coexistence continued to operate on an individual level. The encroachment of tsetse fly bush forced many to evacuate, relegating to wilderness the lands they once had called home.

MOBILIZING SOCIAL LANDSCAPES TO COPE WITH FAMINE AND LEGITIMIZE NEW FORMS OF AUTHORITY

During this period, the social landscapes of the past were still evident in the strategies of diversification, inclusion, and distribution that western Serengeti peoples used to cope with the disasters. Regional networks were mobilized as people went to ask for food (kuhemea) among those who had harvested better that year. Ngoreme elders confirmed the movement to Kerewe Island for food while Ikoma, Ishenyi, and Nata said that drought victims were more likely to go to Sukuma. The extreme severity of the famine is evident because those who went had nothing but their children to trade for food. Elders remembered their grandparents, who would have been children during the famine, telling the stories of lost siblings and walking to Sukuma. The trade of children for food was made socially acceptable as a "bridewealth payment" for girls, leading some parents to dress their boys as girls.[91] The White Fathers also described refugees selling their children into slavery on Kerewe to get food.[92] Gerald Hartwig's reconstruction of Kerewe history shows that from 1850 to 1870 the mainland "Ruri" people from near Musoma brought children, probably kidnapped from neighboring peoples, to Kerewe to sell for food. Then, from about 1875 on, the Kerewe themselves actively searched for children to buy as slaves during the famines by taking boats along the lakeshore and up the Mara River.[93] Most of these children would have been incorporated into Kerewe or Sukuma families, while others were sold on to the clove plantations of Zanzibar or the sugar plantations of Mombasa after the international slave trade had been abolished.[94] Elders said that many who went to Sukuma for food ended up staying a number of years, during which time they found Sukuma patrons who provided support and protection. Sukuma hosts provided the refugees with a plot of land to farm and a place to build their house in return for labor on the host's farm.[95] When the drought was over western Serengeti peoples began slowly moving back home.

The story of first woman, Nyakinywa, becoming the chief of Ikizu is one example of how the social landscapes of oral traditions were reformulated in the context of the disasters. In chapter 1 we saw how the older Ikizu origin story of Muriho establishing a new land came to incorporate a later story of first woman and first man, Samongo, competing for authority over Ikizu. Other traditions legitimize Nyakinywa's authority through her chiefly clan connections in Kanadi,

Sukuma. Because of their connection to the caravans in the nineteenth century, Sukuma people held the advantage in trade relations, sought out as patrons and later as leaders in western Serengeti communities. In Ikizu the stress of the disasters resulted in the need, or opportunity, for a more centralized authority and the creation of an Ikizu identity through a Sukuma-style Kwaya clan chiefship. The Ikizu origin story involves Nyakinywa, a rainmaker and daughter of the *utemi* (chiefly) line, who becomes chief by tricking Samongo, a prophet from the older line that descended from Muriho. Although this is told as an origin story, there is no linguistic or cultural evidence that the Ikizu came from Sukuma. But strong evidence exists that, for the first time, the Ikizu united as one ethnic group under a Sukuma-style rainmaker from the Kwaya clan in the late nineteenth century. The clan stories that established these connections were then recontextualized into the older emergence story of Muriho as a way to legitimize the new power of Sukuma rainmakers. Kwaya clan traditions in Kanadi that claim their connection to the Ikizu chiefship in the late nineteenth century and resemble the Ikizu origin story of Nyakinywa and Samongo also support this conclusion.[96] The Kwaya clan in Kanadi played an ongoing part in the investiture of a new Ikizu mtemi chief by providing ritual items and advice for the ceremony. Just as in Sukuma, the Ikizu inherit the utemi through the maternal line and give the ntemi power to mediate the forces of the wilderness.[97] Nevertheless, the Ikizu adapted the utemi to their own leadership pattern of an obugabho-style rainmaker who distributes patronage and always works together with the local prophet from another clan.

Although oral traditions legitimize the rainmaker's power by establishing continuity with landscapes of the past, the list of chiefs itself provides additional evidence that the Kwaya chiefship dates only to the late nineteenth century. Table 4.1 compares the various lists from Ikizu elders of Ikizu rainmaker chiefs, including anywhere from eight to fourteen rainmakers. In the lists of rainmakers, or *watemi* chiefs in Ikizu, Nykinywa always comes first, linking her to Ikizu origin stories. Although the authors of some of these lists assigned dates to the chiefships, only those from the beginning of the twentieth century can be corroborated by written sources. Among the five versions of this list, no two agree on the names or their order before Gibwege (ca. 1890), in the period of the Hunger of the Feet.[98] This suggests that no unified Ikizu utemi, and thus no unified Ikizu, existed before the 1890 disasters. Rather, many different obugabho-style rainmakers operated throughout what is now Ikizu. The oral traditions of emergence were recontextualized to account for this change. In this process, rainmakers from the Kwaya clan, living in various localities, were incorporated into the genealogical line of Nyakinywa to legitimize the centralization of authority.[99] With the formation of ethnic groups, such as Ikizu, the function of clans changed from that of diffuse pathways of regional knowledge to a consolidated line of power.

Table 4.1. Ikizu utemi chiefs

Informant #1	Informant #2	Informant #3	Informant #4	Informant #5
1. Nyakinywa (1815–25)	1. Nyakinywa	1. Nyakinywa	1. Nyakinywa	1. Nyakinywa
2. Nyakazenzeri (1825–35)	—	2. Wakunja	—	2. Nyekono
3. Hoka (1835–45)	2. Hoka	—	2. Wang'ombe	3. Nyakazenzeri
4. Kesozora (1845–55)	3. Nyambube	3. Kesozora	3. Kisozura	4. Hoka
5. Hoka Nyabusisa (1855–65)	4. Kirongo	4. Nyekono	—	5. Guya
6. Wekunza (1865–75)	5. Kisusura	5. Kerongo (first male)	4. Wekunza	6. Kesozora
7. Nyambobe (1875–85)	—	6. Nyakinywa II	5. Mayai	—
8. Gibwege (1885–95)	6. Gibwega	7. Gibwege	6. Gibwege	7. Gibwege
9. Mwesa Gibwege (1895–01)	7. Mwesa (first man)	8. Mwesa	7. Mwesa	8. Mweda
10. Nyakinywa II (1901–06)	—	—	—	9. Nyakinywa
11. Matutu Mawesa (1906–26)	8. Matutu	9. Matutu	8. Matutu	10. Matutu
12. Makongoro Matutu (1926–58)	9. Makongoro	10. Makongoro	9. Makongoro	11. Makongoro
13. Matutu Matutu (1959–86)	—	11. Matutu Matutu	—	—
14. Adamu Matutu (1986–)	10. Adamu Matutu	12. Adamu Matutu	11. Adamu Matutu	—

Informant #1: P. M. Mturi, "Historia ya Ikizu na Sizaki," in Shetler, *Telling Our Own Stories*, 71–5
Informant #2: Ikota Mwisagija and Kiyarata Mzumari, Kihumbo, 5 July 1995 (Kihumbu).
Informant #3: Maarimo Nyamakena, Sanzate, 10 June 1995 (Kirinero).
Informant #4: Zamberi Manyeni, Guti Manyeni Nyabwango, Sanzate, 15 June 1995.
Informant #5: E. C. Baker, "Notes on the Waikizu and Wasizaki of Musoma," *Tanganyika Notes and Records* 23 (1947): 66–69.

Table by author, 1998

Another example of recontextualizing the social landscapes of clan tradition to create new identities and reconfigure older emergence stories comes from those who claim origins in Sonjo: the Ikoma, Ngoreme, and Ishenyi. While western Serengeti peoples probably lived as neighbors to the Sonjo across the Serengeti since the distant past, the disasters—the most important of which was Maasai raiding—forced them to abandon many of these old sites. They took shelter in hill fortifications both to the east (Sonjo) and the west (Ikoma, Ishenyi, Ngoreme) of what would become Serengeti National Park, abandoning the center to the Maasai. Western Serengeti peoples and the Sonjo were natural allies in their re-sistance to Maasai domination, and, as refugees went in both directions, they called on much older fictive kin, clan, and adoption ties to seek mutual aid. The Sonjo villages had to accommodate Maasai dominance to remain in such close proximity to them, although they remained bitter enemies.[100] Each Sonjo vil-lage, situated on a hillside or mountain and surrounded by a complicated set of fortifications, depended on springs to water irrigated fields on the valley floor below. Today the Sonjo practice Maasai-style linear age-sets and know nothing of generation-sets or the cycling age-set names from western Serengeti.[101] Both the hunter-gatherer Ndorobo in the region and the Sonjo now follow the cere-monial cycle of Maasai age-sets. The first Europeans observed that Sonjo, like western Serengeti peoples, dressed and outwardly appeared like Maasai: they wore red blankets over one shoulder and adorned themselves with beaded jew-elry; the warriors (*murran*) carried long knives.[102] One elder told me that this gear was necessary for safe passage across Maasailand, as they were then indis-tinguishable from the Maasai. In deference to Maasai dominance in livestock, Sonjo who remain in Loliondo District raise only goats and sheep, hunt little, and mainly subsist on agricultural production of millet and beans. Scholars of Maasai history have shown how Sonjo could pass the ethnic boundaries of econ-omy to "become" Maasai if they gained cattle, while those who lost their live-stock became Ndorobo hunters.[103]

Although the linguistic evidence presented in chapter 1 tells us that Sonjo and western Serengeti peoples were in neighborly communication with each other for a very long time, the strongest evidence for the formation of western Serengeti identity with Sonjo origins seems to date to the period of disasters. When ques-tioned about connections to Ikoma, elders from the Sonjo village of Samonge spoke about Tinaga as one of eight villages (Buri, Hajaro, Horane, Hume, Jema, Meje, Tinaga, and Yasi) collectively known as Masabha (the north), located more on the plains than Sonjo villages today.[104] Maasai raiders destroyed Tinaga and the people dispersed, some fleeing to Sonjo villages to the south or as far west as Ikoma. One elder from the Tinaga clan said that the Maasai and Lumbwa fought with the Masabha people for many years until the Maasai took all their cattle and goats and destroyed their villages.[105] Elders from the village of Sa-mongo claim that they can still see the graves, homestead foundations, and grinding stones at the site of Tinaga.[106] Ngoreme traditions mention clans that came from Tinaga and Masabha.[107] One Ikoma version of the emergence story says that the first hunters came from Sonjo Tinaga following the wildebeest migration to get meat during a famine.[108] One elder from the Tinaga clan in Sonjo said that the destruction of Tinaga took place in the time of his grandfather (ca. 1880).[109] These migrations went in both directions from communities that were found in areas later claimed by the Maasai. The stories of the Sonjo prophet Khambageu somewhat similar to the stories of prophetic leadership among the Maasai or Tatoga, say that he came from the west, in Ikoma, and that people went there to propitiate his spirit until only a generation ago. Others say that he came from Tinaga, where many of his miracles took place and which he subsequently cursed, leading to its destruction by the Maasai.[110] Ikoma, Ishenyi, and Ngoreme elders recalled a time when they knew themselves as the Regata people. The Sonjo village of Sale is also known as Rhughata. An elder from Rhughata named origin sites and ancestors with clan names found in the western Serengeti.[111]

Recontextualized social landscapes of clan traditions were the basis for these connections to Sonjo that formed and legitimized the new identities of the Ikoma, Ishenyi, and Ngoreme by reconfiguring emergence traditions. Although the Sonjo immigrants were too few in number to change western Serengeti language or culture, western Serengeti peoples may have valued Sonjo immigrants for their knowledge of Maasai culture and for bringing compelling experience with the warrior ethos and Maasai-type age-sets. Sonjo people would have dealt with raids earlier and more intensely and would have had closer contacts with the Maasai than western Serengeti peoples. If Sonjo knowledge, or simply the proximity of Sonjo to the Maasai, provided the means for resisting Maasai dominance, then as western Serengeti peoples formulated new ethnic identities they would have acknowledged the crucial role of Sonjo in their own emergence as a people.[112] The Maasai themselves might have seen the Ikoma and Sonjo as one people. Similar to the Ikizu case, Ikoma, Ishenyi, and Ngoreme origin stories incorporated the Sonjo as ancestors as a way to legitimate new structures of authority in the region.

Whereas the ecological landscapes of the past began to lose their relevance during this period, the social landscapes of regional networks became even more important as an effective way to cope with famine and to find new forms of identity and authority while maintaining continuity with the past. It is especially apparent in this case that landscapes can be drawn over many times, reimagined to meet the demands of the moment but maintaining a link to past ways of seeing. As a result the interpretation of oral traditions must always take into consideration these various layers put down in different periods and for different purposes with different social groups but retaining the same core spatial images. Interpreting changes in social identity during the time of the disasters means considering that the oral traditions about the disasters are themselves the result of these changes in social identity and therefore do not provide independent evidence about the disasters. Paying close attention to the spatial aspects of oral traditions by identifying previously existing landscape and the ways they were transformed historically allows the historian to identify the various layers of meaning in these traditions.

SACRED LANDSCAPES OF ENCLOSURE AND
NEW KINDS OF AGE-SET TERRITORIES

An enduring vision of the landscape as territories that are defined and healed by ritual enclosure is evident in the move toward concentrated and fortified settlements during this period. The remains of stone fortifications are still present all over the Mara region. Either thick rock walls higher than a man surrounded the entire village (*obugo* in Ngoreme), or smaller stone enclosures (*ruaki* in Nata) protected women, old people, and children as a temporary shelter during the raids. Even a single homestead was surrounded by a stone wall with a barred gate.[113] The German explorer Baumann described an Ngoreme fortified settlement with walls two meters high and almost two kilometers around. One entered the settlement through a gate locked from the inside, finding a large open space inside.[114] Near Ngoreme, the German traveler Kollmann (1899) reported even more strongly fortified villages with stone walls one and one-half meters high, one meter thick, and 250 meters long, and in some places three walls in succession. Inside the walls, a labyrinth of thorny euphorbia hedges divided the individual homesteads.[115] An Ngoreme elder said that each obugo had a guarded front gate and a secret back door for escape, with holes in the walls to look out and shoot through.[116] The fortified settlements themselves were built up on the hillsides or among the rocks with a back door on the upper side of the hill leading out into thick thorn brush and serving as a place to escape if the raiders entered the settlement. The White Fathers, who traveled inland briefly in 1902 and 1904, reported that Ikizu, Ngoreme, and Zanaki people lived in fortified settlements up on the hills among the rocks.[117] The German traveler Kollmann (1899) described "Shashi" (Mara) villages "ensconced in gulleys or high up between fastnesses."[118]

Fortified and concentrated villages were found throughout East Africa during this period, including in Sonjo, where they built substantial fortifications surrounding their hillside villages with thorn thickets accessible only through a stockade entrance of heavy timbers three meters across.[119] Another kind of fortification, a *tembe*, or low log house with a dirt roof in the style of the Gogo of central Tanzania, prevented the Maasai from burning the thatch roof during a raid. They adopted this style from the Tatoga, who brought it from Mbulu during the disasters.[120] The first Nata chief, Megasa, died at the door of his tembe during a Maasai raid in 1914.[121] Scholars have interpreted the move toward fortified and,

Figure 4.1. Remains of stone wall from fortified settlement, Kisaka, Ngoreme. *Photo by Peter Shetler, 1995*

more generally, concentrated settlements throughout East Africa during the late nineteenth century as a response to the insecurities of the caravan trade. Tanzanian historian John Iliffe emphasized the effect of firearms in causing the "ribbon-like settlements along the trade routes to give way to fortified villages" ruled by warlords like Mirambo.[122] Yet this was clearly not the case here, since the western Serengeti was not on the trade routes. Because people distrusted concentrated settlements and described them as unhealthy and potentially dangerous places, powerful reasons must have existed for people to build them. Many colonial officers associated the increase in witchcraft accusations with the concentration of settlement.[123] The fortified structures themselves also represented an enormous outlay of labor for people who were accustomed to building their houses of mud and thatch in a few days.

People began to build fortified and concentrated settlements ostensibly as a way to protect themselves from Maasai raiding but also to strengthen the power of the elders and provide a sense of ritual enclosure and security in time of stress. Scholars elsewhere have postulated that concentrated settlements were associated with the increased authority of political leaders who depended on violence, coercion, and the control of elders over young people. Elders may have used the proximity of concentrated settlements to better control the movement of youth, who had gained some autonomy and prestige as warriors and raiders.[124] Yet the elders had few means, other than insecurity itself, to force people into concentrated settlements against their will. In a situation of expansive land resources, people who disagreed with the leadership could simply leave and be assured of a welcome in any of the other neighboring settlements. People sought protection from the threat of disease, general insecurity, and the need for boundary formation.

The medicine for protection, orokoba, worked against both disease and raids; its power lay in the act of encirclement, or enclosure, of the land against external danger. Fortification was a visually symbolic medicine for protection against threatening external forces.

The cumulative disasters along with western Serengeti incorporation into a Maasai-dominated region became the impetus for recontextualizing the core spatial images of sacred landscapes in order to reorganize the previous age system, which included a dominant generation-set and a subordinate linear age-set more responsive to military needs. The new system divided western Serengeti communities into at least three territorially based age-set cycles, or saiga—Saai, Ngirate, and Rumarancha. The system was cyclical in that each territory would "rule" in turn for eight years before the next would assert its authority to protect the security of the land. In Ikoma, Ishenyi, and Nata, first the Saai cycle, then the Ngirate, then the Rumarancha ruled, after which the rule reverted back to the Saai in a cycle. Nata elders claimed that each saiga territory had three "traditional" names that were used cyclically for each newly initiated set, as well as the unique name that the age-set took in response to events of the time. The age-sets were essentially three parallel linear age-systems, coordinated on the basis of a territorial cycle. Together, the three age-set eras of a cycle, each era eight years long, equaled one rikora generation of twenty-four to twenty-five years. In this way the new idea of a peer group organization, capable of military action, was integrated into the older understanding of twenty-four -year generational cycles to maintain continuity with the past. This age-set chronology is illustrated in table 4.2, in which one can see how the new age-sets were integrated into an older generation-set chronology and into an integrated system that functioned regionwide.[125]

Elders from Ngoreme, Ikoma, Nata, and Ishenyi identified the age-set of Maina (or Ngirabhe), Matara (or Megona) and Masura (ca. 1840–75) as the time when they redefined age-sets into cycles or territorially based associations. A Nata elder said that the first saiga, or age-set, was the Maina (ca. 1840), living at Site, where they divided into the three cycles of Bongirate, Busaai, and Borumarancha.[126] An Ishenyi elder confirmed that the people divided into cycling age-sets when they left Nyiberekira or after they got to Nyigoti, at the time between the Maina and Saai generations.[127] This process of division is clear in an Ishenyi text that reads, "The Amasura [generation-set] gave birth to the Amatara [generation-set] who then gave birth to the Rumarancha, Saai, and Ngirati [age-sets]."[128] In the idiom of the fathers and sons, the generation-set "gave birth" to the three new age-cycles, maintaining the continuity of time. Philipo Haimati used a similar generational idiom in his written history of Ngoreme, where there were five age-set cycles instead of three. He claims that because of the insecurity of raids

> they passed a law that each father should not have all his sons living
> in one homestead in one village. If a war came in one village, then

Table 4.2. Chronology of generations, western Serengeti

Generations	Cycling Age-Sets (Saiga) of Ikoma, Nata and Ishenyi			Cycling Generation-Sets (Rikora) of Ngoreme, Kuria, Ikizu and Zanaki		Praise Names Kuria Nyabasi
	Busaai	Bongirate	Borumarancha	Monyasaai	Monyachuuma	
(c. 1820)	abaNyanyange (abaNyange) (c. 1828)			abaGamunyere	abaGini	1. Gesarwini (1837,39,41)
		abaOrumati (abaHonga) (c. 1836)			abaNyangi	2. Kehanga (1844, 46, 48) 3. Gesambiso (1851, 53, 55)
			abaTing'ori (c. 1844)	abaMaina		
The Generation of Settlement (c. 1845)	abaNgirabhe (abaMaina) (c. 1852)				abaChuma	4. Ngibabe (1858, 60, 62) 5. Machare (1865, 67, 69)
		abaMatara (abaMegona) (c. 1860)				
			abaMasura (c. 1868)	abaSaai		
The Generation of Disasters (1870)	abaSaai (c. 1876)					
		abaNgirate (abaMaase) (c. 1884)			abaMairabe (abaNgorongoro) (abaShirianyi)	6. Getiira (1872, 74, 76) 7. Maase (1879, 81, 83) 8. Nginogo (1886, 88, 90)
			abaRumarancha (c. 1892)	abaNyambureti		
The Generation of Opportunity (c. 1895)	abaKihocha (c. 1900)					
		abaKong'ota (c. 1908)			abaGini	9. Romore (1893, 95, 97) 10. Nginaro (1900, 02, 04) 11. Tamesongo (1907, 09)
			abaKubhura (Romore) (c.1916)	abaGamunyere		
The Generation of the "Tribe" (c. 1920)	abaKinaho (Maina) (c. 1924)				abaNyangi	12. Nyesendeko (1912,14, 16) 13. Kambuni (1920, 22, 24) 14. Kehar (1927, 29)
		abaSanduka (abaMegona) (c. 1932)				
			abaHorochiga (Masura) (c. 1940)	abaMaina		

Table by Peter Shetler, 1998

not all of the brothers would be killed at once. So they combined five circumcision sets in all to be one company of soldiers, one age-set. They called the first children of the age-set whom they circumcised the Saai and gave the Saai land to live on from Maji Moto up to Busawe. They called this land Ikorongo. The Saai called themselves by another praise name that they made up, the Mar'osikeera. They gave them the horn and the drum. The age-set made their own weapons. These were the first company of soldiers. The second year they circumcised other children, they called them the Amatara, to whom they gave the land of Kisaka. They called themselves the Bongirate and were given the horn and the drum and made their own weapons. In the third year they circumcised the next children, to whom they gave the same names of Amataara and Abangirate, but they occupied the land of Kewantena and Bumara. . . . In the fourth year they called the children whom they circumcised Abaga-mutenya and they gave them the land of Ring'wani up to Masinki to live in. . . . The fifth company of soldiers was called the Amasuura. They called themselves the Abarumarancha, living in the land of Iramba. They gave them, too, the horn and the drum, and they made their own weapons. A man who had five sons made this divi-sion, following the circumcision sets. . . . He would spread them out among the five companies as they circumcised them in succes-sive years.[129]

Haimati's story depicts a conscious reorganization of social space, from descent-based settlements to all descent groups joined in an age-set territory.

Given these common understandings of sacred landscapes and the admira-tion of Maasai, western Serengeti peoples might be expected to imitate Maasai age-sets that had so successfully been used as the mechanism for expansion. Age-sets united Maasai age-mates dispersed in separate sections over a wide re-gion and celebrated the ethos of youthful prowess and aggression.[130] And in-deed, the western Serengeti word for individual circumcision sets that would be combined to make up the large age-set, siriti, is a Maasai loanword.[131] In fact, many elders stated that their saiga system was "just like the Maasai" because some of their age-set names and times of initiation corresponded with the Maa-sai, who opened a new age-division every seven years.[132] These common age-set names were the basis for making an alliance with Maasai warriors in the story of the Battle of Ndabaka told previously. The initiation of age-sets was synchronized regionally from east to west, starting in Maasai and ending in Nata. Warriors of the same age-set in different ethnic groups would cooperate with each other in raids, form alliances and adoptions, and provide safe travel and hospitality while they engaged in conflict with different age-sets of their own ethnic groups.

In fact, in this period, when the ethnic groups that exist today were in the process of formation, age-set often took precedence over ethnic identity. Since age rather than descent was now the social basis for settlement, many age-cycle territories became mixed in their ethnic composition. When western Serengeti peoples returned from Sukuma after the famine, Ishenyi and Nata of the Busaai cycle settled at Nagusi; Handajega included western Serengeti peoples as well as Tatoga and Sukuma; and Sibora was a settlement of the Bongirate from the

Ikoma, Ishenyi, and Nata. The Germans assigned separate "tribal" chiefs to age-cycle territories in both Ngoreme and Ishenyi that carried over into the British period. Some elders assert that people were more loyal to age-cycle than to "tribe," getting along better and living nearer to people of the same age-cycle than to others of their own ethnic group in other age-cycles.[133] Placing these territories on a map shows this overlapping of Nata, Ishenyi, and to a lesser extent Ikoma ethnic groups by age-cycle (see map 8).

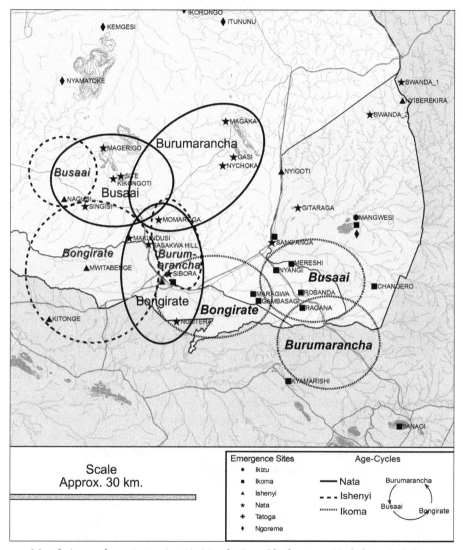

Map 8. Age-cycle territories (saiga). *Map by Peter Shetler, 2005. Underlying GIS data courtesy of Frankfurt Zoological Society*

In spite of this regional coordination and synchronization with the Maasai system, however, the saiga age-set system that evolved in the second half of the nineteenth century was clearly an original western Serengeti recontextualization of past core spatial images. Its structure bears very little resemblance to the Maasai, which uses a straightforward linear age-set pattern, divided into the "right and left hands," without cycling names.[134] The use of cycling names and sets based on the principles of generation, with the outward trappings and ideology of age-sets, was an innovation that respected the older values of the generation while recognizing the need for unity and mobilization of young men. Elders described the saiga as a military device that, for the first time, grouped men of a similar age into organized units; when a raid took place all available able men joined in the chase. But the leader of the age-set was still a man of peace rather than war, and military leadership under a powerful and unifying prophet never developed here as it did among the Maasai, Nandi, and Turkana, except where western Serengeti peoples consulted Tatoga prophets.[135] The western Serengeti saiga did engage in some activities that emphasized their comradeship and equality, as do Maasai age-regiments, but never as armed encampments mobilized for aggression. Age-mates might call a risaga, or work party, for farming or construction, and age-mates might demand a cow to eat together. Yet they more commonly shared meat in communal groups where meat was portioned out according to gender and generation.[136] The fact that women continued to be present at, and crucial to, the important saiga ceremonies as wives is further evidence that the system was based on the concerns of the older generational system with the fertility of the land.

The precedence given to generational principles of elders even in the new age-set system dominated by younger men indicates an ongoing struggle between the interests of elders and youth, men, and women. Some have argued that the Maasai age-system supported the gerontocratic power of senior men over juniors in their control over the disposal of women.[137] Others have seen the adoption of age-set organization as a displacement of elders' centralizing authority by young men gaining power in military activity.[138] In the western Serengeti, although the new age-cycle system offered young men increased autonomy from the constant supervision of their fathers and new channels for gaining respect and authority, in the end the elders maintained their power. Elders testified that during this time young men broke the earlier prohibitions against cattle raiding of any kind. Western Serengeti age-sets imitated Maasai-style raids and dressed up like Maasai to raid on the lake during the age-sets of the Kihocha, Kong'ota, and Kubhura (1900–1916), and perhaps even earlier in Ngoreme. The Lakes people, being afraid of Maasai, would not give chase to the raiders.[139] Livestock populations on the lakeshore were at an all-time high as a result of their better harvests and involvement with the coastal caravan and the Lake Victoria trade; they thus presented an easy raiding target for Maasai and western Serengeti youth recovering from the drought.[140] As a result of this raiding, many young men began to gain their own wealth independent of inheritance from their fathers (see chapter 5).

In times of increased insecurity, East African women often became more vulnerable as the objects of slave raiders or captive wives, and thus more dependent on the protection of men.[141] The celebration of the ethos of male military strength in the aghaso ritual for killing a lion or a Maasai included honoring women who were courageous in childbirth, but not honoring them in a central role.[142] Women were important, as they represented fertility and reproduction in a time when survival was at stake, but that did not necessarily enhance their position in the community. During this time, when men were increasingly mobile and boundary definition became more important, women, as strangers in their husbands' descent groups, came to embody the bounded, enclosed home.[143] A situation in which leaving the boundaries of home was dangerous increasingly restricted women's movements, and elders of both sexes said that women seldom left their home area unless escorted by men. Women did not act as long-distance traders, as they did in Kuria or Kikuyu communities.[144] Although the move of sons away from their fathers' territory may have allowed women to settle nearer to their natal kin, it may equally have deprived them of their mothers-in-law as allies among strangers. The changes of this period do not seem to be have been in the interests of women, although they may have given women the opportunity to gain closer control over the day-to-day management of local community relations, while men focused their minds outside the community.

The formation of territorial age-set cycles allowed descent group elders to diversify their risks by spreading out their sons among various communities and ecological zones according to the older generative principles. The Ngoreme story told by Philipo Haimati expressly talks about age-set reorganization as a way to preserve the descent group, so that not all the sons of one man would be killed in one raid.[145] A man could initiate a younger son into a different age-cycle than his older brother, rather than waiting another sixteen years for the next initiation to come around in his home territory. This allowed each son to enter his saiga with his own age-cohort, when he was most fit for battle. If a father had sons in each of the other age-set territories, the family would always have a place of refuge and expanded sources of security. Other elders said that the purpose of dividing out one's sons to the various territories was to maintain peace within the community by keeping the age-sets from fighting each other.[146] The division of "sons" into different territories was thus a way of expanding the influence, as well as diversifying the risks, of fathers. For example, "sons" in Bongirate would not raid their "brothers" in Busaai, and they would be responsible for feeding their "fathers" in Borumarancha. Fathers, beleaguered by crop loss, disease, raids, and loss of children saw the benefits of tapping into other networks and sent their precious remaining sons away, gambling on their survival. The division of sons was also a way to reconfigure the networks of interdependence after the demise of the old system, when the Asi and Tatoga left.[147] Elders described the territorial saiga divisions in Nata as containing different ecologies and subsistence strategies—Busaai in the hills as farmers and herders, Bongirate on the plains as herders and hunters, Borumarancha in the bush as hunters and farmers.[148] The age-cycle territories allowed for a partial reformulation of an interdependent economy of survival that the events of the disasters had destroyed.

The transformation of age-set structure brought about a dramatic recontextualization of ritual and the social definition of territory. Up to this time, western Serengeti peoples organized their settlements around the idiom of kinship and defined territories as clan land in the walk of the generation-set. With the advent of age-set territories people still settled in small descent groups but did so within the territory of one cycling age-set rather than that of the clan. As the age-sets began settling together, the age-cycle names took on the locational prefix *bu-* or *bo-*, thus producing Busaai (the land of the Saai) Bongirate, Borumarancha, and so on. Because the age-set, rather than the lineage, was now territorially based, it had to take over responsibility for the definition and maintenance of the relationship to the land, just as the generation-sets had done before. The saiga territories, rather than the lineage-based hamate, now became known as the ekyaro, or the land, carrying the sense of ritual control over the land. During this period people still closely identified the descent group with a particular cluster of homesteads within the ekyaro, perhaps now represented by a fortified settlement.[149] Yet for the first time, people also began to identify the united age-set cycle territories with the ekyaro.

In the late nineteenth century, as the age-set began to take responsibility for maintaining a ritual relationship with the land in the larger territory of the three cycling age-sets, rather than individual clan lands, a sense of ethnic identity began to develop. As an Ishenyi elder described it, those in their saiga were responsible for the health of the land in all three of the age-cycle territories within their own ethnic group. It was up to the abachama, or age-set leaders, to do the rituals for protection of the land from raids or disease, to spread the medicine of rainmaking and to consult a prophet when necessary. In the same way that the rikora, or generation-set, had performed these rituals for the ekyaro of the descent- or clan-based hamate, the saiga now began to perform them for all three age-cycles that made up the emerging ethnic group, or the ekyaro.[150] The role of the saiga was to protect the community—a task that more often involved taking up the medicines for "cooling the land" rather than taking up the weapons of war.[151] The ability of the saiga to "encircle the land" in ritual protection defined the spatial limits of the emerging ethnic territory.

Small-scale ethnicity, using the sacred landscapes as the ritual space of the united age-set territories, developed because people connected group identity with a relationship to the land and the medicines of protection to enclose those boundaries. Yet the ethnicity of the western Serengeti was never exclusive, and it formed in relation to other ethnicities within a regional system crosscut by many other kinds of social identities. As western Serengeti peoples found common identities to bridge these differences, they defined the ethnic boundaries that separated them. The sense of ethnicity that evolved out of the space of age-cycle territories was quite different either from colonial ideas about "tribe" or present-day definitions of ethnicity; a larger sense of western Serengeti ethnicity never developed here, as it did in many other places in East Africa. When the Germans came they expected to find completely enclosed and permanent "tribal" boundaries that, in western Serengeti terms, could lead only to death. Western Serengeti peoples had responded to the disasters of the late nineteenth century by using the older visions of the landscape to create forms of identity to cope with change. Yet it was these units, conceived on a very different basis, that the colonial experience molded into "tribes" that continue to have meaning today. These older ideas about ethnicity still underlie a vision of the landscape in which spreading out risk and the ritual health of the land are paramount.

It was in this context of disaster, migration, and radical social transformation that the first Europeans observed a largely "uninhabited," but only recently abandoned, Serengeti at the beginning of the twentieth century. Since they understood Africa to be a place of timeless tradition where firmly set "tribal" cultures remained in place over the millennia, they assumed this to be the permanent, rather than transitory, state. The German traveler Kollmann captures this sense of entering a natural wonderland, devoid of people, when he describes "a broad and mighty steppe of grass and bush" extending "from the south-eastern corner of Speke Gulf, and reaching in the east as far as Ikoma and Elmorow, and in the north up to the Baridi Mountains." He goes on to declare the entire region uninhabited. However he adds that "for a lover of the chase this region is a perfect Eldorado [sic]."[152] The first European sport hunter to enter the area, in 1913, characterized the western Serengeti as useless for agriculture and herding because of poor rainfall, lack of groundwater, and an abundance of tsetse fly, but perfect game country where "indiscriminate shooting during a great many years and by a great many people would hardly affect this marvelous abundance over so extensive an area."[153] From his reading of early accounts of the Serengeti, park warden Myles Turner concluded, "apart from a few nomadic Maasai and wandering bands of Ikoma hunters, the Serengeti had always been an uninhabited game area and was of no interest to the slavers."[154] Bernhard Grzimek's *The Serengeti Shall Not Die* sealed this image of the Serengeti as pristine wilderness in the public eye: "mosquitoes and tsetse fly and lack of water make it uninhabitable by man."[155] As in other places in Africa, the first Europeans legitimized their conquest by sexualizing the landscape as an empty wilderness that must be "penetrated."[156]

Both for local people and the European newcomers, the Serengeti in the early twentieth century had become a barrier rather than a corridor of interaction.

The lack of dry-season water for a stretch of 115 kilometers across the Serengeti plains, the Maasai stronghold in the Loita hills, and the tales of sleeping sickness in Ikoma kept people from moving across the region.[157] In 1915 sport hunter White wrote about his "rediscovery" of the "very last virgin game field of any great size remaining to be discovered and opened up to sportsmen" by pushing west of Lake Natron into Ikoma and Ngoreme territory in the western Serengeti. He said that no one had found this area because natural barriers of mountains and arid wastelands devoid of game protected it as well as the German lack of interest in sport hunting.[158] The Sonjo and Ikoma no longer interacted with one another except as migrant laborers in the colonial system, and soon both would be prohibited from hunting and gathering arrow poison, salt, or honey in the places where they had established historical customary use rights.

～

As the elders and I drove into the game reserves that day back in the summer of 2003 and found the old stone walls of abandoned fortified settlements knocked down to the last three courses, the rocks hauled out to build park buildings, leaving the remaining structure covered with vines in a derelict way, it was amazing to realize that people had indeed lived in this "wilderness."

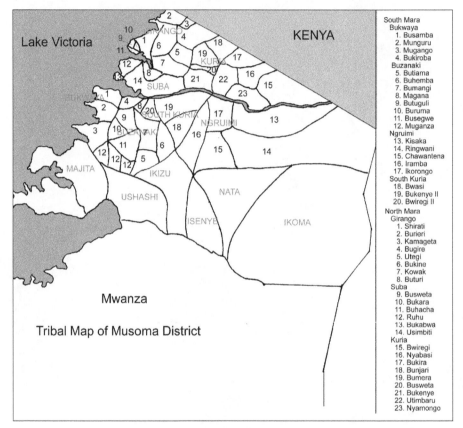

Map 9. Colonial administrative areas. *Map redrawn by Peter Shetler, 1998, with permission from the University of Dar es Salaam, East Africana Library, from Cory Papers, no. 173*

5 ⌁ Resistance to Colonial Incorporation

Becoming "Poachers," 1900–1950

IN A HISTORY OF MEMORY, colonialism was the next crisis that precipitated changes in imagined landscapes just as western Serengeti communities were recovering from the disasters explored in the last chapter. Colonial officials promoted a view of the landscape as a reserve for economic exploitation, resulting in demands for labor and resource extraction as well as radical changes in political authority. Oral narratives about this period use the new core spatial images of hiding and subterfuge as a strategy for resisting these demands. A European view of the landscape that had to be regulated, classified, controlled, and managed emerged in the context of industrial capitalism and the civilizing mission of the empire, which included the conquest of nature. The necessity of "developing" Africa, whose "primitive" patterns of land use were deemed inefficient, unproductive, and destructive to the soil, legitimized colonial rule, particularly during the 1930s depression years.[1] At the same time, the government set aside areas of pristine natural beauty—which were understood as underutilized by natives—for elite hunting, separating nature and culture as the respective sites of leisure and production. As Juhani Koponen argues, amid their contradictory impulses, the Germans sought to develop Tanzania in order to exploit its resources for their own interests.[2] Finding enough labor for these enterprises was a constant struggle throughout the colonial period, and various measures—including taxation, consumer incentives and coercion—aimed at squeezing labor out of the peasant economy.[3] Economic development, with the proliferation of agricultural, veterinary, health, labor, and mining experts, became the overriding goal of the British after World War II, when debt to the United States necessitated an increase in exports from, and a decrease in imports to, the colonies.[4] Although the colonial actors, including administrative officers at various levels, experts, missionaries, and merchants, varied in their advocacy of local native interests, they shared a view of the landscape that was meant to be rationally developed for economic gain and to preserve the productive capacity of the land.[5]

Western Serengeti peoples resisted state demands but their responses were conditioned by past ways of seeing the landscape, which led them to seek recovery from the late-nineteenth-century droughts by trading hunting products to re-stock their herds rather than to join the colonial migrant labor force. Unfortu-nately those same strategies designed to gain livestock wealth, develop new lead-ership styles based on patronage, and avoid both the labor market and cash crop campaigns also brought them into direct conflict with colonial ideas about proper land use, which marginalized them as poachers and unruly natives. Although the social landscapes of the past had ongoing utility during this period, ecologi-cal and sacred landscapes were rendered less relevant as the state took increas-ing control over both political life and the environment. The colonial authorities attempted to transform the appearance of the landscape into consolidated peas-ant farmsteads with large areas of wilderness set aside for sport hunting, while western Serengeti peoples began to disperse their settlements to move further away from colonial control. Colonial demands for rational administrative units as discrete territories led to the creation of "tribes" as elders recontextualized oral traditions to serve their interests in controlling those processes.

Oral traditions about colonial rule begin with the appearance of German military officers in the region soon after 1900 and thus overlap with the dynam-ics described in the last chapter. Once the Germans acquired their piece of the "magnificent African cake" in the 1884–85 Berlin conference they had to prove that they could administer and control German East Africa. Although the Ger-mans established a military post in Mwanza, at the southern end of Lake Victo-ria, by 1891, they did not attempt to administer the area to the north, known as Shashi,[6] until after 1900, when they established posts at Schirati and (Nyabange) Musoma along Lake Victoria.[7] Western Serengeti peoples felt German rule most directly when the Germans built a permanent station called Fort Ikoma between 1905 and 1907 on the western edge of the Serengeti plain.[8] The Ikoma chiefs were compelled to supply labor, both men and women, for building the fort and feeding the soldiers.[9] In 1916, during the campaigns of World War I, the Ger-mans fled Fort Ikoma, their last stronghold in the district, after a battle with the British, returning to recapture the fort before they were finally defeated in 1917.[10] The British ruled Tanganyika as a League of Nations Trust Territory until its in-dependence in 1961.

The new core spatial images of narratives about the colonial period are those of hiding and subterfuge, representing the many ways western Serengeti peoples found to resist colonial demands. Although people took action in their resistance, the mode of resistance was "passive" in that it did not directly challenge colonial authority. James Scott, in *Weapons of the Weak*, describes the form this kind of resistance took as "passive non-compliance, subtle sabotage, evasion, and deception."[11] In these traditions the overwhelming image of the Germans is of those who captured people and instituted chiefs whose job it was to demand labor. They tell about the hardships imposed by the Germans and resistance to those demands. Ikoma elder Mahewa Timanyi told about his encounter as a child:

> Mahewa: When the Germans came they took captives, and people were afraid of them. My father had only two sons, my elder brother and myself. He hid us in the bush and told us not to make any noise so that the Germans would think that there was no one at the homestead and pass by without taking captives. While we were hiding my elder brother started a game of sticks to pass the time. But the sticks would make noise when they slapped down. I got hit in the eye and cried out. My father came to scold us and said that he was hiding us so that the white men would not get us and here we were playing.
>
> Jan: What did they want the captives for?
>
> Samuel: To do work at the fort.
>
> Wilson: They took people as porters to carry their loads.
>
> Samuel: In the year that the English beat the Germans in the war, as they were leaving the Germans rounded up captives to carry their loads. They wanted young men and took them by force to carry their loads. Many ran away.[12]

This response of passive resistance by hiding from the Germans is repeated in a Nata story, told by a member of the chief's family about choosing the first chiefs.[13]

> At that time people ruled themselves. Then one day they were surprised to see a white man, a German, who had come from Mwanza. He came with Sukuma, Takama, and Kahama soldiers. This was at the time when my grandfather was still a young boy. They made their camp at Makundusi, in the bush that is called Mungaraba. The Nata people all came to see, and they cleared a spot inside the bush where they could watch what was going on but not be seen. The German was armed with guns and set up his tents so that he could find and install the sultan. But the Nata people remained hidden in the bush. The German sent his soldier to summon them. He called them in the language of the Sukuma. But they were afraid and refused to come out for the second time.

The German told them to bring out their leader. Among the Nata was an elder named Kiboge who was the big man among them. He was a prophet and had the medicine bundle of war. He dreamed about things that would later take place. The Nata people told him to go and meet the Germans, but he was afraid and refused. So the white man got mad and said that if their leader did not come he would attack them. One young man named Megassa volunteered to go. "I will go so that you all can stay in peace." He had compassion for his people. So he went to the German.

When he approached the white man he was given a stool and a book. The soldier asked his name. He said, "Megassa Nyora Sesera." The German measured his feet, length and width. He gave Megassa boots, a *kanzu* [long white robe], and a tarboosh [fez]. Megassa took off his leather clothing. Then he went back into the bush to his people, but they ran away in fear. He called them and said, "It is only me." So they came and listened to his explanation. They saw that the book looked like a soft sheepskin. When they saw the pen writing they called it *wanag'ora ng'ora* [that which goes around and around]. The white man slept there that night and left the next day. He went to Nyabuta, where he began building Fort Ikoma.

After four days he sent one of his soldiers to the new sultan, Megassa, who lived near the Rubana River at Tarime. He was told to send young men to work on building the fort. They were sent to work near the Orangi River, at a place where there are lots of *ebenturu* trees, which are very heavy. They cut these and carried them to the fort.[14]

In this story the Nata people hid from the German gaze in order to be able to observe and understand the power of the stranger before making their decision about whom to send out. Because there were no preexisting chiefs, the Nata retained their autonomy to choose the colonial chief and thus determine, in part, the terms of the colonial encounter.

This story is representative of western Serengeti peoples' encounter with colonial rule, in which they pursued their own agenda with some success, yet ultimately found themselves in conflict with and forced to adapt to colonial demands and ways of seeing the landscape that were antithetical to their own. Choosing colonial "tribal" chiefs who gained their power from the colonial government rather than from the people undermined the terms of political authority embedded in the sacred landscape as a responsibility to protect and heal the land. Control over the environment continued to erode when the colonial state assumed that responsibility, keeping people from managing the ecological landscapes of the past. Western Serengeti peoples successfully used strategies embedded in social landscapes to recover from the disasters by trading hunting products in Sukuma in order to restock their herds. A new kind of leadership style of wealthy cattle patrons, the *nyangi* elders, developed from this trade. These strategies, however, ran into direct opposition to the colonial government, who began to see western Serengeti peoples as lawless and unruly renegades, uncivilized and backward natives, and, ultimately, cruel and indiscriminate "poachers." These dynamics created a subculture of passive resistance to government initiatives that is still prevalent today.

THE EMERGING THREAT TO SACRED LANDSCAPES AND POLITICAL LEADERSHIP

Ways of seeing the sacred landscapes of the past that equated political authority with the ritual protection of the land were severely threatened by the appointment of colonial chiefs. Ethnic groups—as united age-set territories or clans dependent on one rainmaker—that had only begun to emerge in the period of the disasters were now solidified by colonial administration. Because both personnel and finances were limited, both the Germans and the British employed a system of indirect rule in which they had to induce already established local leaders by coercion or benefits to carry out their orders. Local leaders, as the face of colonialism, would also provoke less resistance to unpopular demands. Based on evolutionary assumptions about migrating Goths and Huns in pre-Roman Europe, colonial officers expected to find discrete "tribes" ruled by chiefs.[15] Yet in the western Serengeti there had not been a system of chiefs before colonial rule, and people maintained some autonomy in determining who would become chief, among the various positions of leadership. However, ultimately these chiefs were accountable to their colonial masters rather than to the people. With this divorce of political authority from ritual control over the land, the sacred landscapes had less relevance. The chiefs had to carry out the orders of the colonial government rather than do what was necessary to heal the land.

As a result, the core spatial images of hiding and subterfuge inherent in early colonial stories represent passive resistance to the authority of the chiefs with disobedience or by quietly moving away.[16] In 1906 the Germans instituted civilian authority and needed local chiefs to rule on their behalf.[17] So they sent out a German officer from Mwanza guided by a small troop of native soldiers to visit the so-called Shashi and identify the "tribal" leader that would then be installed as the sultan, or chief. The German-appointed chiefs were required to enforce colonial rule, and, in return, they gained material advantage as well as the backing of a powerful patron. If the chief or the people did not comply with German orders, they would be punished by military action.[18] The Zanaki, just west of Ikizu, resisted numerous German military expeditions before finally succumbing.[19] When the chief appointed by the Germans in Ikizu was overthrown by another contender, a German military force of native soldiers attacked the village, burning houses, killing people, and taking cattle and women back to Shirati.[20] In addition to these violent confrontations, the German military reported that the Shashi and Ngoreme people were "not favourable to German rule, sometimes offering passive resistance."[21] The colonial authority's perception that western Serengeti chiefs were corrupt and its people unruly was created by severing the connection to between political authority and the landscapes of memory.

Because the western Serengeti had not been governed by hierarchically structured chiefdoms and their recently formed ethnic groups were flexible, the people were often at a loss to identify the chief that the German requested to install. As the Germans themselves put it, the Shashi, "acknowledge no chieftain, but every village community is autonomous."[22] In fact, there is no word in local languages for chief, and the British later concluded that the closest thing to governance was the "rule" of the age-sets, those who walk to encircle the land for protection.[23] Precolonial political authority was heterarchically based in a number of different positions of leadership, each with its own separate sphere of authority based on the older generative principles of distribution, descent, and generation. This provided each territory with various leaders who might be considered candidates for the German "chief." In Ikizu the Sukuma-style rainmaker of Nyakinywa's line seemed the obvious candidate. But, because there was more than one rainmaker who claimed chiefship, the Germans orchestrated a rainmaking contest to determine the Ikizu chief.[24] In Ikoma and Ishenyi the rainmakers were Tatoga prophets, so the first Ikoma chief was not Ikoma at all but the Tatoga prophet Gambareku, succeeded by his son Kichaguchi. However, before the Germans left, an Ikoma age-set leader, Rukini, became chief. The Tatoga refused to be chief in Ishenyi, and in Ngoreme a variety of prophets, age-set leaders, and opportunists became chiefs over the different age-set territories.[25] A frustrated British officer lamented that in the whole district there were "only two Sultanates in which the Sultan was established prior to German occupation . . . all other areas being ruled by rain doctors, war doctors or councils of elders."[26]

The Germans gave chiefs authority that they did not possess under the previous heterarchical system and sometimes appointed chiefs with little to no authority among the people. A European hunter traveling in Ngoreme in 1913 encountered young Chief Myeru, whose father said he had "made *shauri* with [advised] the Deutsche [Germans] that my son be Sultan [Chief]." He also talked to Chief Missambi, "a bright intelligent boy of twelve or thirteen educated by the Germans to read and write Swahili." These chiefs distinguished themselves by wearing clothes and knowing the ways of the white people but had little legitimacy with their own people.[27] Throughout the larger Mara region there are many stories of how young men without standing became chiefs by fooling the Germans through creative translation or ingratiating themselves by providing valuable services such as forging nails or constructing boats.[28] Clearly, western Serengeti peoples did not give much respect or value to the colonial chiefs.

Narratives from this period make it clear that the authority of the German chiefs was separate from preexisting forms of authority rooted in the landscapes of memory and that the chiefs ruled through the "medicines" of the colonial power. In the story told in the introduction of choosing the first Nata chief, their first choice was a prophet who kept the medicine of protection during war, followed by Megassa, who finally agreed to go out to save the people. In many versions Megassa appears as a person of no particular authority when he became chief, but others say that he held the highest rank in the eldership titles and was the spokesman for his age-set who possessed powerful medicines. However, it is clear in another version of the story that once Megassa was made chief, his authority rested on the German medicines of power, embodied in the symbols of office given to the new chiefs—the fez, kanzu, boots, pen, and book. According to a narrative, employing similar core spatial images of hiding and subterfuge, told by Nyamaganda Magoto:

> When he [Megassa] went back to the people they were all scared of the medicine that he had gotten from the Germans; he possessed authority. The other institutions of authority in the community continued to function, except for the relationship with the colonial authorities. Neither Megassa nor Rotegenga, his son, would interfere with them. The chief's job was seen mainly as collecting taxes and labor conscription; all else was under traditional authority.[29]

Megassa's authority as chief, even as it was legitimized by the top eldership titles, was clearly determined by the Germans. He was feared, but his authority was restricted to those areas related to the state, not to the things that really mattered to the community.

In addition to these oral traditions about colonial rule employing new core spatial images, earlier narratives about farmers, hunters and herders, clans, descent groups, and age- and generation-sets came to be understood as ethnic stories about origins, migration, and settlement. The core spatial images of these narratives were recontextualized and elaborated as Nata or Ikoma stories, as they appear today. In his investigation of the "political organization of the Musoma tribes in pre-European times" for administrative purposes, government anthropologist Hans Cory called two elders from each chiefdom for a meeting at Musoma, lasting two weeks, in which they discussed "the histories of the tribal units and their racial inter-relationship."[30] In this context the elders were asked to tell the story of their "tribe" in relation to other "tribes." The Musoma district books provide the history of each discrete "tribe" with distinct territorial boundaries. The Nata ethnic story obviously incorporates clan narratives of first man and first woman, while the Ngoreme story is little more than the amalgamation of narratives about age-set territories uniting to fight the Maasai. These histories were necessary to support the authority of the colonial chief over the ethnic unit, yet their core spatial images point to fundamentally different sources of legitimacy.

Although the lack of local respect for the chiefs, and the resilience of older forms of authority, became apparent during the war, when colonial control weakened, the incoming British continued to support the chiefs. When the Germans left, the people overthrew most of the chiefs, because, as one early British officer noted, they "had no tradition to support them and, moreover, they were chosen often without consultation of the wishes of the people."[31] The British also reported that the people looted the property of the oppressive chiefs, who no longer had the Germans to keep them in power.[32] When the first British political officer, Major Coote, arrived in Musoma in 1917, all the German chiefs except one in Sizaki had fled or been driven out, and governance had reverted to what they understood as "elders' councils."[33] Significantly, the elders that revolted against the chiefs are designated in the records as Bagini, the name for the generation-set under the Chuuma cycle. This meant that the generation-set had reasserted control and protection over the land. However, the German administrative structure survived amazingly intact, and, out of the need for stability, the British quickly reinstalled the German chiefs or held elections to choose between the chiefs' relatives. In 1919 the district meeting in Musoma brought an end to formal generation-set leadership by ruling that the Bagini "will in the future have no power in the District as they have caused much trouble with witchcraft." British-appointed headmen, who answered to the chiefs or to imported functionaries from Buganda (*akidas*), replaced the Bagini, who held the rituals to heal the land.[34]

Colonial power was thus limited because the chiefs, with little to no authority over the community and the land, were often ineffectual or at least restricted in their influence, and thus prone to corruption.[35] Although the Germans eventually established the district headquarters in Nyabange, and then Musoma in 1912, to more closely administer the eastern lake area, the western Serengeti was still relatively isolated from direct colonial supervision even up through the end of the British period.[36] A German report noted that the Shashi people were always causing trouble or running away, while their chiefs were powerless to do anything about it.[37] One of the first British reports on local politics stated that in Musoma District the people were "somewhat ill-disciplined and lawless," while the chiefs were uniformly "illiterate and inefficient."[38] Because the weak authority of chiefs provided a shaky basis for indirect rule, Musoma was a difficult district to administer, and when the British took administrative control of the area they already saw the people as unruly, "backwards tribes" who were prone to hostility. German attempts at categorizing peoples, used by the British administration as an orientation to the area when they took control, indiscriminately classified the

Ikizu, Sizaki, Ngoreme, and Ishenyi, among the Shashi, or circumcised people as, a "vile people who recognized no authority and accepted no refined culture, prone to rebellion and trouble."[39] One British officer described Nata as "the home of the tax-dodger and border jumper," where an alleged community of outlaws lived "a nomadic existence and went from chiefdom to chiefdom hunting, living in temporary camps and not registered as tax payers."[40] These unfortunate characterizations set the terms for an antagonistic relationship that was not in the long-term interests of western Serengeti peoples.

One way to measure the depth of western Serengeti disaffection and resistance is to look at population movements within the chiefdoms, where people continued to vote with their feet as they moved away from chiefs that were not seen to be acting in the people's interests. Although the colonial officers liked Nata's Chief Rotegenga, calling him "trustworthy, keen, energetic and just," he began to lose his constituency when a large part of the Bongirate age-set territory membership left the chieftaincy to live elsewhere. In 1927 the Nata population stood at 1,870 while between 1933 and 1934 it dropped from 1,581 to 1,028. It was during that year that an influential elder in Nata, Magoto Mossi along with others, left for political reasons or personal disagreements with the chief, taking their many cattle to the plains in the Ikizu chiefdom.[41] The British reported that by 1941 the Nata population, under Chief Rotegenga, had dwindled to almost nothing and the Ishenyi population was scattered under Chief Sarota.[42] In contrast, the district office complained that, "instead of ruling," Chief Nyambeho of Ikoma, who was unanimously elected in 1925, "is ruled by his [headmen] and in their turn they are guided by the wishes of their villagers which seldom coincide with the requirements of the law."[43] Yet Nyambeho's people stayed. Western Serengeti peoples passively resisted colonial rules that did not meet their requirements for good leadership and tolerated colonial chiefs only insofar as they did not interfere with previously existing patterns or make onerous labor or tax demands. The lack of effective local political leadership made it difficult for the British to regulate and control the efficient use of natural resources as they wished. Sacred landscapes survived and continued to be recontextualized during the colonial period, in spite of the threat from a political system based on a very different kind of authority.

COLONIAL CONTROL OVER ECOLOGICAL LANDSCAPES

Throughout the colonial years, western Serengeti peoples' view of an ecological landscape, in which people were an integral part of its management as interdependent hunters, herders, and farmers, was also threatened as the state assumed more authority over the regulation of the environment as a method for controlling people. Colonial policy was influenced by a view of the "natural" landscape that, just as the people, must be tamed by the British civilizing influence.

Colonialism operated on an economic logic that demanded profit, or at least sustainability, for the empire. Both the German and British colonial governments carried out this mandate by attempting to shape and control the landscape to fit their vision. People were to farm and herd in "rational" ways on fields and pastures that resembled the British countryside and should not be dependent on hunting for their livelihood since that represented an earlier evolutionary stage. The colonial governments instituted hunting regulations and wildlife reserves in response to the loss of game in the rinderpest epidemics, even though recovery was underway and the decimation of wildlife nowhere near what it had been in central and southern Africa. Rather than managing disease through controlled immunity, colonial policy called for its eradication by measures that would radically disrupt communities if necessary. Human settlement, as the zone of production, had to be separated from wilderness, as the zone of nature alone. These policies led to resistance because they violated the standard of the moral economy implicit in past ways of seeing the landscape.[44]

Because local ecological control had broken down during the period of disasters, colonial governments attempted to institute their own controls to prevent sleeping sickness, which did more to promote other colonial agenda and upset community life than to eradicate the disease. Although previous western Serengeti strategies had maintained immunity against sleeping sickness with controlled contact, both German and British approaches attempted to separate humans and tsetse fly completely by prohibiting all native movement in the wilderness, clearing tsetse-infested bush in densely populated areas, and, in sparsely populated areas, forcibly resettling whole communities. In Nata and Ishenyi, people were resettled from Nyigoti, and in Ngoreme from Mwibara up to Maji Moto.[45] Aside from preventing disease, the policy of resettlement and concentration outside of tsetse-infested areas was part of a larger colonial development scheme to induce a certain standard of civilized behavior and economic production.[46] The game warden, Swynnerton, argued that concentrated settlement would bring about protection, ease of administration, supply of services, increasingly moral behavior, and prosperity.[47] Although the massive clearing and resettlement schemes carried out in Shinyanga were not repeated in the western Serengeti because there was not enough labor for such vast areas, forced labor groups in Nata and Ikoma cleared tsetse bush between 1929 and 1934.[48] While the Germans had outlawed fires altogether, in 1921 the British colonial game department made controlled burning part of its tsetse fly eradication scheme and ordered that no grass fires be set before the middle of September (when the short rains began), or if possible every other year, to ensure a hot fire that would destroy the bush.[49] In contrast to the traditional system of burning that depended on an orderly progression of cool burns controlled by the elders, they now had to wait for a cue from the political officer for a hot burn that threatened grazing areas.[50]

Again, western Serengeti peoples passively resisted many of the sleeping sickness measures imposed on them by the colonial government because they were incompatible with former ways of seeing the landscape. Even back when the Germans tried to isolate sleeping sickness patients in camps, the Schirati project was abandoned for lack of compliance.[51] In 1928 the district officer reported that the Ikoma headmen were reluctant to round up sleeping sickness patients to take them to the hospital.[52] Many people were afraid of the injections and were reluctant to go to hospitals, where people died. The Horochiga age-set (ca. 1940) sang a song that said, "the needle has finished us off," forcing the chiefs to go door to door to induce sick people to go to the hospital.[53] In fact early drug therapy did not always cure the disease; it sometimes reappeared and the drugs caused side effects like pain and blindness.[54] People resisted sleeping sickness measures not because they were backward and uneducated (as the colonial officers surmised) but because the measures had larger implications for control over their land. A number of colonial officers themselves questioned the sleeping sickness measures, concluding that the natural course of population growth and development would eventually push back the bush without draconian measures. Others ventured that vehicle traffic, rather than hunters and fishermen, spread the fly in the bush.[55] These comments suggest that western Serengeti peoples who resisted sleeping sickness measures had rightly perceived that a deeper colonial agenda was at stake.

The correlation of the outbreaks of sleeping sickness with times of famine suggests that when drought came people abandoned their caution about limited exposure to the disease environment and took increasingly dangerous risks as they went into the wilderness to hunt or find grazing areas for their livestock. Yet the government used sleeping sickness prevention as a way of controlling access to the wilderness, even though the outbreaks were temporary and related to famine conditions.[56] The biggest outbreak of sleeping sickness came to Ikoma during the famine years of 1929–33. Before that, a 1927 medical report showed an incidence of sleeping sickness of less than 1 percent, with no tendency to epidemic spread at that time. People did not live in prolonged contact with fly populations and became infected only when they went out for extended periods of hunting and fishing.[57] Yet by the next year, with the onset of famine, sleeping sickness cases began to appear in Ikoma, and in 1932 the epidemic began to spread at "an alarming rate," with sixty-five new cases in the first two weeks of January, as opposed to seventy-five for all of the previous year. In his report the district commissioner said that the outbreak was caused by "the widespread hunting of game by natives in infected areas as the result of food shortage."[58] Drought drove livestock owners deeper into the bush to find adequate grazing for livestock and dangerously close to tsetse-infested areas. One Nata elder said that the sleeping sickness outbreak was the result of a Tatoga prophet calling everyone out to chase Maasai cattle raiders. They followed the raiders a long way but it turned out to be a false alarm, and when they returned they had contracted sleeping sickness.[59] Another outbreak during the 1941 famine filled the Ikoma hospital with sleeping sickness patients.[60] Sleeping sickness became a legitimate reason for implementing the colonial agenda of restricting access to wilderness areas that were to be set aside as "natural."

Access to wilderness resources was also controlled through hunting and Game Department laws that preserved these areas exclusively for European sport hunting. Both German and British colonial officers often came from an aristocratic, sometimes military, class that enjoyed hunting as an elite privilege and mark of civilization rather than a means for providing meat. A hunting ethos developed among these elite hunters during the late nineteenth century in both Britain and Germany that asserted the need for sportsmanship, including a fair and humane hunt. European hunters were also the first conservationists, as they sought to preserve the wild animals from the near extinction that bison in North America and wild game in South Africa had already met as a result of the advance of "civilization." Evoking a racist orientation, European hunters viewed themselves as uniquely able to protect the animals against what they saw as the cruel and indiscriminate slaughter carried out by Africans, who, according to Darwinian thinking of the time, were retrogressively hunting with primitive weapons for meat rather than farming. In fact, a clean and humane kill was made possible only with the invention of the high-velocity rifle in the 1880s. Although game in East Africa was not under pressure of extinction, as it was in central and southern Africa, the same approach was followed because it fit the larger colonial agenda of reconstructing the African landscape.[61]

A British view of the landscape as a reservoir for economic extraction gave white hunters the moral right and responsibility to harvest this resource but prevented native hunting. Coming into the unmapped northwest Serengeti, sport hunter Steward Edward White exclaimed that these "virgin game fields" were the last unexplored territory in Africa where game stopped and stared at men.[62] Early in the colonial years, officers began to call for more protection of Serengeti wildlife from Ikoma hunters and save it as "a resort for bona fide sportsmen, collectors and naturalists."[63] A party of professional Italian hunters and their three clients from Nairobi took home the following trophies from their hunt in 1929: six rhino skulls, including a baby, and upwards of eighty trophies, including three lions and twenty-four tails.[64] Another white hunter, Paul Hoefler, bragged about the uncountable number of hyenas his party shot and used to bait lions for photographing.[65] Although early European exploration and conquest had been subsidized by wild meat, harvested to pay labor, by the early years of colonial rule it was restricted to elite sport hunters.[66] These hunters still claimed the moral right to control hunting in the Serengeti because of their sportsmanship and humane treatment of animals.

In order to take control over the environment, the British criminalized native hunting by imposing their own views about civilized forms of hunting through the game laws.[67] While the first German game ordinance was proclaimed in 1896, a similar British Game Preservation Ordinance of 1921 laid out the basic rules for game reserves, game licenses, professional (white) hunters, and trophy dealers. No one could hunt unless he or she held a game license or killed an animal in "defense of any person or for the protection of property," and no game license could be issued to a native without the consent of the governor, who could arbitrarily impose "such special conditions as he may think fit." The effect of this ordinance was that virtually no natives could hunt legally, and if a native killed an animal in defense of his property, he was still required to report it to the game officer.[68] In order to address British concerns about cruelty and indiscriminate hunting, the 1921 law further laid out the permissible methods of hunting, making it illegal to use or be in possession of nets, traps, snares, pitfalls, and poisoned weapons, including fish poisons or the use of dogs in hunting and setting fires for the purpose of driving game.[69] A series of game law revisions ruled out all forms of indigenous hunting, leaving guns as the only permissible weapon, as is the case today. Since few local people owned guns, hunting was essentially restricted to the wealthy.[70]

In addition to hunting regulations, the declaration of game reserves imposed severe restrictions on land that western Serengeti peoples considered part of their extensive and diverse resource base. The idea of the game preserve, even though an ancient idea, was developed in Britain and Scotland during the eighteenth and nineteenth centuries, during the height of the aristocratic hunting craze, when productive crop and pasture land was converted to carefully managed deer parks and later shooting estates separated from human settlements. These new landscapes of "planned wilderness" created by Britain's hunting elite in fact became the image of nature itself in European paintings and literature.[71] The Germans set aside two game reserves in 1896 and eleven more in 1903, with a total of fourteen by 1913.[72] The British continued this policy with the 1921 ordinance that defined a complete game reserve as a place where no one was allowed to hunt, fish, camp, cut trees, or burn grass. Although the British declared the Serengeti (Complete) Game Reserve in 1929, only one year later it was downgraded to a "closed" reserve, where only hunting was prohibited in recognition of the inability of the Game Department to enforce other restrictions. The Game Ordinance of 1940 again restricted residence and entry in the Serengeti as a complete reserve until the area was declared a park in 1951.[73] The boundaries were also amended a number of times; at first they encompassed only 230 square kilometers; from 1930 to 1932 the reserve included all of Musoma District and part of Arusha District.[74] Very much in keeping with the precedent of a private British deer park, local people knew the Serengeti Closed Game Reserve as

Shamba la Bibi (lit., grandmother's field; i.e., Queen Victoria). Many elders said that although some continued to hunt there, they recognized the danger of doing so and respected the boundaries. Others said that the Germans kept them out of the reserve by telling them that the wild animals had a sickness.[75]

In reality the 1921 Game Ordinance and those that followed were both practically and morally unenforceable, especially when local people were outwardly obedient. District officers and even game wardens found it difficult to prosecute men bringing home meat for their families, using weapons that they had inherited from their grandfathers, or living on land they had occupied for generations. The chief secretary in Dar es Salaam issued a circular in 1926 stating that outside of the game reserves, "natives were not pursued for killing game unless they drove them or used cruel means of destruction."[76] In spite of the 1932 ordinance prohibiting possession of poisoned arrows without a permit, western Serengeti peoples could not be prevented from carrying their bows and poisoned arrows everywhere they went for protection.[77] Ikoma and Nata elders said that in the colonial years one could hunt openly anywhere but in the game reserves without fear of reprisal.[78] Some colonial officers objected to this discontinuity between law and practice, and in 1934 the chief secretary suggested, "We must legitimize legitimate hunting by natives before we can prevent illegitimate hunting."[79] The 1933 attempt of Sukuma and later some Musoma chiefs to challenge the legality of racial distinction between native and European hunters by applying for minor game licenses was subverted by making a license contingent on gun ownership.[80] Finally, in 1951, Africans were able to obtain a general game license under the Native Authority, but with the similar prohibitions, enforced by local game scouts.[81]

The same discontinuity between law and practice appeared in the governance of game reserves, which were too big and isolated to patrol effectively. A 1923 letter from the chief secretary, in response to the problems of enforcing the game ordinance, stated that because the intent was not to remove already settled people from the game reserves, the inhabitants should be "considered as having received permission to reside therein."[82] Yet very soon the Game Department consistently lobbied to remove squatters who were building, cultivating, and grazing within the game reserves, making the case that human settlement and wildlife were incompatible.[83] Yet people kept on quietly pursuing the economic benefits of hunting within the reserves in spite of the new rules, with frequent reports similar to that of 1921 stating that the Shashi natives annually burned the grass on the Mbalageti and Ruwana plains in order to hunt game.[84] Although the establishment of the Banagi ranger station with a permanent European game ranger in 1929 resulted in much closer supervision of the Serengeti Closed Reserve and an increase in convictions for illegal hunting, it was not until 1959 that game posts built within the park actually made an impact on poaching.[85]

Although there was ambiguity in the game laws, the Game Department was determined to wipe out the commercial market in bush meat or any other wild product except for trophy hunting. Though the government did not stop all hunting in Ikoma, it did try to eradicate the frequent "organized and constant game drives."[86] District commissioners were instructed to fill in all pitfalls, destroy springs and snares, and stop the trade in game meat and skins.[87] Yet they could not enforce even this liberal interpretation of the hunting laws, given that the western Serengeti was a long way from Musoma and covered a huge, sparsely populated territory with very few staff members. A 1942 law stated that "no person shall, except with the written permission of a Provincial Commissioner, sell or offer for sale the meat derived from any wild animal."[88] This act was especially aimed at western Serengeti peoples who used commercial hunting as a way to avoid the labor market.

Since western Serengeti peoples kept on hunting by practicing passive resistance, Musoma officers periodically renewed the call for erecting more game posts, increasing staff, and posting a European game official, in order to stop "the very considerable and wholly unjustifiable slaughter of game animals."[89] In 1923 the district office, frustrated that everyone in Ikoma seemed to be involved in illegal hunting, tried a communal punishment scheme whereby each descent group in each of the three age-set territories was required to leave two cows with the chief for six months "as a pledge of good behaviour." They then had to fill in the game pits and destroy the fences, 450 of which were filled under the district officer's supervision.[90] However, after an initial period of compliance, the Ikoma soon reverted to hunting. As the concern about illegal hunting mounted, game scouts, who had been used to eradicate vermin or rogue animals, were increasingly trained for military or police action. Western Serengeti peoples knew the scouts only as those who arrested or shot them, since the Game Department specifically sought to minimize contact between the scouts and the community.[91] Some game scouts took advantage of their position of authority in isolated circumstances and preyed on local villagers, resulting in some violent incidents.[92] Most western Serengeti peoples avoided game scouts and viewed them as enemies.

The government also used the strategy of concentrating and settling people in villages in order to control environmental resources more closely. In 1924 the game warden concluded that only when the Ikoma, "who now hug the woods for the sake of hunting are induced to settle and grow economic crops, and making money, to buy cattle, . . . they also will be anchored and become useful members of the community."[93] However, since German times western Serengeti peoples had passively resisted government control and forced labor by moving out into the wilderness.[94] The British reported that the people refused concentration in villages, living instead "a somewhat nomadic existence," for fear of witchcraft, the desire to escape government control, and the need to be near hunting grounds.[95] Persistent colonial reports referred to the lawless native in western Serengeti who settled in the wilderness to "avoid tribal authority" and "all useful work and cultivation of economic crops."[96] Although the government worked to shape the landscape in their own image, they met with constant resistance, as people continued to use resources extensively, recontextualizing the core spatial image of diversification.

What the colonial officers rightly observed was a much larger movement away from the stone forts and concentrated settlements in age-set territories of the disaster period to homesteads more spread out and removed from the centers of colonial control.[97] A European hunter traveling through Ngoreme in 1913 noted that "the country must have been at peace a long time" because isolated homes were scattered everywhere, while the "ruins of old villages perched high and fortified in the rocks" testified that it was "not always so."[98] Dispersal brought an end to age-set territories, and people began to regroup according to clan rather than age-set. A British officer said, "The Waikoma [Ikoma] have apparently decided to reorganize themselves on a clan basis instead of according to Sega [age-set territory] and this is all to the good."[99] People moved out of the hills and onto the plains to take advantage of the grasslands for their growing cattle herds and for proximity to hunting. Nata, Ishenyi, and Ikoma Bongirate age-sets occupied the interethnic plains settlements of Sibora and Mugeta, while the Ngoreme moved to the Mwibara plains from Maji Moto. The introduction of the ox plow also allowed innovative farmers with wealth to exploit the thick clay soils of the plains.[100] People moved away from concentrated settlements to avoid both colonial and community control in more compact settlements. The weakening of indigenous political authority, including the power of elders to control juniors, led to settlement dispersal across the colony.[101] People resisted German colonial rule by building their homesteads on the peripheries, in direct proximity to hunting areas, where colonial officers and chiefs could not observe the wilderness harvest and where it was more difficult to collect taxes and conscript labor. The British, faulting German policy for the dispersal of settlements, reported a "tendency before the war, owing to oppression by German *askari* [soldiers], for natives to break away from larger villages and establish small family villages in the bush."[102] They also noticed that people avoided large settlements for fear of witchcraft, which may also have reflected the social tensions of this period of shifting authority.[103]

Because of the increasing control over spheres of everyday life critical to survival, western Serengeti peoples were in constant tension with the government, which also precipitated some cases of active and violent resistance. The Ikoma came close to rebellion as a result of the game-related disputes. In a 1936 trip to Ikoma, the chief secretary warned the people that the government had said they must stop hunting in the game reserve and killing forbidden animals or they would be prosecuted to the full extent of the law.[104] By 1938 the game warden reported that, with the increase in arrests and supervision, the Ikoma were "boasting that they intend to hunt in the reserves and that they will fire with poisoned arrows if stopped." In fact they, along with their Asi friends, went on to burn parts of the complete reserve, even around the Banagi ranger station, shoot poison arrows at and otherwise threaten scout patrols on a number of occasions, and blatantly kill zebras within several kilometers of the station. The game ranger warned that the Ikoma were "becoming thoroughly out of hand" and that unless the government took "immediate and drastic action" the situation would "develop into guerilla warfare." His report concluded that although they had plenty of places to hunt outside the reserve, "they abuse this privilege and openly defy the government."[105] The provincial commissioner met with Ikoma and Nata in public meetings and cut both the chiefs' and headmen's salaries until they took action. He also asked the game ranger at Banagi to find ways to establish good relations with the Ikoma, since the work of the scouts would be impossible without local help. The chiefs aided in making some arrests; they brought in additional staff and a crisis was averted.[106]

Passive resistance, as the primary western Serengeti response to colonial demands represented in oral traditions as the core spatial images of hiding and subterfuge, were understood in a very different way by colonial authorities. Early in the colonial period western Serengeti peoples had been called poachers and lawless renegades, which the British thought resulted from their innate moral character and timeless traditions that would be difficult to break. Yet the hunting practices and resource use that led to these labels stemmed from previous ways of seeing the landscape that gave authority to those that healed rather than harmed the land, managed the environment to protect the people, and depended on hunting and other wilderness resources when famine threatened. They were not deliberately resisting colonial authority but rather responding to their own agendas and ways of living on the land. The colonial authorities, too, were responding to the challenges they faced through their own views of what the landscape should look like and were working to make that a reality. The landscape that the colonial officers represented as "natural" was in fact a tool for reshaping human society for the economic benefit of the colony.

SOCIAL LANDSCAPES IN THE RECOVERY FROM DROUGHT

The agenda of the peoples of the western Serengeti is most clearly seen in the ways they used a view of past social landscapes with considerable effectiveness in their recovery from the late-nineteenth-century disasters. Much of what brought western Serengeti peoples into conflict with the colonial government was their efforts to restock their herds after the disasters rather than direct resistance to colonial control. The introduction of colonial rule allowed for a period of relative peace and stability when the number and intensity of Maasai raids was considerably reduced. Since the vast majority of western Serengeti livestock had either been stolen in raids or killed by disease and drought, people were eager to rebuild their wealth using the resources of the wilderness that developed into a commercialized hunting trade in the beginning of the twentieth century. It was this cattle wealth, to a large extent, that allowed western Serengeti peoples to successfully resist labor migration and cash crop production that so disrupted other regions of colonial Tanganyika and to develop a new leadership style based on the older generative principles of distribution and feeding the people.

When the Kihocha age-set (ca. 1900) were ready to marry and found that their fathers' livestock had been devastated by the string of disasters, they did what generations before had done: they hunted and took their dried meat and skins to trade on regional trips (orutani/obutani).[107] As a result of their famine-related connections to Sukuma, they discovered a large market for wildebeest, zebra, and giraffe tails with which to make bracelets and anklets (budodi), ornaments for dancing, and fly whisks for prophets and rainmakers. Although the trade in wildebeest tails predominated, western Serengeti peoples also exchanged lion skins, elephant tusks, ostrich eggs and plumes, arrow poison, other animal products used in medicines, as well as the traditional dried wild meat and cured skins.[108] In return they received tobacco, salt, iron hoes, all of which came to play a significant role in western Serengeti culture; but, most significantly, they received livestock.[109] Since Sukuma people lived at the terminus of the overland route at Lake Victoria, they had access to trade goods and wealth from their work as porters for the caravans, giving them a disposable income for wilderness products.[110] In the stories of obutani, or long-distance trade trips to Sukuma, elders told how, as young men, they had formed small groups and armed themselves against raiders to make the three-day trip of hard walking to Sukuma. The Ikizu, more removed from access to the wilderness products, traded gourds instead, taking on foot as many as fifty gourds tied together as far as Geita.[111] The German explorer Baumann reported seeing a "worn path" in Ikoma that "was the trader's trail to Sukuma."[112]

Age-set groups often organized these hunting and trading trips, while direction of the process was still in the hands of the elders, on the basis of older views of the landscape; the elders made sure that young men gained their bridewealth through legitimate means like hunting, rather than cattle raiding.[113] When they went to hunt together, elders taught the youth the historical geography of the land and the moral economy of hunting. As soon as they killed the first animal, they made a camp where the elders would stay with the carcasses and eat the first division of roasted meat. The youth made sure that the elders had water and firewood before they left to continue the hunt, fearing a fine if they did not pay their elders proper respect. All the game was brought for butchering and drying back to this camp, where they stayed for days or weeks at a time. Everyone in the community benefited from the hunt, which began after the harvest in July, when the wildebeest migration arrived.[114] Although hunting trips were exclusively male, women came to the camps to bring food and to help carry the meat home, where the hunters were greeted with cries of joy. Everyone shared in the distribution of fresh meat, consistent with the older landscapes of a reciprocal economy.[115]

The use of bows and poison arrows remained the most common method of hunting, now supplemented by commercial methods on a much larger scale for the trade. Hunters chose a place that was naturally constricted by hills, augmented it by building brush fences, and drove the wildebeest herds into pitfalls they had constructed on the other side.[116] German traveler Kollmann reported counting over two hundred deep and narrow pitfalls for game covered in brush on a half-hour's walk in the extensive woodlands of the Ruwana plain and in Nata.[117] Each pit had been dug by a particular family, and the man who owned the pit claimed all the game that fell into it, even though all had participated in the drive equally. According to a 1934 report, thirty pits were found together in one spot.[118] Yet in spite of this trend, in 1913 a European hunter found that the longbow was still the "weapon of choice" and was amazed to observe a dog and a man with a bow run down a wildebeest on foot.[119] The evidence clearly shows a shift toward large-scale harvesting of meat and other wilderness products for commercial purposes.

Although ivory from elephant tusks had become a staple in regional trade, most western Serengeti peoples claim that they did not have a tradition of elephant hunting, although they used ivory for ornaments and the emblems of eldership titles.[120] Elders said previous generations did not hunt elephants because it was so difficult and dangerous when one hunted with poison arrows. Elephant hunting was also ritually prohibited because elephants are like people—the females have breasts and they bury their dead. If a hunter killed an elephant, he performed a complete funeral and observed a period of mourning with a false grave for the elephant. An Ikoma elder told the story of Tatoga prophet Masuche, who turned his wife, Nyabhoke, along with her house, into an elephant when she would not cook for him. When elephant hunters returned home, they sang a song saying that they were bringing back Masuche's wife.[121] Neither was there a political monopoly over the ivory trade in this region, as there was elsewhere in East Africa where chiefs extracted tribute by claiming from each kill the "ground tusk," the one that first touched the land they controlled as the elephant fell.[122]

Yet by the end of the nineteenth century ivory had become the most lucrative trade item to Zanzibar, and western Serengeti peoples sought ways to benefit. Ivory from East Africa went to India for bride's bangles and to Europe and the United States for billiard balls, piano keys, and combs.[123] One Ikoma elder said that the Maase age-set (ca. 1884), and to a greater extent the Kong'ota (ca. 1908), were the first to break the prohibition against elephant hunting, although the wealth it generated was never used to build up a homestead but was rather squandered on personal indulgences.[124] The Nata sang songs about famous sharpshooters, called the Abaronda and known as great elephant hunters, like

Maincha from the Kihocha age-set (ca. 1900) or others from the Rumarancha age-set territory. They began to use muzzle-loading rifles that came in with the caravan trade.[125] Still, these hunters were the exception. Most western Serengeti peoples who traded ivory in Sukuma obtained it first from Asi hunter-gatherers. The Asi continued to be recognized as the expert elephant and rhino hunters.[126] Ikoma elders tell stories of particularly close relations with Asi, forming friendships with individual families who brought ivory and other wild products to trade at their homes.[127] The Ngoreme and Ikizu, on the other hand, claim never to have hunted elephants or rhinos.[128] Commercial elephant hunting reduced populations in many of the intensively hunted areas throughout the colony; in 1890 alone 209,000 kilograms of ivory were exported from German-controlled ports.[129]

The colonial government found this strategy to regain livestock through the commercial trade of hunting products morally repugnant, as it went against their sensibility of the hunting code of sportsmanship developed by British aristocrats. In 1934 the Mwanza senior commissioner elicited sympathy by showing photos from Ikoma of dead animals around pitfalls with "the head and neck of one poor giraffe still alive . . . showing above the pit." The British were concerned with the suffering of animals not killed immediately, the cruel way in which they were killed, and pitfalls that did not discriminate in regard to sex, number, age, or species, but killed all alike in a wasteful manner.[130] However, they also objected to native hunting because it was considered primitive, only the first step in human evolutionary development and not something to which Nata or Ikoma farmers should "revert." One officer questioned why people who were "ostensibly agricultural" participated in extensive hunting operations and "make it for much of the year their prime work, camping in the bush for months together." He found it abhorrent and at odds with the urgent needs of the empire that "appreciable sections of the population and such strapping fellows . . . should be devoting their attention to the chase instead of playing their part in the development of this very rich country."[131] The British concluded that western Serengeti peoples hunted out of laziness, contrasting them with the cotton-growing, cattle-owning Sukuma, who were "not greatly implicated in the lawlessness and idleness of the area."[132] In this view of the landscape, civilizing the people and civilizing the landscape were integrally connected.

The large cattle herds amassed as a result of this trade in hunting products in the early colonial years found an excellent home in the productive grazing lands of the western Serengeti. One veterinary officer called it "the finest cattle district in Tanganyika."[133] The provincial commissioner noted that, although he found the Musoma people to be much more "backward and isolated" than the Sukuma, they inhabited "a superior country" and possessed "great livestock wealth" that was "outstanding in size and quality"—surpassing any that he had yet seen in the province.[134] One European game hunter observed in 1913 that Ngoreme cattle kept the "fattest humped cattle I have seen out here" and that Ishenyi cattle herds covered the entire Ruwana plain.[135] In 1935 the western Serengeti chiefdoms had the highest proportion of cattle per person in Musoma District, ranging from Ikoma (2.2) to Ishenyi (6.0), compared to the renowned Kuria cattle keepers (only 1.8).[136] The first census in Musoma District during the British period, in 1926, showed 332,010 cattle, increasing to an all-time high of 465,084 in 1933.[137]

As in other places throughout the territory, young men now had the ability to build up their own wealth, marry, and establish their own homesteads without their fathers' help. One elder lamented that the men of that generation got intoxicated with their own wealth and ceased to respect their elders.[138] As a result of increased cattle wealth, and the desire of elders to control bridewealth, the price to marry increased dramatically, and laws to restrict bridewealth began to appear beginning in the 1920s.[139] Young men often hunted and traded for about five years and then returned home to establish a homestead and begin the climb to become "big men" themselves. Ikoma, Ishenyi, and Nata elders said that it was during the age-sets of the Kongota (ca. 1908), Kubhura (ca. 1916), Kinaho (ca. 1924), and Sanduka (ca. 1932) that men gained significant cattle wealth, while Ngoreme named the Gini generation-set (ca. 1900).[140] During the previous generation of the Mairabe, a wealthy man had ten to thirty head of cattle, and most people kept only sheep and goats before the disasters, but this generation began to number their cattle in the hundreds and even thousands.[141]

The wealthy new cattle owners converted their economic wealth to political power as cattle patrons (*omunibi*) by recontextualizing the core spatial images of distribution and reciprocal networks.[142] In contrast to previous forms of distributive leadership, these patrons were almost exclusively men. Women, without access to cattle, were devalued as the product, rather than the producers, of wealth. Patrilineal rapidly replaced matrilineal descent systems as men sought to consolidate their wealth in their immediate families. "Big men" had to develop widespread lateral relationships of patronage throughout the region to maintain their position. Because of the threat of cattle disease and raiding, a man with thousands of cattle could not risk keeping his entire herd in one area, nor could he secure pasture and water for that many cattle at one place. Thus the practice of cattle trusteeship took on a new prominence during this era. In addition, the big man had to gain the respect and trust of the local youth who would recover his stock, through regional age-set networks, in the case of a raid. However, class differentiation did not develop from the accumulation of wealth because, according to the older landscapes, keeping a wealthy man's position of respect demanded the redistribution of his wealth by "feeding" the people.

This new kind of leadership, recontextualized from older core spatial images during the postdisaster era of trading in Sukuma, was institutionalized in the nyangi (eldership titles).[143] Around the turn of the century, peoples of the western Serengeti transformed the celebration of the passages between the basic life stages into a system of eldership titles attainable only by the wealthiest men, each of whom had to give a large feast in order to become an initiate. Each ethnic group continued to mark a different set of life stages for both men and women, including, for example, naming, cutting the teeth, circumcision, setting up a homestead, and the circumcision of the first child, but now the eldership titles for wealthy and prestigious men were added as additional stages.[144] Nata elders explain the origins of the nyangi system in its present form with the story of a prophet named Kikong'oti who gave people the secrets for joining the nyangi when they went to Sukuma during the famines so they would not "forget Nata and die in a foreign land."

> At that time there was a great famine, and the people dispersed and went to Kreti [Sukuma]. As the Nata went, they all passed by the place where this Nata, who was like first man, an Asi, named Kikong'oti lived, on a little hill. As they passed, he warned them not to forget Nata and not to all die in a foreign land. He was concerned that they would forget the things of Nata. So he showed them the Nata nyangi. There was a big *mragawa* tree in front of his house. When people passed by, he would ask them to come and then he would show them the nyangi so that they would not be finished off in Sukuma. He would take the fruit of that tree and cut it into four parts, and with each part [as the symbolic feast] he would initiate them into one of the nyangi secrets. Then they went west, and when they returned they came from the south, from a place called Getongi. . . . So they came back to Nata and found that Kikong'oti had gotten very old. He was the beginning of the Nata nyangi. Those that have cattle, if the cow dies, they do not want to eat by themselves. Those who farm and harvest lots of millet, they wonder why their neighbors do not come and visit them. So they make a big pot of beer. The different nyangi arose around the names of different pots of beer.[145]

Some elders credit Kikong'oti with the beginning of the Nata nyangi because he showed them what do when they have lots of cattle but "do not want to eat by themselves" or harvest lots of millet and "wonder why their neighbors do not come and visit them."[146] In one elder's version of the Kikong'oti story he speculated that the nyangi started because people were looking for a way to get fed by wealthy men during the famines and wealthy men were looking for a way to enjoy themselves and pass the time after the harvest.[147] An Ikizu elder said that the nyangi celebrations with big feasts were created in the early colonial years, after a time of hunger, when "the poor man figured out a way to get some food to take to his children."[148] Some elders indicated that the Ikizu and Nata learned some of the powerful medicines (*masubho*) taught to initiates taking the titles from the Sukuma, who had a reputation for dangerous witchcraft.[149]

Wealthy men gained a new kind of recontextualized regional power by wielding "medicines" that they acquired when they were initiated into the nyangi titles, each medicine specific to a given ethnic group.[150] The color of an elder's wildebeest or cow tail flywhisk—black, red, or white—was a regional indication of his status.[151] Ideally, the nyangi elders put the good of the community first and used their medicines to identify thieves, punish criminals, or enforce local custom, but people feared incurring the wrath of the titled elders.[152] Nyangi leaders in Nata could not live outside the boundaries of the ethnic territory, and elders understood the secrets of the nyangi as the heart of what it was to "be Nata."[153] Titled elders became the guardians of tradition and the principal narrators of ethnic history.[154] Even though the nyangi were ethnic-group leaders, they built regional networks between themselves. The Ikizu and the Nata shared eldership titles and eldership secrets and attended each other's ceremonies, while the Ikoma Mwancha clan shared eldership titles with the Nata Mwancha clan, and the Ngoreme, Ikoma, and Ishenyi shared some nyangi titles and attended each other's ceremonies.[155] Both Nata and Ikizu tell the story of how the Tirina River flooded and kept the messengers from either side from crossing and inviting the other to attend their nyangi with a bundle of tobacco. So they threw their bundles across the river, saying, "You go do your ceremony and we will do ours, but we are still one." The story is told to explain how Nata and Ikizu people, who are essentially one, were separated.[156]

Because, according to the older social landscapes, a leader was someone who fed the people and had a large following, the nyangi initiate had to put on a huge community feast in order to assume an eldership title.[157] Hundreds of people at a time attended the nyangi ceremonies, which lasted up to eight days, while other titles required simply dividing out meat. One man described his *aguho nyangi* feast in which he prepared sixty goats, ten cows, a thousand sides of dried wild meat (ebimoro), numerous barrels of beer, and a store of grain for elders to divide.[158] Eldership titles forced the "big man" to distribute his wealth to the poor so, as one elder claimed, people could eat at the rich man's house. He thought that the elder's medicines, or masubho, were only later added to give the men incentive to take the titles.[159] The nyangi elders became a powerful behind-the-scenes political presence in the colonial period. As an indication of the waning power of the age-set with the prominence of titled elders, the British reported that age-set governance was undermined by the power of prophets and rich men, whose dependents supported them in the council.[160]

This increased cattle wealth, and the power that it brought as a result of hunting, allowed western Serengeti peoples to have more control over their level of involvement with the colonial economy. Colonial officers noted that because the western Serengeti had some of the best cattle country in the territory, they had no need to join the wage labor force.[161] They were much better off than agricul-tural people because they could pay their hut or poll taxes by selling cattle for much less total labor than cash-crop production.[162] Others commented that there were no cash crops in the western Serengeti because people got the money they needed "by selling livestock, skins and tails of game."[163] But avoid-ing wage labor through cattle was not part of the colonial economic plan. The colonial government considered traditional forms of livestock husbandry through-out the territory to be wasteful and uneconomic in its concern for the display of wealth rather than production for market.[164] The job of the veterinary and agri-culture departments was to bring peasants into the cash economy and teach them "that wealth is not necessarily the possession of cattle."[165] Despite the gov-ernment's efforts at education, western Serengeti peoples continued to resist in-corporation with the livestock wealth gained by hunting.

Cash-crop production was at the center of the colonial view of the economi-cally exploitable landscape and necessary for the viability of the colony but was once again resisted. From German times on, Musoma District authorities pro-moted a number of different crops, including sansevieria (for rope fiber), peanuts, sesame, groundnuts, and, most important, cotton.[166] Both during the 1930s de-pression and WWII, "increased production campaigns" aimed at remedying the serious economic crisis and lack of revenue due to price falls, drought, and the political uncertainties of war.[167] According to a district report, the natives were given "a definite warning, if they do not either obtain work or grow something that can be sold they may find themselves unable to pay tax and become de-faulters."[168] When peasants refused to grow cash crops as a result of depressed prices, the government decided that each native was "due some guidance" to "choose the manner of his livelihood," and it made growing cotton compulsory in some districts.[169] Throughout the territory the government appealed to the peasant's sense of duty "to make contribution to the needs of the empire and al-lied countries for raw materials necessary for the war" and to Britain's need for cotton from nondollar sources.[170]

In Lake Province the British put their hopes for cash-crop production in cot-ton, introducing it first in 1924 along the lakeshore and in Ikizu, and beginning the push for more cotton cultivation in 1933.[171] Musoma District crop statistics for cotton fluctuated wildly, starting at 3,435 hectares in 1937, dropping to only 818 hectares in the following year, rising to 8,180 by 1941. The area plummeted again, to 695 hectares in the drought year of 1943, then reached a record 16,156 hectares in 1945, before plunging back to 2,945 hectares in 1946.[172] These statis-tics reflect the Musoma people's acceptance of the crop only as they were forced to or benefited from it in the good years, while rejecting it or practicing passive resistance during the hard years. A British agricultural officer reported that the natives who wished to avoid cotton cultivation were "forever discovering on their lands a certain shrub, the existence of which is claimed (with a small degree of truth) to militate against successful planting of cotton." A story told by many and reported by an agricultural officer in 1936–37 relates how Zanaki people, through the leadership of the elders, gained exemption from cotton cultivation by cooking the cottonseeds before planting them so that when the agricultural officer inspected the fields he concluded that the area was unsuited to cotton.[173] Cotton did not become a significant cash crop in the western Serengeti until right before independence.[174]

Although the colonial government saw cotton resistance as obstinacy, there were logical reasons why western Serengeti peoples decided not to grow it. In the midst of a drought in 1930–31, cotton prices fell dramatically due to the global depression. Most farmers had no choice but to put their effort into food rather than cash crops.[175] Although many were forced to grow cotton during the war, the 1943 drought again turned many back to an emphasis on food crops.[176] Another good reason for relying on hunting instead of growing cotton was distance from markets and difficulty in transportation.[177] Elders said that cotton created labor bottlenecks because it needed to be weeded three times and then hand cleaned before market. People had livestock or could hunt and so saw no reason to plant cotton. One Nata elder said, "We did not need cotton to make us rich, we were already rich [in cattle] then; cotton was too much trouble."[178] While agricultural officers promoted cotton, "as a drought resistant crop with a long maturing period," people told me over and over that "cotton caused hunger" and exhausted the soil.[179] Western Serengeti peoples did not resist growing cotton because they did not want to become part of the market economy but because they gained more profit through livestock. This is demonstrated by the huge success of the ghee industry in Musoma District, which became the largest producer in the territory in 1934 through the use of low-cost home cream separators, but saw its demise, along with the hides industry, as a result of government regulation.[180]

If western Serengeti peoples did not readily embrace the cash-crop economy, based on other ways of imagining the landscape, they were just as hesitant to become migrant laborers or mine workers. The colonial government never allowed heavy labor recruitment in Musoma District because it was often struggling with food sufficiency or quarantined against sleeping sickness and also because the gold-mining industry in the region claimed first priority for local labor.[181] Thus Musoma District never became a labor reserve for plantations; only 450 workers left the district in 1927.[182] In 1936, out of a population of about two hundred thousand in Musoma District, 2,967 worked in the mines (1.5 percent).[183] While the government aimed for 10 percent of a district's adult male labor, Musoma District had only 6.7 percent of adult males enrolled as mineworkers even at the height of the mining period, and only 4 percent of those that actually came to work at the mines on any given day.[184] Many men preferred work in other places to the mines nearer home because they did not like mine work. The men who entered the colonial labor market were usually those whose fathers could not provide them bridewealth to marry or those seeking the adventure of leaving home and proving their manhood.[185] Many went to Nairobi or Magadi Soda in Kenya or became drivers.[186] Musoma District had a reputation for filling the ranks of the army and police. In 1942 fifty-five hundred soldiers came from Musoma District and served in places like Burma, Somalia, or Sri Lanka during WWII.[187] Most laborers worked outside the region for only one or two years, until they had earned enough money to buy cattle, still the measure of success in the older social landscape of patronage.[188]

But gold mining was still the most important wage-labor option for western Serengeti peoples. Gold was discovered in Ikoma in 1902 and soon after in Ngoreme, but during the 1920s the Musoma mines produced little and had trouble attracting enough labor.[189] Then in the early 1930s a boom period began as new mines opened, and by 1938 the Mara Mine in North Mara and Buhemba Mine in Zanaki were the number one and two gold producers in the territory, with twenty-four mines total, about half of which were in the western Serengeti. Buhemba Mine employed fourteen hundred native workers and twenty-two Europeans.[190] The boom was over by 1942 and the last mines closed in 1965. Labor shortage in the mines, reflecting worker resistance, was the most serious problem for the gold mines. Although much of the labor for the Musoma mines came from within a thirty-kilometer radius, and 90 to 95 percent came from within the district, western Serengeti peoples are barely mentioned in the list of worker ethnic groups, except for the Ngoreme in mines within their own territory. The largest percentages of workers were Luo (60–80 percent) from North Mara, Sukuma, or Jita peoples; Zanaki, Kerewe, Ikizu, and Ruri were also mentioned, as well as skilled workers from outside the district. The Musoma mines used the *kipande* labor system, common throughout the territory, in which each worker that registered at the mine would receive a thirty-day labor card. The worker could choose when to work off those thirty days in order to be paid, and the mines had no way to coerce them to show up each day. Many stayed away for months at a time and then returned to finish their cards, seldom appearing during the farming or hunting seasons. The 1935 mine inspection reported that only 60 percent of the workers showed up for work on any given day.[191]

As much as the mine owners complained about the kipande system, Musoma District workers would not work for contracts, and the high demand for labor, along with their alternative sources for cash, allowed them that choice. As one labor officer rightly surmised, these were "target workers," that is, people who had a farm and livestock at home but entered the labor market to make money for a specific need, like a bicycle, bridewealth, or paying their tax, and left when that was achieved.[192] The British, however, saw them as lazy and idle, and hoped to teach these "primitive people rich in stock who have no incentive to work . . . the value of money and more of personal comfort."[193] Mine owners surmised that if they paid higher wages, the workers would simply leave sooner. Although many mine owners continually called on the government to force natives to work, the British had abolished compulsory labor, except for cases of specified government service when full market wages were paid. They were reluctant to even discuss the possibility, given the resentment toward forced labor in the German times.[194] The one time that western Serengeti peoples remember being forced into the labor market in British times was during the war years (1944), called *manumba*, when wartime needs for sisal took priority and Musoma District, among others, received a quota to supply labor to the sisal plantations in Tanga.[195] Western Serengeti peoples still remember the manumba work with great resentment. To get their laborers to stay, the labor officer suggested that the mines provided better conditions, including accommodation for families, creating an understanding and respected management and improving camp conditions.[196]

Yet in spite of the cattle wealth amassed during the early colonial years, which allowed western Serengeti peoples some choices in the colonial economy, periodic droughts continued to occur in the region, and state restrictions kept people from responding to them by using the core spatial images of diversification embedded in the social landscape. Throughout the 1920s and 1930s East African rainfall was below average, with territorywide drought reoccurring in 1943. In 1946 food shortages throughout East Africa resulted in a dramatic decrease in crop area.[197] With the increasing restrictions on hunting and movement, western Serengeti peoples were unable to recover their wealth and grew increasingly poorer. While the colonial government recommended planting cash crops like cotton and food crops like cassava as a famine prevention strategy, western Serengeti peoples pursued hunting and the trade in wild animal products as their logical preference. The colonial officers increasingly saw the western Serengeti as a "famine land," concluding that, "Ikoma is an area in which semi-famine conditions may be considered to be endemic."[198]

The first big drought during the British years in western Serengeti began in 1927 and was at its height in 1931, when the chiefs' *baraza* (court or council) in Musoma reported a famine situation in which locusts had invaded and most places had gone four to twelve months without any rain.[199] While in 1928 the territorial statistics reported some 42,000 hectares under cultivation in Musoma District, by 1932 that area had been reduced to 29,000 hectares.[200] The government prohibited the export of native foodstuffs from the district and gave some food relief in 1931 and 1934 as a response to famine conditions.[201] Drought also exacerbated disease and led to a severe reduction in Musoma District livestock. An outbreak of rinderpest in the western Serengeti in 1927 killed 1,281 cattle in Issenye and 1,136 in Ikoma.[202] In 1931 and 1932 the veterinary department reported that livestock in Musoma District were suffering from locust, East Coast fever, rinderpest, and trypanosomiasis.[203] Between 1933 and 1934 Musoma cattle were reduced by 25 percent. Even more dramatic, during the same period 57 percent of all goats were lost.[204] During the 1940s, similar outbreaks of cattle disease occurred, resulting in quarantines and compulsory dipping.[205]

The overwhelming government solution to famine was the promotion of cassava as a famine crop that could stay in the ground for a number of years, would not be destroyed by locusts, and was drought resistant. The first German traveler in the region found the Ikizu growing cassava in 1892, and a German description of the region in 1912 reported that the Ikizu grew cassava almost exclusively, with some sorghum for beer.[206] The Ngoreme say that they readily accepted cassava as a famine crop.[207] Yet the first British agricultural statistics for the territory show Musoma District growing only 1,020 hectares of cassava, as compared to 37,000 hectares of millet in 1927. Millet and sorghum remained the biggest food crop throughout the British period; not until 1945 did cassava became a major crop.[208] Cassava was promoted, and made compulsory, in the western Serengeti during the famine of 1929–34, especially as an antilocust measure.[209] The district officer first told the chiefs in 1931 that it was their responsibility to get the people to plant cassava.[210]

Yet the Ikoma Federation—consisting of the chiefdoms of Ikoma, Nata, and Ishenyi people—refused to plant cassava, planting it only when they were forced to do so.[211] In 1933 the provincial commissioner reported that the Ikoma, Nata, and Ishenyi were in danger of famine not only because of rain but also because of their "lack of response" to "repeated exhortations to plant cassava as a reserve food supply."[212] In these chiefdoms very little or no food was planted in 1932.[213] The government blamed the people's "laziness" or "apathy" for not planting cassava and did not take seriously their explanation that the soil was not right for cassava, since trials had already grown well in Ikoma. In fact the Ikoma elders themselves blamed famine on the "laziness of their sons," who were out hunting instead of farming cassava, leading the government to conclude that this was another example of the "age-old custom of women to the field and men to the chase" that retarded progress among the Ikoma.[214] The government saw this rebelliousness as a sign of the lawless nature of the people and the ineptitude of the chiefs. They compared the vast amounts of cassava being planted in Ikizu to the resistance in the eastern chiefdoms and concluded that the difference was in the strength of the chiefs' leadership. The district officer did commend Chief Rotegenga of Nata for his communal cassava fields while noting that Chief Sarota of Issenye not only failed to plant cassava but also wrote to ask for a supply of millet seeds to plant.[215] The salaries of the chiefs were reduced as punishment for their neglect of the famine.[216]

However, once again there were logical reasons, within a different view of the landscape, why they resisted cassava. Nata elders said that the Bongirate age-set territory, living out on the plains near Sibora for better livestock pasture, refused to grow cassava because the mbuga, or black-cotton soil, was not sandy enough for cassava, which prefers to grow in the hills. These communities had begun to reap the benefits of growing grain crops on the black plains soil with the introduction of ox plows and were not inclined to return to the hills. Some people left Nata in 1932 to escape the fines for not planting cassava. However, the government used cassava as a way to get Nata to move back to the hill ecologies and out of the plains areas, which were gradually being set aside for game reserves and were out of the range of government surveillance. Chief Rotegenga's Saye age-set territory, living in the hills, accepted cassava, and those who suffered most from the hunger were resettled there.[217] People did not like the taste of cassava, saying that it resembled hyena feces.[218] Cassava also took a number of years to mature and thus would keep people from moving on to new land when problems arose.[219] Many elders said that there were simply other ways of coping with hunger, mainly that of going to neighboring areas to trade hunting products or cattle for food.[220] Even though they noted the shortage of food in Ikoma, Nata, and Ishenyi in 1933, the government did not recommend importing food because, as well as having large livestock herds and plentiful grazing areas, "the people are great hunters and game is plentiful." In fact, when they were offered free grain in Musoma, people preferred to "barter with their livestock for grain from neighboring chiefdoms" rather than going into Musoma for relief food.[221] After the famine ended in 1935, the district officer concluded, "The people of Ikoma are indolent and look upon meat as their staple diet" so the government should "avoid giving them relief." In fact, the South Mara chiefs' council then decided not to give further relief food to Ikoma because some individuals had sold it to the Maasai.[222] Even in the face of famine, western Serengeti peoples remained skeptical of government controls that threatened older ways of seeing and using the landscape.

Although there were still a large number of cattle in the western Serengeti, government control over cattle movement and numbers reduced western Serengeti peoples' ability to use livestock to prevent famine and diversify their risks. After the outbreak of rinderpest in the region, the Veterinary Department issued strict restrictions on cattle movement that kept people from finding pasture during the drought and trading their cattle across the Kenya border for a higher price.[223] Ngoreme people asked only to be left alone, since they perceived that more cattle died "under the treatment of the Veterinary Officers than was the case when they treated them themselves." In Kisaka they even offered to pay double taxes if the veterinary officer would leave. This was after a 50 percent, and sometimes 100 percent, mortality rate for inoculated cattle was reported in Nata, Ishenyi, and Ikoma as well. The officer stated that Ngoreme resistance took the form of "hiding the sick cattle, breaking quarantine, and spreading malicious rumor."[224] In 1928 the largest number of convictions in Musoma District Native Courts was for veterinary offences (226 of 347).[225] The Musoma Cattle Rules, issued in 1933 and 1936 to prevent cattle theft, prohibited people from depositing livestock, even bridewealth, with another person without first obtaining a permit.[226]

The policy of destocking as a way to prevent erosion in overgrazed areas like Sukuma never got very far in Musoma and was finally dropped in 1947 because there was so much resistance and little objective evidence of overstocking.[227]

During the colonial years, hunting was used increasingly as an emergency famine strategy rather than a commercial enterprise.[228] The government was somewhat sympathetic to this necessity, and at the beginning of the famine in 1928 the district officer announced to the Ikoma that, "no administrative action would be taken against hunters provided that they did not go in for wholesale slaughter."[229] The government turned a blind eye when people relieved their hunger by eating game meat but not if they would walk to Sukuma and trade that meat for grain. Western Serengeti peoples do not consider meat a meal by itself; to have eaten, one must have grain in the form of ugali. Although the grain one could get in return for meat would feed many more people than the meat itself, this was not legal. At the height of the Ikoma famine in 1929, reports of increased poaching in Ikoma included descriptions of "wildebeests killed solely for their tails and beards, carcasses being left to the vultures and beasts of prey," and the chief secretary instructed district officers to "take action against increased poaching in the Serengeti region."[230] By 1933 the game warden warned of an impending crisis if poaching did not stop, while the provincial commissioner denied that poaching had increased and asserted that, with a provincial "population of 200,000, a list of 26 offenses committed by an aggregate of 211 natives is not an impressive one" and represented only the increased "zeal of the rangers."[231] Yet hunting still made most sense as a response to famine.

Although illegal hunting was surely an ongoing practice, it is significant that reports of increased poaching show up so prominently in times of famine. By 1939, when the drought was over and the production of food and cash crops in the western Serengeti had increased dramatically, the Game Department ceased its dire reports of poaching in Ikoma. Predictably, the government attributed this to "an increased interest in their responsibilities."[232] However, in 1941, another dry year, reports of increased poaching began to emerge again from the Banagi station.[233] With increasing impoverishment, hunting became a strategy not to get wealth but just to survive. In 1941 a district officer noted that one reason for "the problem of the professional poacher" in Ikoma was people's relatively poor economic position in which "hides and tails" were one of the few resources available to generate cash for "daily necessities in the local dukas [stores]." [234] The long-term effects of the loss of control over the environment and epidemic outbreak took a toll on western Serengeti peoples. Whether by death or migration, the Ikoma lost 20 percent of its taxpayers between 1933 and 1934, and the Nata lost 35 percent.[235] Venereal disease, often spread in the mines, was prevalent during the famine years of 1927 to 1933 and affected a whole generation of women who became infertile.[236] By the late colonial years, as people became poorer in cattle and their old networks of reciprocal responsibility based on the social landscapes of the past had broken down, the government began to bring in relief food during times of famine. The Ngoreme remember that the Indian traders also brought food to sell during the famine of 1942–44.[237] As the colonial years progressed the reputation of western Serengeti peoples went from wealthy cattle owners to poor recipients of food aid.

Older western Serengeti views of the landscape facilitated the use of hunting and trade, first to regain their wealth and then simply to survive and keep their communities intact during the colonial years. New traditions about the colonial period using the core spatial images of hiding and subterfuge conditioned a subculture of passive resistance in the region. What this meant in practice was that they constantly ran into conflict with colonial ways of seeing the landscape embedded manifest in law and sensibility about "civilized" behavior. Whether by disregarding hunting laws, game control areas, growing cotton or cassava, working in the mines or obeying sleeping sickness measures, western Serengeti peoples resisted. It is no wonder then that western Serengeti peoples were "marked for their unruly and rather lawless attitude."[238] Yet that resistance was not a result of opposition to the cash economy but rather their own agenda to gain cattle wealth using strategies embedded in the moral economy of past landscapes. Increasing government control over the environment kept western Serengeti peoples from managing the ecological landscapes of the past, while the colonial backing of chiefs eventually brought a separation between leadership and the health of the land. The commercialization of hunting certainly affected wildlife numbers and habitats, changing the relationship of people to their environment. While western Serengeti peoples developed creative solutions to the challenges of colonial rule based on older ways of seeing the landscape, during this period they experienced increasing impoverishment and marginalization in the colonial world.

⌣

When I traveled through the game reserves to find the old Ikoma settlement sites, I heard stories of how these elders moved from Robanda to open up new settlements in Mugumu in their youth because of the famines. I was keenly aware that these respected elders knew this area because they had hunted here, presumably illegally, in their youth. These men who had shown me such kind hospitality and generosity were certainly not my image of the "poacher" so often demonized in the popular press. The tragedy of the colonial years is, at least in part, a result of these fundamentally different ways of seeing and using the landscape. The resulting dynamics of resistance were to play a fundamental role in the next historical period with the establishment of Serengeti National Park.

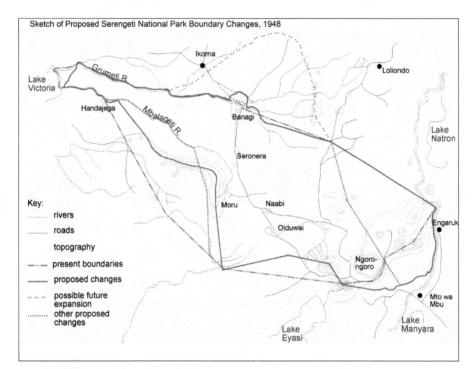

Map 10. Shifting boundaries of Serengeti National Park. *Map by Peter Shetler, 2000, from sketch by G. H. Swynnerton, Game Ranger, 19/9/1948, National Game Parks, 215/350/II, TNA*

6 ⮑ The Creation of Serengeti National Park

Voicing Global Concerns, 1950–2003

THE CREATION OF Serengeti National Park in 1952 was the third challenge faced by western Serengeti peoples, coming in short succession after the late-nineteenth-century disasters together with Maasai hegemony and the imposition of colonial rule at the turn of the century. The park finally set in place boundaries and rules that imposed a hegemonic conservationist view onto the Serengeti landscape in ways that made it difficult for any competing visions to survive. The core spatial images of constriction and restriction characterize the stories western Serengeti peoples tell about this most recent historical context as they try to make sense of their loss. The global conservationist view was different from, and often in conflict with, a colonial view of the landscape as an economic resource.[1] Colonial officers on both sides of the issue argued over whether human rights for economic development or wildlife rights to survival should take precedence. The conservationist view was represented initially by outside organizations like the Society for the Preservation of the Fauna of the Empire and inside Tanganyika by the Game Department (later the National Park Service), hunting and tourism companies, as well as sympathetic administrative officers. Soon after independence, many Tanzanian nationals, some educated outside the country or at Mweka, the premier African Wildlife Management College in Tanzania, also shared this vision.[2] The government profited from outside funding and tourist dollars to run the park and finance other basic services, as well as the global prestige and national pride of its commitment to wildlife preservation and running a world-class research institute in the park. This powerful and compelling way of seeing the Serengeti as a global resource to be preserved for future generations has been supported by both the colonial and independence governments, including those of the African Socialist ujamaa years, the liberalization era of recent years, and new programs under the rubric of "community conservation."

The global conservationist view of the landscape developed in Europe and the colonies as a critique of the destructive impact of industrialization and the colonial economic view of the landscape. Yet this view, too, embraced the ideological separation between nature and culture represented in European landscape paintings of the eighteenth and nineteenth centuries as an unpeopled wilderness reserved for elite contemplation of the picturesque, in contrast to the dirty cities, where poor workers labored in the factories. The seeds of environmentalism, in fact, evolved hand in hand with imperialism as its negative impact on the tropics became apparent.[3] The new science of ecology also grew out of this vision from the colonial periphery of a natural system in equilibrium, becoming a popular paradigm for all kinds of knowledge.[4] Back in Europe, Romanticism brought a critique of industrialization, which, along with increased population and development, identified humans as destructive agents in the natural world, demanding the pro-active intervention of the state. Americans or Europeans—like the Grzimeks, whose film on the Serengeti was featured in the introduction—came to work in Africa disillusioned with progress and in search of the idealized landscape of Eden, including its "noble savages."[5] In this view nature is understood as a global resource, further removing local people from a voice in its preservation. Parks everywhere have taken on the spatial image of a walled fortress to keep out human development and let in only those consuming the wilderness scenery as an aesthetic leisure, rather than a productive activity.[6]

When western Serengeti peoples talk about this last and most recent period of history and their relations with the park, it is often a story of loss, using the core spatial images of constriction and restriction to speak to those in power. With the creation of the park, its game control buffer areas, and the state's increasing ability to enforce law and restrict land rights, western Serengeti peoples finally lost control over the ecological landscapes of the past and were denied customary user rights to the common lands on which they had depended. They could no longer enter the game reserves to hunt; gather wood, thatch, or salt; propitiate ancestral spirits; or even walk the land without permission, a vehicle, and an escort. While western Serengeti peoples have had to adapt to new ways of seeing the landscape as the state assumed control over ecological and social landscapes, they continue to debate the legitimacy of local leadership as defined by older landscapes of oral tradition and their role in shaping wildlife policy. One Nata elder's narrative about the park uses the core spatial images of constriction and restriction:

> In the German days you could hunt in your own land but not in "Shamba la Bibi" [the Queen's Field], the game reserve. The British used the same boundaries, but with independence they

expanded the boundaries and moved people out. These were bad changes. You can't fish anymore and are arrested if you fish in the old places. . . . There are lots of fish, . . . but you can't do it anymore. . . . The Bongirate age-set lived along the river, and wild animals were everywhere. Everything was bush [*pori*] or wilderness then. Many of them were resident animals, but now they go and come with the seasons. There are less wild animals now than before. There used to be lots and lots. . . . Our biggest problem is loss of land. The park constricts us. We can't even cut down trees. We live in the bush, but it is like we live in the city. The huge open bush is for nothing. We can't hunt or graze there. We used to share it with the animals. The animals ran if they saw a bow and arrow, but if you were just herding livestock with a stick they were not scared. Now we don't have a place to graze and must go far away to gather firewood. It is a problem to get charcoal because we can't cut the trees and we can't get wood for building. How can we keep people from hunting illegally? Well, how will we ever be able to go back to the places we used to live? Could we live with the animals again? If there were no guns and no dried meat to export, the problem would be solved. We need to have a place to graze.[7]

This narrative combines expressions of the loss of the extensive wilderness resources integral to the social landscapes of the past, the newer core spatial images of constriction and restriction, and the language of global conservation landscapes. Western observers would not understand this sense of constriction in a land so vast and seemingly empty. Yet it reflects increasing impoverishment as the western Serengeti resource base is increasingly restricted and subsistence farmers find themselves dependent on a cash economy. This narrative also makes it clear that western Serengeti peoples too have begun to use the language of wilderness landscapes in order to make their own voice heard in the global debate. In romanticizing their own past, they argue that their people hunted only to get meat for their families and that they were the natural conservationists, living in harmony with the animals. Accepting the designation of a separate space of nature apart from culture, they make the claim that "everything was bush or wilderness then." The narrator uses older generative principles of the moral economy when he refers to the government's lack of a distribution ethic, to give people what they need to live, indicating a stingy leadership that does not feed the people. The animals are disappearing because the land has not been healed by the walk of the generation-set or the propitiation of the ancestral spirits, as it was when "the Bongirate age-set lived along the river and wild animals were everywhere." Although there are some attempts like these at inserting the older generative concepts embedded in the core spatial images, people increasingly employ the new core spatial images of constriction and restriction in order to speak to the hegemonic structures that control the landscape and leave them feeling powerless to effect change.

Serengeti National Park was created through a contested process involving various levels and branches of the colonial government, international conservation organizations, the European public, and some native authorities in Sukuma and Maasai, both drawing on and reacting to this new vision of a wilderness landscape that must be separated from people. In the years just before and after independence, global interests forcibly curtailed western Serengeti peoples' control over the management and use of ecological resources through the massive antipoaching campaigns of the 1950s and 60s. After World War II poaching itself had shifted to become a commercial enterprise in the context of a global market economy and increasing impoverishment. In the postindependence years of *ujamaa* socialism and beyond, evictions from the park and land tenure laws that restricted people to small plots of land under village or state control affected past strategies growing out of a view of the social landscapes to diversify and spread out risks in the use of wilderness resources. In the liberalization era from the 1980s to present, the new move toward community conservation has provided an opening for a western Serengeti voice and sparked a debate over the legitimacy of leadership in relation to older spatial images of generation-set ritual protection of the land and feeding the people, as well as newer images of restriction and constriction. Although the conservation landscape vision has become hegemonic, western Serengeti peoples struggle to preserve a history and a way of interacting with the environment that is embedded in the landscape of their ancestors. It is the creative recontextualization of core spatial images from past landscapes that can provide more just and sustainable, richer and deeper, more hybrid ways of imaging Serengeti for the future.

COMPETING LANDSCAPE VIEWS IN CREATING SERENGETI NATIONAL PARK

With the establishment of the park, western Serengeti peoples confronted a new and powerful conservationist view of the landscape coming from the imperial periphery that challenged the destructive consequences of the colonial economic view. This new view demanded active state intervention for a larger global good that took precedence over local needs.[8] The debates over the definition of Serengeti National Park and its boundaries, beginning in 1929 with the establishment of the Serengeti Closed Reserve and ending in 1958 with the setting of the final park boundaries, were shaped by a number of conflicting interests in rights over common land, including those of the colonial government, the peoples surrounding the park, and the international conservation community. Although the Serengeti was gazetted as a national park in 1940 under the Game Ordinance, it was not until 1951 that it was proclaimed a national park with revised boundaries under the National Parks Ordinance.[9] The government first established game reserves with an understanding of the landscape as an economic resource to be efficiently developed; the landscape was a place where European sport hunters harvested wildlife resources in areas that did not seem suitable for human habitation, and native hunting was restricted as a way to control labor and to order the rural landscape in a rational economic way.[10] The new way of seeing a natural landscape began to be taken seriously with awareness of the decimation of game and soil erosion in southern and central Africa, bison depletion in the American West, as well as ecological collapse as a result of the American dust bowl in the 1930s. Conservationists argued that impending crisis required radical solutions in which human economic development would take a backseat to the survival needs of wildlife in at least a few restricted spaces like parks.[11] Roderick Neumann argues the conservationist view imagines parks as "landscapes of consumption" for tourists, justifying the need for state intervention and surveillance of rural life.[12]

A global conservation view of the landscape produced blind spots for preservationists reading the environmental evidence. Universalizing ecological crisis on the basis of a few limited cases meant that those who looked at Serengeti generalized anomalous snapshots in time to indicate historical trends.[13] The deliberate steps toward the creation of a park in 1930, 1940, and 1951, pushed forward by a sense of impending doom, coincided with peak poaching years as a result of famine. Although the famine was noted, officers did not see that increased poaching might be a temporary local solution to famine, rather than a long term trend. At the same time, as conservationists were raising the threat of extinction, others in the administration expressed concern that the wildebeest herds, in fact, were too large and causing erosion.[14] Even today, while some sources claim that the wildebeest population is near collapse, with the estimated rate of illegal offtake perhaps as high as two hundred thousand animals per year, perhaps involving as many as thirty thousand hunters, the latest findings show that the total wildebeest population of 1.3 million has not changed appreciably since 1977. Yet at the same time commercial poaching has virtually eliminated rhinoceroses from the park, while elephants, buffalo, roan antelope, and wild dogs suffered from severe decreases during the 1980s. Of these, only elephants have made a dramatic recovery, with the world ban on ivory trade in 1989.[15] The reasons for these uneven shifts in wildlife numbers are clearly more complex than a linear trend in increasing destruction due to local population pressures.[16]

The first global alert of danger to Serengeti wildlife came as a result of European hunting abuses, even though native hunters bore the brunt of the ensuing restrictions. The first Serengeti game reserve was created in 1929 after the international press exposed excessive European lion hunting by unsporting methods. In 1920 an American sport hunter constructed a vehicle road, following a route similar to one that now exists from the Kenya border to Seronera, and professional European hunters soon began to erect semipermanent camps to entertain their wealthy clients in what hunters described as "the last virgin hunting grounds left in Africa."[17] After 1925 European film crews came into the Serengeti, making it particularly famous for its lions, which were baited behind vehicles in order to give tourists a good view. This also made the lions easy shots for hunters in vehicles who sometimes killed up to one hundred at a time, since game laws classified lions as "vermin" for which a bounty was paid.[18] On finding this out, the Society for the Preservation of the Fauna of the Empire wrote a strong letter to the undersecretary of state for the colonies in London and printed an article in the *Times* exposing the destructive habit of "pursing game in motorcars." The SPFE also mentioned "the slaughter of game by natives for reasons other than the protection of their persons or property." What was not mentioned was that 1929 was a year of severe famine in the western Serengeti. Although even the Germans had not seen the Serengeti as threatened enough to warrant action, the British declared a game reserve in 1929 and a lion sanctuary in 1935.[19]

In spite of increasing international pressure for wildlife preservation, many colonial officers continued to give priority to native people's rights, still operating out of the view of landscape as a resource for economic development. In 1922 an agricultural officer pointed out the enormous crop loss to game each year, which, because of the game ordinance, people were "helpless to control" even as they faced "starvation." Colonial officers recommended changes in game laws, prohibiting game reserves near agricultural populations, and giving some of the revenue from hunting fees to farmers for seed and implements in the areas most directly affected by game damage.[20] Medical officer J. B. Davey objected to game reserves in areas like Ikoma, where game brought sleeping sickness and rinderpest, and called for the restriction of game reserves to areas unsuitable for human development.[21] In the postwar era of economic development, many colonial officers often saw game as an unfortunate but inevitable casualty of "the onward march of progress and economic development." In 1935 the chief secretary in Dar es Salaam declared that when the interests of game and humans come into conflict, "human requirements have prior claim."[22] Those who advocated for larger and more parks to preserve wildlife constantly ran into conflict with colonial officers, who put native peoples and economic development first, calling for moderation in land allocated for wildlife. One district officer claimed, "Once an Administrative Officer spends any length of time among the Masai, he becomes more Masai than the Masai, and his whole outlook is jaundiced by the unreasoning affection which he had for the Masai and his complete inability to see any other point of view."[23]

As the debates over park boundaries continued, international conservation groups increasingly defined concerns for human development as antithetical to the needs of wildlife. Major R. W. G. Hingston, who visited Tanganyika on behalf of the SPFE in 1930, was the first to suggest that the Serengeti Game

Reserve become a national park, since without permanent protection it would give way to increased population and economic claims. He concluded that the game would disappear unless all humans were removed from the park, except perhaps hunter-gatherers, who "continue to hunt with their primitive native weapons."[24] In 1939 the SPFE was already pushing to connect a park in Serengeti to a Kenyan park that together would encompass the area of the wildebeest migration.[25] They claimed international rights over common land so that visitors might view some of the last of the earth's wild animals.[26] The society later lobbied London to live up to their responsibilities outlined in the International Convention for the Protection of the Fauna and Flora of Africa, which Britain signed in 1933, calling for wildlife areas that were permanently set aside, large enough to encompass animal migration routes and different from the earlier reserves, which served the needs of white hunters.[27] In concert with the game wardens, the SPFE objected to customary hunting rights of any natives and advocated ultimate authority over the park from London, rather than the colony's legislative body, because colonial officers would be influenced by pressure for native people's rights.[28]

In the tough negotiations to define park boundaries the government did not consult western Serengeti farmers because their low population, lack of a unified political voice, and reputation as poachers meant that they were not a viable political force. The government recognized the Maasai as the only ethnic group with legitimate traditional rights of occupancy in the park and considered "the western zone" to be "undisputed as regards vested human interests."[29] The 1931 report proposing the establishment of Serengeti National Park suggested that the western Serengeti contained "few native inhabitants," and thus "there will probably be no serious objection to those few being transferred outside the boundary."[30] Small groups involving only two hundred families of Sukuma or Maasai, however, gained land concessions on the basis of future land needs for agricultural expansion, in the case of the Sukuma, or because they "neither hunt game nor cultivate the soil," in the case of the Maasai.[31] The Musoma district officer himself dismissed western Serengeti peoples' claims in the park when he reported that, although many considered it their "traditional hunting area," their "traditional method of hunting" involved "cruelty and indiscriminate slaughter."[32] In Musoma District, neither the district officers, the chiefs, nor the people themselves spoke on behalf of western Serengeti interests in the park.[33]

Implementing a vision of the landscape as pure nature or unpeopled wilderness ran into conflict with hard realities on the ground, where people lived. One of the most controversial issues in the debate over the park's boundaries was whether the already resident Maasai could continue living within the park indefinitely, as promised by the Game Ordinance of 1940 and National Parks Ordinance of 1948.[34] When the park was proclaimed, the government reported a total of five thousand Maasai with one hundred fifty thousand cattle living within the boundaries of the park, constituting about one fifth of the Maasai in the whole territory.[35] Those supporting the Maasai argued that they did not traditionally hunt and lived in harmony with the animals.[36] Others opposed continued Maasai presence in the park because it would exacerbate cattle raiding and poaching in the western Serengeti.[37] Yet Maasai rights within the park were only durable as long as international visitors wanted to view them along with the game. In 1950 the governor expressed his concern that in order to stay in the park the Maasai had to become "a museum exhibit, living in a kind of human national park."[38] While the commissioner of the Northern Province urged the park board not to evict the Maasai from Ngorongoro, stating that "they are the most interesting feature of the crater for tourists to photograph," he was clearly disturbed by the implicit inability of Maasai to develop and improve their economic position if they wanted to stay in the park.[39] The park trustees eventually hoped to attract the Maasai to leave the park by offering improved water and grazing outside the park, while refusing to develop those resources in the park.[40] Many felt that the park rules were unjustly "designed to maintain the Maasai in their present primitive state" and therefore would have to be revised to give the Maasai more autonomy.[41] The Maasai themselves were not happy with park restrictions, stating that "Maasai value personal freedom above anything else and in the National Park we are not free."[42]

Mbulu district officer Gordon Russell's 1950 report made the first proposal for the park's boundaries, suggesting that they begin with a nucleus in the northern crater, highlands with the rest remaining as a national reserve until such time as the park could be expanded. Others pushed for, and eventually gained, a much larger park area, although extensive areas were also excised because of Maasai and Sukuma concessions, while Ikoma land in Banagi was added to the park.[43] In 1954, as pressure from the Maasai increased, district officer Grant of Monduli put forth a proposal supporting the Maasai's legitimate rights to life in the park and recommending new boundaries that would allow for their development, based on the fact that the Maasai used the Serengeti plains only during the dry season, when the wildebeest herds were not present. Maasai numbers in the park had only increased due to the difficult droughts in the 1940s and 1950s. Grant proposed giving the Maasai the proceeds of tourist income as well as excising the central plains from the park, in return for their submission to park rules and giving up the right to farm or to move freely.[44]

Needless to say, conservation advocates were appalled at these suggestions for carving up the park; they were determined to "fight this proposal by all means at our disposal including, if necessary, an appeal to public opinion throughout the world to whom the Serengeti Plains means plains game and lions and not cattle, small stock and Maasai."[45] Many in the government feared for the integrity of the park, complaining that the Maasai harbored Ndorobo hunter-gatherers and "undesirable traders" and that more and more Maasai were moving into the park.[46] They evicted some one hundred Ndorobo families from the Moru kopjes, along with 21,780 head of livestock, leaving 927 "lawful inhabitants" with 20,000 cattle and 16,000 sheep and goats in the western Serengeti portion of the park.[47] The park wardens were convinced, however, that even the pastoralist Maasai presence was detrimental to the park, citing the long-term results of the destruction of trees for building materials and erosion from overgrazing.[48] According to an official memorandum, by 1955 it was becoming clear to those on both sides of the debate that in the long term "the policy of coexistence of Masai stock and game was not going to work," and that perhaps the vast area of the park was overly ambitious.[49]

In 1956, after getting prior approval from both the park's board of trustees and the Maasai Council, the Legislative Council of Tanganyika issued the controversial white paper on Serengeti National Park as a compromise solution that mediated between the needs of people and the needs of wildlife. The publication ruled that "the maintenance of human rights in a national park is not compatible with the interests of wild life therein," but at the same time asserted that the existing rights of the Maasai that were protected by law would not be infringed upon.[50] As a result, three areas were declared for total exclusion of people in a national park: the western Serengeti (3,625 square kilometers), the Ngorongoro Crater (1,165 km²) and the Embagai Crater (260 km²), with a corridor for game to pass between the park areas. The new plan removed the central plains, along with the Moru kopjes, and returned them to the Maasai, reducing Serengeti National Park from 11,550 square kilometers to only 6,730. The plan also sought to add land in Sukuma Maswa District, the northern extension, and a small piece of land on Lake Victoria, along with suggesting three new controlled areas, including the Grumeti Game Control Area.[51] As expected, there was outcry from those who advocated global land rights over those of the Maasai.[52] The *Sunday News* from Dar es Salaam representing the English-speaking population of Tanganyika, ran an article accusing the Tanzania national park trustees of "failing in their trust" by bowing to pressure and engaging in "secret negotiations from which people were excluded." The writer made a case for the Serengeti as a symbol of national life and common pride.[53] The *East African Standard* called "the dismemberment" of the park "a tragedy" that "threatens to sacrifice one of Nature's greatest heritages of wildlife for fewer than 100 Maasai families who entered illegally into the Ngorongoro Crater in 1946."[54]

International conservation agencies—including the wildlife societies of Kenya and Tanganyika, the International Union for the Conservation of Nature (IUCN), and the Fauna Preservation Society (FPS)—lobbied the colonial office in London for an independent inquiry and legitimized their claims to the Serengeti by mobilizing ecological science to their cause. The FPS sponsored an ecological survey conducted by W. H. Pearsall in November and December 1956. Pearsall's subsequent report argued that the park should include all the land within the annual wildebeest migration route but should not include pastoralists, whose livestock were destroying the environment. Furthermore, on the basis of ecological science, the report recommended that the Moru kopjes and the crater highlands should be included in the park as critical water catchment areas. Pearsall gained his underlying political agenda because his report was accepted on the basis of the authority of ecological science. Yet even park managers admitted that at the time the report came out little was known about the extent or path of the wildebeest migration, nor of the park's ecology in general.[55] The park trustees supported the Pearsall report, forcing the government to appoint a committee of enquiry to reevaluate the matter yet again, a process that produced another ecological report by game warden G. H. Swynnerton and a bitter statement from the Maasai.[56] As park warden Myles Turner stated, setting the park's boundaries had "become the hottest conservation issue in the world."[57]

The government was sympathetic to the conservation views represented by the Pearsall report but was also keenly aware of the competing economic view of the landscape, necessitating political support from the Maasai and Sukuma and outside finances to support the park.[58] They were additionally pushed by European public opinion, aroused as a result of reports from the new antipoaching campaign in the park that generated extraordinary stories, aired by the press and witnesses before the legislature, about the confiscation of thousands of kilograms of dried meat and steel-wire snares.[59] Ironically, Turner, who wrote the original report on poaching, heard for the first time on British radio the estimate, attributed to himself, of two hundred thousand animals killed illegally in the park each year. Turner reported that he laughed and called it "good propaganda for wildlife conservation." This unlikely figure, exaggerated from a dubious estimate made by the park's director, would have amounted to a full tenth of the park's animals.[60] In 1957 the chief secretary wrote a circular to all provincial commissioners alerting them to the strong "public opinion against the illegal destruction of game."[61]

The final decision in 1958 set the boundaries of Serengeti National Park nearly as they appear today, including the whole area of what was assumed to be the wildebeest migration route and excluding human settlement. The new national park included the western and central plains, the Moru kopjes, and by 1969, land on the Duma River in Sukumaland and the land northward to the Kenya border that became known as the northern extension, for a total of 14,500 square kilometers. The Ngorongoro Crater highlands were removed from the park to become a special conservation unit where Maasai could go on living as a concession for the loss of the Moru kopjes and the central plains, while the floors of Ngorongoro and Embagai craters had special status as nature reserves.[62] After that decision the park systematically evicted Maasai, Sukuma, Ikoma, and Ngoreme people who still lived in the park and achieved its goal by 1960, before independence. There was no debate about including western Serengeti peoples in the park, since the committee of enquiry declared that it was of "only mar-

ginal value for human use and will rapidly deteriorate under continued occupation."[63] But the debate over Maasai rights continued as the IUCN, the Nature
Conservancy, and other conservation groups, particularly in the United States,
demanded that the Ngorongoro Crater be kept within the park. This took place
as the colonial government was realizing that it would soon have to turn over the
colony to national leadership.[64] Additionally, the government did not know how
it would pay for the park when it then faced the need to cut social services for
lack of funds. The government could only hope that the worldwide concern for
the preservation of the crater would somehow produce the funds necessary to
administer the park.[65]

On the eve of Tanzanian independence, those closest to Serengeti National
Park were working hard to ensure its continuity under national leadership and
gain the support of its citizens, afraid that they might do away with the park altogether.[66] It was in this climate that Dr. Bernhard Grzimek, director of the Frankfurt Zoo, published *The Serengeti Shall Not Die*, the book and subsequent film
that popularized the view of the Serengeti landscape as pure nature. The book
aimed, as his publisher said, to "try to stir up public opinion so that the Colonial
Government does not . . . diminish the area of this sanctuary for wild animals."[67]
Grzimek and his son pioneered the aerial census and monitoring of wildlife.
However, their research was not based on disinterested scientific observation but
rather on the political agenda of proving both that wildlife migration patterns
could not be accommodated if the park was split in two and that humans must
be separated from wild animals.[68] Conservation organizations like the Nature
Conservancy and the Frankfurt Zoological Society were involved in fund raising
for a research project in the Ngorongoro Conservation Area and a research fa-
cility in the park itself, which became the Serengeti Research Institute in 1966.[69]
But park director John Owen also knew that in order to survive, the animals had
to "pay for themselves" by bringing in tourist dollars. This they did; between 1956
and 1972 the annual number of visitors to the park rose from four hundred to
fifty-two thousand.[70]

The story of the SNP demonstrates how this global conservationist view of the
landscape came to be hegemonic not only in the Serengeti but throughout Tanzania. It is now the unquestioned orthodoxy; the question is no longer *if* there
should be a park on the global model but how big and under whose control.
Creating the SNP was not a straightforward exercise in protecting the wildlife. It
involved competing views of the landscapes and interests of many kinds, both
within Tanzania and without, and decisions that were often made on the basis
of political rather than ecological concerns. As the colonial era came to a close,
Tanzanians looked forward to a time when they could rule their own land. Yet,
oddly enough, it was only during the era of independence, once the SNP was
fully funded and staffed, that the state was able to firmly control the wilderness
landscapes of the Serengeti and to exclude western Serengeti peoples.

Both the efforts of the state in controlling wilderness resources and the response of western Serengeti peoples must be understood within the global context of African independence and the unquestioned dominance of a conservationist view of the landscape.[71] Tanzanian independence came in 1961, on the heels of one of the worst droughts on record in the Serengeti. Then in 1962, during the first farming season after Mwalimu Julius Kambarage Nyerere became president of the United Republic of Tanganyika, the heavens opened up and it rained and rained and rained. Throughout Tanzania, rain is a symbol of fertility and blessing, and to local peoples the overabundance of rain symbolized a divine blessing on independence.[72] In the SNP it rained so much that Naabi Hill became an island and the central plains a sea. The rivers shut off any movement in or out of the park during that rainy season.[73] Thus, independence brought hopes for the promises of self-rule, economic development, and equal justice. Tanzanian citizens assumed leadership for the parks after 1970, with Solomon ole Saibull becoming director of Tanzania's national parks and David Stevens Babu becoming chief warden of Serengeti National Park.[74] President Nyerere—the son of a rainmaker chief from Butiama, in the Mara region, and holder of an MA from Edinburgh—represented those dreams.[75] He, more than anyone else in power, knew what it was like to live on the edge of the Serengeti, among the wild animals, and was aware of the global concerns that surrounded them.

Given the expectations for independence, the most astonishing thing about the administration and regulation of the SNP in the postcolonial years has been its continuity with both the colonial policy and global conservation views of the landscape, including the complexity of the debates surrounding the park. In fact, united Tanzania expanded the protected areas to include 32 percent of the nation's total land area. Mwalimu Nyerere, the new republic's first president, declared at the 1961 Arusha conference that Tanzania would accept responsibility for conserving this world heritage, while looking for monetary support and expertise from outside. The impetus for the conference had come from the international conservation community, which was concerned that with independence Africans had neither the training nor the commitment to conserve wildlife in Africa.[76] While the discussion about Maasai occupation of the Ngorongoro Conservation Area continued, the SNP spurred little debate about its legitimacy but ever-increasing regulation and enforcement. The independent government developed its own interests in preserving the SNP to gain foreign currency, finding that it could not run the park without support from outside conservation organizations. Therefore, it was increasingly in the interest of the government to promote a conservation view of the landscape over the interests of Serengeti peoples.[77]

While hunting had been a strategy embedded in the social landscapes of the past to diversify local economies and prevent famine through wide regional networks, in the post–World War II context both a new kind of poaching and a new antipoaching response from the government emerged that subverted the older landscapes. This new form of poaching may be compared to Kuria cattle-raiding gangs in Michael Fleisher's analysis; both should be seen as modern, rather than primitive phenomena. While both cattle raiding and hunting in earlier times supported networks of exchange in the community, the new forms involved commoditized products and tore apart rather than unified those same communities. Cattle raiding has been a part of local practice since at least the nineteenth-century Maasai expansion but, according to Fleisher, took on its modern form after World War II as a "highly organized, cash-market-oriented enterprise, carried out by heavily armed, multi-clan and multi-ethnic gangs."[78] Stolen cattle were destined for a global market via a chain of accomplices and middlemen across the border in Kenya. The proceeds from raiding were no longer meant to restock the family herd but to supply cash income. A tradition of interclan and intergroup conflict allowed raiding to acquire legitimacy and acceptance even in its new form. Both cattle raiding and poaching developed as part of the informal, illegal economy that flourished in the global market.[79]

As the western Serengeti became increasingly impoverished within a world system in which they were at a disadvantage, poaching gangs, like cattle raiders, attracted young men who had no other opportunity for advancement. Fleisher found that cattle raiders were young men in their twenties and thirties who tended to come from the same families, those that had more sons than daughters. Young men normally received cattle to pay bridewealth for marriage from the bridewealth paid to their family for their sisters' marriages. Young men from a poor family without sisters had to find another way to get the necessary livestock and saw cattle raiding as a way to get rich quickly without much effort. The same profile may well apply to poachers in the Serengeti. Few local opportunities exist for young men there to make money for bridewealth or to gain an education for nonagricultural jobs.[80] A 1999 survey showed that in the western Serengeti villages the elementary school with the highest number of students passing the national exams was Mbisso with six, while none in Makundusi passed. Attendance in primary school has been notoriously low and for those who go on to secondary school, even lower. Many young people do not want to farm; they leave the area to find unskilled jobs in the city. The same survey shows an exodus of young men, rather than their attraction to the area as poachers; in the five prime poaching villages along the park borders included in the survey, 33 percent of all residents were adult women and 23 percent adult men. Farming and herding were still the main occupations of western Serengeti peoples, with 65 percent of caloric consumption produced locally.[81] Poaching was clearly a tempting option for the young men who stayed at home. Poachers continue to claim their hunting use rights to common land now set within the context of a modern global economy where the informal sector and growing market for wild-animal products flourished in the face of increasing impoverishment.

A conservationist view that sought to control wilderness resources and protect animals at all costs, even violence, led to the militarization of the park, which included antipoaching campaigns.[82] The case is somewhat similar to one analyzed by E. P. Thompson: the 1723 Waltham Black Acts in England, which inflicted capital punishment on deer poachers and wood and turf gatherers in the royal forests. Thompson argues that these excessive laws were brought about by a new capitalist merchant class who extinguished the customary use rights of poor people in the royal forests by fencing them off for direct exploitation and for the display of wealth in the form of deer parks and manicured gardens. The people rebelled by blackening their faces and raiding the forests to defend their customary rights to land for grazing, farming, or gathering. Although people had always hunted in the royal forests, the powerful now saw "poaching" as an affront to their authority and individual interests.[83] In the western Serengeti case, poaching represented a challenge to a state that sought to control the processes of modernization and development. Bureaucrats of the independence era took on much of the attitude of their colonial forbearers in seeing rural people as backward and primitive, in need of the benevolent guidance of the state.[84]

Although hunting in what is now the park had been prohibited since the early colonial years, according to both park wardens and local people, antipoaching efforts in the Serengeti only really began to be effective when the park's boundaries were set and resources came in from outside to fund personnel, transport, and communication. SNP warden Myles Turner reported that when he took over the park in 1956 he had ten staff members, one Land Rover, and one Bedford truck at his disposal to establish a "presence" by manned ranger posts in isolated areas where "poaching had gone on virtually unchecked."[85] In 1969 the Frankfurt Zoological Society made their first of many gifts to antipoaching campaigns with three new Land Rovers, and the government began to inflict steeper fines and prison sentences on convicted poachers. By 1971 antipoaching efforts in the Serengeti had developed into a full-scale military campaign. The Serengeti Field Force consisted of seventy-two noncommissioned officers and rangers with six Land Rovers, one truck, and two Cessna aircraft. The peak years for poaching convictions were 1969 (364 convictions; 2,715 snares confiscated) and 1971 (346 convictions, 2,725 snares). The rangers, in army fatigues, carried rifles and practiced drills each morning. The ranger posts began to resemble stonewalled, inaccessible forts rather than wilderness camps.[86] Yet the success of the antipoaching campaign depended as much on the psychological illusion of pressure and presence as much as it did on force. Poachers, in turn, relied on a widespread belief in their ability to turn themselves into animals or administer a curse on the rangers. In Myles Turner's sixteen years as deputy park warden in charge of field operations, he confiscated twenty-two thousand snares and captured one thousand prisoners for conviction. By 1972 the administration announced that poaching within the park had been brought to a "near standstill."[87]

The park rationalized the huge outlay of expense and the use of coercive force in the antipoaching campaign by defining illegal hunters as a criminal subculture of cruel professional gangs rather than poor farmers. The park warden characterized the poachers he pursued as "gangs" who practiced cruel and unskilled methods of hunting, distinct from the "traditional" hunters of times past, whom he admired.[88] Grzimek popularized this demonized image of the poacher in his reports of coming upon poacher's camps where the ground was "dyed red with fresh blood."[89] The director of national parks, Lt. Col. P. G. Molloy, referred to the hunters' "fiendish system" of wire snares and thorn fences as one that "demands the minimum of effort from the hunter"—its "ruthless efficiency only matched by its hideous cruelty" that will "only result in total extermination of wild life."[90] The image of poachers as wealthy outsiders rather than poor local people became a popular, politically correct way to legitimize antipoaching in an era of independence and self-determination for Africans. A children's book called *Poachers in the Serengeti*, written in 1965, portrays the leader of the poaching gang it describes as an Arab named Mohammed Ali, with "a blood-chilling grin that revealed long discolored teeth."[91]

Poachers were classified as criminals because of their cruel and indiscriminate means of hunting according to the European "hunting ethos." Their new method was to set hundreds of steel-wire snares along openings in fences constructed on game trails in the passes between hills or near watering points. The steel cable was sold in three-meter lengths in the market for a few shillings and was often stolen from gold mines in the area. Sometime after World War II, steel snares had replaced the traditional sisal snares, used mainly by young boys for capturing small game near home. The "gangs" usually included twenty to thirty men, divided into groups of fence makers, hunters (snare setters), and porters, each with their own part in the hunt. They set up camps in the dry season and hauled out dried meat and skins until the camp was either raided or the game depleted. Rangers often encountered animals deeply wounded by the wire snares, with their flesh cut to the bone, dragging the broken snares behind them.[92] Kay Turner writes in her memoirs, "No one who has seen the results of wire snares can remain indifferent to its use."[93] The West has become sensitized to this issue by the many films about the Serengeti, including those of the Grzimeks, showing in graphic detail the suffering of snared animals alongside thousands of captured snares being destroyed by game rangers.

Despite the government's efforts to demonize poachers, a significant shift *had* taken place after World War II—from the age-set based, communitywide hunts in times of hunger or in pursuit of replenished livestock herds to the appearance of organized, professional poachers. Reports from this area before the war do not mention the presence of gangs, though they do note that the Ikoma valued the tails of zebras, wildebeests, and giraffes as well as dried wild meat.[94] Characteristic of the earlier period, in her memoirs Audrey Moore, wife of the Banagi game ranger, refers in her 1938 memoirs to illegal Ikoma hunters as "law-breaking hunters" or "tribesmen," who no longer used pitfalls but "always have a smile and a word for us, and bear no malice at all even when caught red-handed amongst their game snares."[95] But as demand on the world market brought wealthy dealers in trophies, meat, and hides, park warden Myles Turner reported seeing a new brand of poachers. Many mention that Ikizu, Zanaki, or Luo hunters were the first professionals with guns and vehicles who delivered game according to advance orders. In 1969 the market for illegal skins and trophies increased all over East Africa, leading to "motorized gangs" who killed zebras, leopards, and other animals whose skins would sell for a high price in Nairobi or Arusha.[96] In the local market the price of wildebeest tails also skyrocketed due to increased pressure on hunting. Kay Turner writes that the "greatest threat to wildlife" was not the local hunters, who killed for meat, but the "commercial poachers" with rifles and muzzle-loaders working at night in vehicles for "material gain."[97] Another type of poaching that appears in the park warden's reports was "motorized poaching by police, army and government officials and others in the Serengeti." Myles Turner reports poaching raids in which he apprehended government employees, including Game Department personnel, driving government transport vehicles.[98] The park soon became another state resource that could be pilfered by those with access to power who also suffered under a declining economy.

Yet in spite of the dominance of the new kind of poaching, other incidents reported by the antipoaching patrol indicate that some of the old-style hunting was still being practiced and that people continued to claim their customary use rights to common land. In contrast to the image of a hardened criminal poacher, Turner also reported capturing and releasing two nine-year-old Ikoma boys hunting with miniature bows, quivers of poisoned arrows, and knives. In the northern extension in 1969 a patrol came upon a community wildebeest hunt in which men, women, and children armed with bows, axes, and picks, along with their dogs, drove the animals into gulleys in broad daylight. Similarly, they once observed large numbers of Sukuma, including many children out hunting when the migration came through during the day. The Ikizu village of Hembe reportedly had more than half its adult male population in prison on poaching charges in 1965, suggesting that more than criminal gangs practiced hunting. Turner's account also seems to indicate that poaching was worst in drought years such as 1967 and 1969, the two worst poaching years on record.[99] Although professionalized criminal "gangs" do seem to have been operating and increasing the number of animals killed, communities around the park also participated in commercial hunts and protected the gangs. The popular acceptance of hunting and a willingness to defy the law by passive resistance goes back to colonial experiences with hunting laws. In fact it seems hard to distinguish between traditional hunters and criminal poachers, since most men participated in hunting at one time or another.

A subculture of resistance to game laws, including animosity and mutual resentment, was firmly in place long before the antipoaching campaigns began. In his memoirs, Turner described Ikoma, Kuria, and Ngoreme men as little more than a "heavy poaching fraternity." He also recalled the resentment he felt from the other side as he flew the park's airplane over the Mugumu area and a young boy threw a rock at the plane, while a little later a man shot an arrow at it. He commented in his memoirs, "We have no friends here!"[100] The Kuria were particularly resentful of losing land in the park's Lamai Wedge and would throw rocks down over the escarpment at game patrols.[101] This animosity was a result of arrests and harassment but also of the intrusion of government restrictions and rules into every facet of life. At one point the game rangers were even confiscating weapons from herders, who successfully argued that one could not be out in the wilderness without a bow and arrow or spear to protect against both wild animals and Maasai raiders.[102] Of course, the game rangers also expressed their own resentment toward the poachers. Bernhard Grzimek recorded his son's words as they buzzed poachers from their airplane at three to four meters: "I'd like to have a machine-gun which fired through the propeller." These same "poachers" fired back arrows at the plane, one of which struck the wing.[103] Clearly, there was no love lost between western Serengeti peoples and the park's representatives. Mutual enmity and a subculture of resistance still infuse interactions between the park and the surrounding communities, even when the park tries to accommodate local needs.

SOCIAL LANDSCAPES AND THE DEFINITION OF LAND RIGHTS IN THE UJAMAA ERA

At the same time the park became increasingly conservative and fortresslike, the society around it was experiencing profound political and economic shifts as the nation moved from colonialism to independence and then from socialism to free-market capitalism. During this entire period, however, the state continued to undermine customary rights to the land and access to wilderness resources. While the older social landscapes defined land rights in terms of clan or descent group membership and use rights to larger wilderness resources, the postcolonial government accepted a European definition of property rights from the colonial era in which the state controlled all land. In 1966 President Nyerere declared that Tanzania would follow its own path—called African Socialism, or ujamaa—to achieve its goals of egalitarian development, which included nationalizing industry and moving rural people into centralized cooperative villages. The government promoted national unity through successful Swahili-language adult literacy campaigns. The postcolonial Tanzanian elite who ran the park had their counterparts in the political elite who ran the nation. Both were committed to the modern development project of their colonial predecessors, though now using the rhetoric of traditional African concepts of family and cooperation in the face of decreasing resources from abroad. While this may have been good for building the nation, it was not good for the economy, and Tanzania's debts mounted.[104]

The most profound effect of the ujamaa legislation on life in the western Serengeti was the further erosion of land rights in the villagization campaign of 1973–76. While at first ujamaa villages were promoted as voluntary experiments in cooperative living, they soon became mandatory. All rural peoples now had to move away from their homesteads and fields to designated village centers, where they were promised better social services like schools, clinics, and economic opportunities. In 1975 these villages received legal recognition as the entities for regulating rural life from that time forward. Rural people all across Tanzania resisted relocation and tell tales of the enormous suffering it engendered. One account of "Operation Mara" describes how armed men came to move people without warning, tearing off thatch, removing doors and windows, and taking them in trucks to new places that had not been prepared. The early December rains started before they could build new homes, and agricultural production suffered, forcing people to walk back to their old fields, three to four kilometers away, to harvest anything.[105] With the concentration of people and their farms in villages, pressure on the fields and grazing lands led to erosion, soil degradation, and an increase in epidemics, while food production decreased and the bush spread. People also had their own reasons for distrusting the new sites, often in relation to the location of ancestral land spirits. When the government realized the enormity of its mistake and allowed people to gradually return to their old sites in 1981, the damage had already been done.[106] For rural people, ujamaa ultimately meant additional dispossession of land and state regulation.

Western Serengeti narratives about villagization that furthered the government's goal of moving people away from the park's boundaries were expressed through the core spatial images of constriction and restriction.[107] Nyamaganda Magoto was teacher at the primary school at Sibora from a year after it was opened in 1957 to a few years before 1974, when the village was moved because of ujamaa. He described Sibora, within what is now a game control area as a large town with many shops and houses, bigger than present-day Mbiso where the people were moved. It therefore seemed that the government had had another agenda when it moved people out of already concentrated settlements like Sibora. During ujamaa people watched as the game reserve buffer zone around the park expanded into areas that had just been emptied of people for villagization.[108] Sibora, along with a number of other villages, was built out on the plains, where wealthy cattlemen moved in the early part of the century to have more space. It was also a place where men experimented with ox plows in the black-cotton soils. With villagization these plains settlements were moved up into the hills, where cassava and cotton grew better. One elder described Sibora this way:

> Everyone was moved out of Sibora forcibly in 1977 with ujamaa and the game control area. Sibora was an incredibly productive place and fed all of Nata. The soil doesn't dry out. There was no cattle disease; when the colonial government brought in cattle medicines, they were not used, not even for Ndigana. Sibora was the biggest settlement anywhere out here. Cassava does not grow well at Sibora, but millet and sorghum do great. The soil was farmed with ploughs because they were mbuga [black-cotton] soils. Those who settled here were the age-set of the Sanduka, from the Bongirate cycle, Nata, Ikoma, and Issenye all settled here together.[109]

Another elder said that when people were forced to move back into the hills they settled by smaller descent groups rather than by age-sets, as they had on the plains.[110] The main Ikoma village of Robanda, out on the plains and in close proximity to the park gate, successfully resisted evacuation with villagization because that would have meant taking the ritual elephant tusks, the Machaba, across the Grumeti River. Although many moved out of ujamaa villages after 1981, they have not been allowed to move back to the plains settlements, now part of the game control areas.[111]

Villagization also brought significant changes in mobility; western Serengeti peoples were now tied to their village residence and so were not free to move every few years as they chose. In addition, the move to the hills to plant cassava, which takes a number of years to mature, meant that they had a longer-term investment and could not leave as easily.[112] Many people who had not been moved to a village too far from their fields continued to walk back to those fields to farm. Larger-scale migrations were considerably slowed or halted after ujamaa. Ikoma people began to move away from Banagi (now within the park) and Robanda into the open wilderness area around Mugumu, known as the land of the Asi, just after World War II, while immigrants from Kenya (Maigori, Nandi, and Luo) and North Mara (Kuria and Luo) moved there in the 1950s because of population pressure. Although park authorities saw the move into Mugumu as a simple desire to increase poaching activities, Mugumu elders say that they came searching for better farmland because of the drought.[113] Many Kenyans returned home with the villagization of the ujamaa years. Violent interclan and interethnic cattle raiding became endemic in the Mugumu area and into Ngoreme from the 1970s on, in part as a result of the dynamics of villagization and the impoverishment that came with restriction of resources.[114] Loss of settlements now part of the park were doubly hard for Ngoreme and Ikoma peoples to take because at the same time the government was giving large areas of Mugumu, Tabora B, and Ikorongo to the newcomers as "empty land."[115] The Tanzania census shows a net increase in population of 5.6 percent between 1957 and 1967 due to immigration into the districts around the western side of Serengeti National Park.[116]

The state's right to take land from people, both to create a park and to form ujamaa villages, is based on British colonial law, which declared that the king owned all land by right of conquest. Although the 1928 Land Ordinance recognized customary land rights, peasants had no title or defensible protection for their claims, which could be alienated to settlers or foreign corporations without legal recourse. A dual land tenure system evolved in which some land under statutory tenure was bought and sold on the market with secure private title, while the vast majority of land remained under insecure customary tenure.[117] In 1995, as a result of conflicts over land resulting from villagization and the new push toward privatization and market economy, a new land law was passed. Land remains the property of the nation, under the direction of the president, who controls it on behalf of all citizens through the Department of Land. In order to change the category of land, the president must first give notice of the change, get the consent of those involved, and recognize land rights obtained during ujamaa villagization. All land in Tanzania falls into one of four categories: private land, conservation land, village land, or common land. A process is now in place to deal with land allocation and dispute resolution, and a decision by the village assembly required to change land rights. One can obtain a certificate (*hati ya kimila*) that confirms customary land tenure rights as a form of ownership for a specified period of time, anywhere from five to ninety-nine years.[118] However, Tanzanian scholar Issa Shivji concludes that, in spite of the rhetoric of a "bottom up, participatory process," the new land law actually disempowers common people and does not give them secure tenure. The system for titling land is overly bureaucratic and impossible for rural peasants to negotiate. It gives the commissioner of land total authority while excluding the village assemblies and the National Assembly. In line with the present trend toward liberalization, the market determines who gets land, which will mean the further alienation of customary land rights.[119]

Western Serengeti peoples express a sense of restriction and constriction in the implementation of the new land law. If land is registered they will have to give up much of the farmland that they control for rotation, fallow, or lending to others. Customary land that is registered under the new law will be taxable and few will be able to pay for all the land they now claim. This will destroy the extensive system of shifting cultivation and mobility, demanding expensive inputs in either labor or capital for fertilizer on much smaller plots of land. Without customary land tenure, people can no longer move homesteads and fields to take advantage of soil, climatic conditions, or shifting social alliances, nor return to lands they once considered their own.[120] Hunting, fishing, cutting trees, and

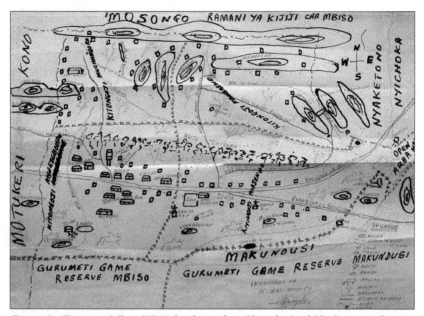

Figure 6.1. Tentative Mbiso WMA land use plan. *Photo by Paul Shetler, 2003, from map provided by Mbiso village officer*

gathering other forest products is now against the law in the game reserves. Since the law gives them no recourse to guard their customary rights to common resources, western Serengeti peoples continue to find ways to defend those rights by their own means, either by continuing to cut wood and fish or by active protest.[121] Today, villages participating in the wildlife management areas are required to create a management plan in which they designate which parcels of village land will be used for what purposes. Although these plans fit the government's need for a clear land use strategy, it does not take into consideration the older system, based on social landscapes, in which land use shifted as people moved between various homesteads, fields, and grazing areas over the years. They allow for no flexibility as new conditions arise or when land is closed off for tourism and hunting companies to make adequate provisions for activities elsewhere. When elders today claim that they are feeling constricted on the land and suffering from land shortages, they are not referring to "crowding" but to the lack of access to a wide range of resources necessary for the extensive agropastoral system.

Many of the evictions from park land and incorporation of new land into Serengeti National Park were aimed specifically at disrupting resistance to the government's authority. One of the last areas to be incorporated into the park was the small piece of land that connected the SNP with Lake Victoria so that migrating herds could reach the lake in times of drought. Of the 800 people who were evicted from this area between the road and the lake and in Handajega, 256 were eventually compensated.[122] While some questioned whether this extension was really necessary for the wildlife, a bigger reason behind the evictions was the fact that Handajega had long been one of the most troublesome spots for poaching and resistance to park laws. In 1950 when the warden visited the growing base of the notorious poacher named Sonda at Handajega, he found wild meat drying everywhere, "festooned from trees and every rock was covered with it," as well as the "skins of the zebra, wildebeest, grants [Grant's gazelle], topi, reedbuck and buffalo."[123] Another troublesome area for poaching was north of the Mara River, now known as the northern extension, which connects Serengeti National Park with Kenya's Maasai Mara Park. The Musoma district officer protested that he and the Ikoma and Ngoreme chiefs had been part of a survey to fix the western boundary of the partial game reserve, only to find out that it was to become the new park boundary.[124] The expanding Kuria population had been increasingly moving into this area when it became park land in 1966 and some families refused to go, resulting in opposition and near violence.[125] Turner reports that at a meeting attended by field force rangers and 250 people armed with spears, the crowd declared, "We don't want National Parks, Game Department, or anything. This is our country." The issue was brought up again in 1972, resulting in the removal of 180 square kilometers of park for grazing.[126] Court cases concerning evictions from the park or game reserves continue to appear today.

No sooner were the park boundaries settled than the park began to make demands to "buffer" the boundaries with additional protective zones in the form of a partial game reserve to the south and a controlled area to the west.[127] Game warden Swynnerton first broached the idea of "buffer zones" in 1952, suggesting that the change of park boundaries was made with the needs of humans rather than game in mind, and that the former needed to be protected when they left the park.[128] Although people were promised that they could continue to live, cultivate, and graze their livestock in this land so long as they did not shoot the protected species, those promises did not hold.[129] After the park was created, the Grumeti was declared as a controlled area (hunting by permission of the game warden) and the Ikorongo as a partial game reserve (hunting of particular species), neither of which restricted entry or residence.[130] Since 1994 the Ikorongo-Grumeti Game Reserves have been upgraded from game control areas to game reserves where no residence, farming, or hunting is allowed, and the boundaries have been modified.[131] In 1964 administrators in the newly formed Mara region reported "considerable opposition" from Ikoma, "who seem to derive much profit from poaching and grazing" in the areas from which they were being "evacuated" on the western boundary of the park.[132] Elders remember that the Maragwe community of Ikoma was still left within the park boundaries, while they had previously abandoned Banagi when the game post was established there. The Nata had already moved out of Girawera and Senagora because of anthrax spread from

Thomson's gazelles and the threat of Maasai raids.[133] Some of the last formal evictions of western Serengeti peoples from the buffer zones took place in the Grumeti extension from 1967 through 1969, when twenty-six families that the park claimed "merely used it as a poaching reserve" refused to move until the Musoma police finally forcibly removed and settled them in Nata.[134]

The core spatial images of restriction and constriction are evident in stories about the park in recent years as people search for a way to express their discontent at what has happened to their land. The districts that include the Serengeti-Maswa conservation area have the highest percentage of district land under some form of conservation in Tanzania, with 36.5 percent of the total land area.[135] Since the government did not understand land rights as extensive access to various wilderness resources and customary use, they saw no contradiction in declaring that the SNP "did nothing to affect the existing rights of any person in or over the land included in the park."[136] The overwhelming perception among western Serengeti peoples about the park is expressed by an elder who said that "it gradually crept up on us." First there was a small area designated as a game control area under the Germans; then, under the British, that area grew bigger and bigger until the park took its place and a new game control area appeared on its boundaries. New designations like "open areas" or "wildlife management areas" seem like one more step in the encroachment process.[137] Another elder commented, "Before the park came the game control officers let us hunt in our own areas, they gave us jobs as guards and we were all happy, but now game reserves have become part of the village and we can't hunt anymore."[138] A third complained that the influence of the park is steadily growing: "Before, you would be arrested if you were caught within the park boundaries; now, they can come to your house and get you."[139] A common sentiment among those I interviewed was that "each year the park would get more land until now people are really constricted and upset about it."[140] One elder said, "This is our land but we can't even go into it anymore."[141] Western Serengeti peoples have made extraordinary sacrifices for the park without getting anything in return. They employ the core spatial images of constriction and restriction, rather than the older landscape images, to express their concerns because this is the only language that the people involved at all levels of the powerful structure that controls the land can understand.

SACRED LANDSCAPES AND COMMUNITY CONSERVATION
IN THE LIBERALIZATION ERA

In the most recent period of liberalization, new community conservation programs have provided an opening for western Serengeti peoples to debate the meaning of good conservation leadership within a moral economy by recontextualizing older core spatial images, even though the dialogue is dominated by global landscape images. These debates take place within a national context of structural adjustment policies imposed by the World Bank in the 1980s to make Tanzania responsible for its debts by increasing revenues through exporting larger quantities of raw materials, creating a business climate conducive to outside investment, privatizing industry, cutting back on social services and decentralizing government structures. In spite of the relative weakness and poverty of the African state, it has played a disproportionately large role in the economy, making the system vulnerable to corruption.[142] Whereas in the past Tanzania prided itself on the services and infrastructure provided by the state, private companies from outside the country are now taking over many of those functions. Companies owned by South Africans and other expatriates control many of the nation's gold mines, its brewing company, and its airlines. Structural adjustment was good for the tourist industry, as it became relatively cheap for foreigners to take a Serengeti safari, allowing a few Tanzanians to cash in on the influx of foreign capital. But structural adjustment has not been good for the majority of people, who found themselves poorer than they were ten years earlier. Funds for health care, education, and social services have been cut, and many have lost their jobs. Oddly enough, structural adjustment and the involvement of international organizations have also created a new openness, at least in rhetoric, to participatory local control while at the same time increasing government management and surveillance of rural life.[143]

Since the 1990s and into the new millennium the park recovered from cutbacks, resulting from much larger regional and global events, leaving it with wealth that could be used to support national development. The border with Kenya was closed from 1977 to 1986, reducing the number of foreign visitors to the park from seventy thousand in 1976 to ten thousand in 1977. During this period, the park's operating budget often exceeded its revenue, showing a net loss that restricted antipoaching efforts. Part of the structural adjustment package also included devaluation of the shilling, which accounted for an additional rebound in foreign visitors in the 1990s. In 1986 revenues began to exceed the operating budget. By 1990 the park's income had reached $972,581, while the operating budget was $445,656—near where it stood in 1982, before the cutbacks.[144] The increase in revenue has allowed the government room to consider new ways of approaching conservation and antipoaching as well as to share some of that revenue with other parks and communities surrounding them. The number of dollars visibly rolling into the Serengeti each day during tourist season, as well as the increasing poverty of rural Tanzanians, has made the people around the park question how that money should be spent.[145] The irony is that this wealth has not benefited the people displaced to allow for its production.

Two other factors determined the direction conservation activity took in this new era. First, although difficult to measure, park authorities concur that poaching has increased markedly over the past twenty years. This was the combined result of population increases along the western boundaries, as much as 15 percent per year, and of decreases in antipoaching patrolling during the 1980s as a result of budget constrictions. While the migrating wildebeest population has not decreased significantly, resident grazing animals outside the park declined by 95 percent between 1988 and 1992.[146] Rhinos, roan antelope, and wild dogs are still considered endangered, but elephants have made a startling recovery and are becoming pests in Ikoma fields in Robanda once again.[147] At the same time, the western Serengeti has experienced increasing impoverishment over the last century. A recent survey of Nata and Ikoma villages reported that people harvested an average of one to five sacks of crops per person annually and had a monthly income of two thousand Tanzanian shillings (about two dollars). Education, healthcare, transportation, and communication all are below the national standard. Malnutrition was also high in the area, with 19 percent of children under five years, 12 percent of men, and 15 percent of women malnourished. In 1998–99 ten people died in Makundusi as a result of malnutrition and four of complications in birth. The most common diseases treated at local clinics included malaria, diarrhea, gonorrhea, bilharzia (schistosomiasis), TB, and typhoid, many of which are related to poverty. An additional sign of severe stress on these communities was the decrease in livestock herds as people sold their cattle to buy grain or lost them to disease. Once the main livestock center, Serengeti District now has less than 2 percent of the region's livestock.[148]

Both in response to the real needs for development of people around the park and the failure of militarization to protect the park, the liberalization era is characterized by a shift in conservation philosophy from one of defending the park as a fortress to one of involving and benefiting the community, a change that has become known as integrated conservation and development, or simply community conservation. The international conservation community looked to models in Botswana, South Africa, Zambia, and Zimbabwe that gave communities a stake in protecting wildlife. It was clear to all that the park would not survive the next century unless people living on its boundaries got benefits from, rather than paying increasing costs for, wildlife preservation. If people found other ways of making a living, they would not need to hunt, and if their living depended on the prosperity of wildlife, for example in tourism, then they would protect the wildlife. The goals of these programs are to decentralize control over natural resources with participation at the local level and to change the local subculture of resistance and antipathy to park policy. Community conservation philosophy holds that economic development will also result in more sustainable conservation of the park's ecology. The complementary link between development and conservation has yet to be demonstrated, however, either here or in other places in Africa.[149] If the park wants to reduce resistance, those in charge must address the long-standing issues behind that resistance, which are grounded in a fundamentally different way of viewing the landscape.

In spite of their laudable goals and concerted work, community conservation projects throughout Africa have not achieved self-sufficiency; they remain closely tied to donor support and have not shown significant success in either development or conservation.[150] This is largely because these programs have provided "development" in the form of community projects like schools, clinics, and wells, which do not substitute for income that would allow individuals to stop hunting while economic hardships persist.[151] Since community conservation programs were the result of outside initiatives from donor agencies and pressure from structural adjustment policies toward decentralization, many wonder whether communities are really committed to the process. The ownership of land and the control of natural resources remain the most contentious issues.[152] Even inside evaluations of community conservation conclude that the "participatory approach" has not been effective and cannot substitute for enforcement.[153] Ultimately the state still owns, controls, and makes all important decisions over natural resources, and local people can only choose to go along with the program or not. Despite the rhetoric of decentralization, the programs are accountable to donor agencies and government agencies rather than to local people, who are unorganized and poor, leaving the programs open to corruption and manipulation. Western Serengeti communities have not been able to take advantage of the possibility of voicing their concerns in the context of community conservation because they have not been convinced that the programs can make a difference and because they must find the language to do so within a global conservationist discourse.[154]

However, the Tanzania National Parks (TNP) and SNP *have* made enormous strides in improving relations with the surrounding communities through various programs like the Ujirani Mwema (good neighborliness) program and the Community Conservation Service.[155] In 1985 the park initiated the Serengeti Regional Conservation Project (SRCP) to work with fourteen villages, seven in Bunda District and seven in Serengeti District, along the western park boundaries.[156] The main work of the SRCP is to organize village conservation committees for the oversight of natural resource development, selling meat and skins from legally harvested game, and using money generated from natural resources for village development. The SRCP staff worked at education through village workshops, coordinated selling wild meat, helped in the oversight of private companies, and empowered villagers through capacity building. The SRCP also initiated the village game scouts chosen by the village conservation committees, often from among former poachers, and trained by the SRCP to protect natural resources on village land, including looking for poachers, supervising the harvest of natural resources such as forest products, beekeeping and fishing, monitoring hunting licenses, taking care of problem animals, overseeing campsites for visitors and stopping fires.[157] The local elders' council sentences poachers caught by village game scouts to village development work rather than prison.[158] According

Figure 6.2. SRCP headquarters, Fort Ikoma. *Left to right:* author, unknown, Wambura Edward Kora, John Muya, Wilson Shanyangi Machota. *Photo by Paul Shetler, 2003*

to the SRCP plan, village conservation work should allow the local projects to become self-sufficient and even generate income for village development projects like schools, dispensaries, offices, bridges, loans for income generating projects, and for ongoing conservation work.[159] Yet the projects are not yet self-sufficient and the SRCP continues to subsidize them by providing training, humanpower, vehicles, equipment, and other supplies.

In the discussions generated by community conservation work, local people express their concern about the legitimacy of conservation leadership using the core spatial images of sacred landscapes that promote an egalitarian leadership of the generation-set that heals the land along with patrons who distribute resources and feed the people. The biggest complaint from the villages involved is that revenue from these programs does little more than pay for itself. This leads many to believe that village conservation committee members are enriching themselves, rather than "feeding" the community within the moral economy of reciprocity between patrons and clients.[160] The SRCP investigated a number of villages for the misuse of funds, and at least one chairman admitted to me that the villagers were upset about the infrequency and unreliability of wild-meat sales.[161] When the SRCP's first phase came to an end in 1997, their review for the second phase uncovered many problems, including misuse of village funds, ineffective village conservation committees, an increase in poaching, lack of follow-up at the village level, lack of education, and lack of commitment to the project.[162] A grant from the Norwegian government allowed the project to go into a second stage, with more emphasis on local participation.[163] Yet that participation continually fails to materialize when the rules of the moral economy are broken and benefits are not forthcoming.

The new 1998 Wildlife Policy opened up the possibility for more local control of natural resources by declaring a new type of land designation called the wildlife management area (WMA). The goal of the WMA was to get communities involved in conservation by allowing them to oversee wildlife and natural resource use on village land for their own benefit, if they followed the rules involved with setting up a land use plan.[164] In December 2002, WMAs were legalized, including new regulations for private-sector companies on WMA land, contingent on approval by the wildlife director. Revenues collected by the villages from private companies, hunting fees or fines, and the sale of meat would be divided for natural resource development, to benefit the village, and to put back into the project. While a general meeting of the village had to approve all major decisions, it was still not clear who will have ultimate control over these resources.[165] The first WMA, called IKONA, was proposed in Robanda village and open area, and was expanded in 1999 to include four villages: Robanda, Nyichoka, Makundusi/Nyakitono, and Mbiso, with a total population of 7,444, of which 33 percent lived in Mbiso, and 15 percent lived in Robanda.[166] The IKONA planning group has included members from the Frankfurt Zoological Society, Serengeti Regional Conservation Project, Ikorongo-Grumeti Game Reserves, the district game officer for Serengeti, village chairmen, and some village representatives.[167]

Planning for the WMA further expanded the debate in the western Serengeti about the legitimacy of leadership by speaking the language of global landscapes while also inserting the new core spatial images of restriction and constriction. Donor agencies such as the Frankfurt Zoological Society (FZS), who have played an important role in the planning and implementation of the WMA, are still viewed with suspicion and hostility by many local people who doubt their motives because of the FZS's association with antipoaching over the years.[168] Many also question the powerful influence that the FZS exercises in the SRCP as a result of their access to funding. Five years into the project, villagers were still concerned about the uneven distribution of power and the breakdown of reciprocity within the moral economy. At an FZS workshop, village participants questioned the power of the wildlife director in deciding policy for IKONA, worrying that this was a scheme to take away their land.[169] Other leaders more pointedly asked why there had been no follow-through on the already promised resources and conservation education from TNP, or on monitoring transparency in recruiting park employees. In general, the village leaders complained that the government had not been open to participation of local people in either the declaration of the Ikorongo-Grumeti Game Reserves or the new wildlife act.

Frustrated by their loss of control over the landscape and increasing impoverishment, village leaders often retreated from taking responsibility for these programs, demanding that the government or outside agencies play the role of patron to feed the people. In the WMA planning meetings, village leaders expressed their feelings about wildlife conservation, noting that they have always cared for and lived with the wild animals but that this is one of the poorest regions in Tanzania and help should be provided for the transition from poaching to sustainable use. What the village leaders feared most about the prospect of joining the wildlife management area was the abuse of power by government and other experts, cost overruns and lack of funds, loss of land, changes in laws, and conflict between villagers.[170] The FZS village surveys noted an overall "sense of powerlessness and resentment" among villagers, who said that in spite of their "strong traditional stewardship and responsibility for wildlife and environmental quality," they had been "discouraged from contributing their knowledge and skills" by the colonial regime and were accused of poaching and environmental destruction.[171] While western Serengeti leaders, working from an older landscape vision, demanded that those with power play by the rules of the moral economy, they were accused by the government and development agencies, working from a landscape vision of global conservation, of lacking responsibility and integrity.

The experience of western Serengeti peoples in both the SRCP and a WMA so far suggests that although the efforts of these groups were aimed at greater community participation, they actually served to increase state intervention and control while providing few benefits for the surrounding communities.[172] The assessment of the new wildlife policy by the Tanzanian Lawyers' Environmental Action Team (LEAT) concludes that it extends the jurisdiction of the park outside its boundaries in unprecedented ways that will not protect the tenure rights of rural communities.[173] Almost twenty years later people complain that the benefits that they get from the park still represent a token amount (less than 1 percent) of the revenue that the park generates, and still less than what people have had to give up.[174] Many believe that, although Tanzanians run the park, it seems to be controlled by outside interests. They say that the politicians get their share and the poor get nothing, not even jobs. The only tangible result has been improved relationships between the park and the people.[175] One elder concluded that his people want to conserve the animals so that their children and grandchildren can see them around their homes instead of in the park, which is now only for the rich people.[176] The perception is that government and other agencies do not trust western Serengeti peoples, and many elders told me that the government valued wild animals more than people because they bring dollars.[177] Others said that the common people, not just the leaders, must be informed about and brought into the decision-making process, which is now dominated by self-interest.[178] Many people that I talked to in 2003 did not know what a wildlife management area was or whether their village was a part of it. The WMA advisory team making the real decisions is made up of experts and professionals, with very little representation by village people. Although the WMA gives villages some control over their natural resources, they have no infrastructure to manage it properly — including vehicles, communication equipment, or training.[179]

LEAT concludes that the procedures for WMAs are "complex, cumbersome, time-consuming and expensive" for rural communities to carry out.[180]

Yet the discourse about leadership, although sometimes grounded in the core spatial images of sacred landscapes, also seeks to approach those in power through a new discourse built around the core spatial images of eviction and constriction that, so far, has not been conducive to good leadership. Local people blame the park and the government for lack of a democratic process and the lack of benefits. Yet they have not found effective ways to take ownership for these projects, to organize the communities in ways that would give them a voice in the process, or even to challenge the existing park paradigm altogether.[181] During the colonial years, people told stories using the core spatial images of hiding and subterfuge that encouraged a subculture of resistance to government initiatives. In the independence era community conservation projects have not been successful and therefore do not warrant the investment of time and energy taken away from other work. In addition, the region's experience with colonial chiefs lacking local legitimacy did not provide good models of state leadership. People who are disgusted with leadership, now and in the past, simply avoid it rather than confront it or work to change it. A number of studies of the Mara region during the ujamaa era undertaken by Tanzanian scholars lamented the lack of positive leadership. One study concluded that ineffective leadership, specifically the use of coercive methods that provoked resistance and pilfering common funds, is what kills self-help projects. Another noted that as a result of leadership, women have not been politically mobilized, yet they do all the work while the men sit around and talk. Both studies noted a subculture of resistance to and distrust of any government initiatives.[182] Given the history of displacement and the terms of discourse dominated by the structures of power, the spaces that have opened up in community conservation for a western Serengeti voice in decisions about the environment have not been, nor perhaps could have been, utilized. A new kind of relationship with the state needs to be forged that empowers, rather than discourages, accountable local leadership.

Another important piece to community conservation that also generates discussion about leadership is the private sector tourist companies, who are supposed to provide many of the benefits that villagers will realize from wildlife. Robanda has the most experience in hosting tour companies, including Sengo Safaris at Ikoma Bush Camp. The VIP tour company, located at Sasakwa Hill, near Mbiso and Nyakitono, is a good example of both the problems and promises of private-sector involvement in conservation. In 1995 a private deal was made between the foreign owner of VIP, whose Tanzanian partners included some prominent national political figures and a few leaders from Nyakitono, to alienate some two thousand hectares of village land. The villagers protested when they found out about this secret deal on a ninety-nine-year lease, saying that the owner of VIP did very little for the community and had bad relations

Figure 6.3. Well funded by VIP, Mbiso. *Photo by Paul Shetler, 2003*

with the people around him because he had not involved the community in his plans, was not friendly, and did not talk to people or care about their concerns.[183] They expected him to act like a patron in the moral economy. The tension between VIP and villagers became so intense that the original VIP owner decided to leave.[184] In recent years, Paul Tudor Jones, a wealthy American conservationist, bought VIP and has done a lot to help the community. He built a secondary school in Nata, improved dispensaries and schools, dug wells, provided transport and ambulance services, paid for scholarships, accepted proposals for small-business loans, sponsored the five-village soccer league, trained village game scouts, and gave people, especially poachers, jobs at the camp. Local businesses are also thriving because more people have cash-paying jobs.[185] VIP is building a large hotel and a tented camp for ecotourism, but in 2003 the emphasis was on nonprofit activities through the Grumeti Trust.[186] An additional 113,000 hectares in hunting concessions were recently added to the land controlled by VIP under the name of the Grumeti Reserve.[187]

While in 2003 local people were glad that the first VIP owners left and appreciated the work of the new owners, some suspicions remained about them and other local NGOs. Western Serengeti peoples still used older landscape visions to think about VIP as a new patron but also worried about constriction and restriction if development was left to private companies who were ultimately concerned about profit and did not understand the reciprocal terms of patronage. Rumors circulated that the park no longer brought wild meat for sale in the villages because VIP had bought all the hunting quotas.[188] Villagers were suspicious of what kind of deals village leaders were making when VIP provided the leaders with radios to call for transport whenever they needed it and invited them for parties and meals at the camp, where future plans were discussed. Others were concerned that VIP was only making people happy so that they could quietly grab the land and restrict them from even walking across it. One elder said, "I still have doubts about the tour companies, they will not leave easily once they come; they must be given stiffer rules."[189] Many people were worried about losing their land to foreign companies, so that the only future for their children in this area will be to work as hired hands for the tour companies.[190]

Another potentially hopeful opportunity for community involvement is local NGOs, which have begun to proliferate as a result of new government registration procedures. There are now organizations in many of the small towns in the region, for helping AIDS orphans, community development, livestock breeding, and environmental awareness. Predictably, some are legitimate while others find a skeptical reception in the community, which sees them as opportunists looking for outside donors. In both cases there has been no unified response, nor little sense of community ownership in these projects. Western Serengeti peoples have been astute in observing that these new private initiatives significantly change the terms under which they will have access to the land. Yet local NGOs also hold out the critical yet unrealized potential for locally controlled community organization and effective leadership for environmental action.

The discussion about leadership and the direction of community conservation is thus an eclectic combination of recontextualized core spatial images of past landscapes, the core spatial images of constriction and restriction, and the global landscape of conservation. For example, some elders conclude that the diminished number and quality of animals in the area is a sign that the land is not healthy and needs to be healed. They speak nostalgically about childhoods spent herding livestock among the wild herds and lament that some species are only rarely seen anymore; seasonally or in places where migration patterns have shifted, the large herds are no longer visible.[191] The extensive plains stretching between Ring'wani and Kemegesi or the Nyambureti hills in Ngoreme used to be covered with migrating herds in the dry season. But ever since the mid- to late 1960s almost no animals appear there, with the last rhino spotted in 1964.[192] As population is increasing in this area, putting pressure on the resources that are left, some local people follow conservationists in concluding that the only realistic solution is increased antipoaching enforcement against the big operators.[193]

The solutions to conservation conflicts that some elders suggest based on older visions of the landscape often sound unrealistic and romantic in the face of the hard realities of conservation politics today. Some have suggested that the elders, nyangi eldership titleholders, need to be given more authority in the community again. This approach worked with the cattle theft in the past; perhaps it will work with hunting if young people could be interested in taking the expensive titles again.[194] The Rogoro Museum and Cultural Centre in Nata is one of the new local NGOs trying to earn income from tourist camping while support-

Figure 6.4. Rogoro Museum and Cultural Centre, Mbiso. *Photo by Paul Shetler, 2003*

ing a museum to preserve local history and culture, even calling together the remaining saiga age-set leaders to discuss the problem.[195] In the face of increasing impoverishment and loss of control over resources, western Serengeti peoples continue to search for creative ways to respond to these challenges by using the landscapes of the past. Yet it is clear to everyone that there is no going back and that the park is here to stay.

IMAGINING SERENGETI HISTORY

Although some echoes of past landscapes are evident in these eclectic discussions about leadership in community conservation, what is striking about this last period in western Serengeti history is the relative lack of a tangible response to the issues surrounding the park on the basis of past landscapes. During the Time of Disasters, when western Serengeti peoples were hard pressed by famine, epidemics, and increased raiding, they used the core spatial images from the ecological landscapes of the emergence stories to rethink the relationships between farmers, hunters, and herders. It was during this period that the emergence stories were recontextualized as ethnic origins in Sonjo or Sukuma to take into account the shifting spaces of power. The social landscapes of clan networks were also reconfigured as a means of finding food and security. Finally, the sacred landscapes of protection through encirclement were recontextualized as age-set territories and fortified settlements, forming new kinds of ethnic identities out of the older core spatial images. With the challenge of colonialism, too, the older landscapes were used to find ways to recover from the disasters by exchanging hunting products for livestock. This in turn allowed western Serengeti peoples to maintain some control over their landscape by successfully resisting labor migration and cash crop production. In each case, western Serengeti peoples met these challenges by recontextualizing the older core spatial images and maintaining contact with the landscapes of the past.

In the last challenge faced by western Serengeti peoples with the establishment of Serengeti National Park, people have suffered from their inclusion as marginal players in a global economy. Yet a response that would successfully recontextualize past landscape images in order to confront these problems has not been as evident. There are a number of possible reasons why that is so. First, perhaps these discussions and processes *are* underway locally and either I did not have access to those conversations and sources, or it is simply too early to see the results. Western Serengeti peoples speak to those in power through the core spatial images of constriction and restriction. Perhaps in the context of stories about the park they spoke to me in similar ways. Another possibility is that the current conservationist view of the landscape is so absolutely hegemonic that people cannot even think outside those images anymore. Or if they can, there are no forums in which to do so that would understand this form of discourse and way of seeing the landscape. Another possibility is that the oral traditions themselves are no longer living traditions. Many young people are leaving the region for opportunities elsewhere, while few have taken the time to learn oral traditions from their grandparents or have found those traditions irrelevant for solving today's difficult problems.

While each of these explanations contains some element of truth, a more convincing theory is that because the park and game reserves have physically cut people off from the landscapes of past, they can no longer imagine new ways of using those memories. Because western Serengeti peoples have written their history on the landscape, to deny them access to these landscapes is like burning their history books. The living interaction with oral traditions by recontextualizing the core spatial images of oral traditions in order to meet new challenges is impossible outside living contact with the particular landscapes in which they are rooted. This is why on the walk I took with Pastor Wilson Shanyangi Machota and Wambura Edward Kora over the old Ikoma settlement sites in the Ikorongo-Grumeti Game Reserves in 2003, they were so concerned that their children and grandchildren learn about these places and see them again for themselves. The elders wanted to make sure that the maps and place names in my book were accurate so that, even if they died, their descendants could find their way back again someday. This is the tragedy of saving seemingly undifferentiated wild space for the animals alone. For people whose history is rooted in the landscape, the particular places must be revisited and known in the routines of everyday life in order to pass on the stories.[196]

Yet daily life and memory goes on in western Serengeti. Social landscapes remain current as people keep on forming networks with their relatives who work in Dar es Salaam and provide their children access to education and jobs, or with friends who will let them bring vegetables to sell at the park lodges. Wealthy men continue to fulfill the role of patrons in helping youth to find opportunities outside the community. Elders still sit around pots of homemade beer and tell stories from the past to explain the dynamics of current events. Although western Serengeti peoples cannot easily get back to the land within the park and game control areas there is land outside them which they continue to interact with and enjoy. Theories about collective memory state that oral societies that bank their knowledge in the "memory of things"—in this case landscapes—will change with literacy to a "memory of words," or semantic knowledge.[197] That might be the case as western Serengeti peoples are increasingly alienated from the specific landscapes where memory resides and as the younger generation continues to rely on book knowledge. Yet because of that shift their story is critical to hear and document at this juncture, to understand their tragic experience and search for a response to global conservation, as well as to explore the possibility of reimaging the park through these radically different landscape visions.

Although this book has covered a long sweep of western Serengeti environmental history, it is fundamentally a history of landscape memory and imagination over the last two millennia. Like a sixteenth-century pottery shard that resurfaces as a dish to water chickens or a plaything for children, core spatial images remain the most durable fragments from the past in oral memory. Even though elders use origin stories to legitimize their own authority in ethnic groups that might not have existed four hundred years ago, these narratives still preserve a memory of how linguistically and culturally diverse farmers and hunters meeting on the frontier worked out their relationship to the land and to each other many millennia ago. These origin stories are often combined with exclusive clan narratives, which can nevertheless still be recontextualized in an earlier time when constructing regional networks for aid in times of hunger took precedence. In the context of park restrictions, reasserting demands to practice rituals at the sacred sites of the emisambwa is one of the few ways that western Serengeti peoples can claim a legitimate right to return to the landscapes of memory. Yet they also keep alive the names and histories of specific rainmakers, prophets, and ancestors of the not so distant past. Because of the way that our brains spatialize memory, references to place and spatial patterns in the stories we tell about the past are the key to unraveling a history that is otherwise obscured by the layers of archaeological sediment built up over hundreds and thousands of years. These layers can be identified by the established method in African history of comparing the data with other evidence from other kinds of sources that can be reliably dated. Rather than as an inconsequential backdrop for the plotline, historians should now see ways of imaging the landscape, the core spatial images of oral tradition, as a central concern of historical inquiry.

A spatial analysis of oral tradition for reconstructing landscape memory provides a test case for application in other places and times. It is now obvious that this methodology cannot be applied mechanically and that many contradictions in the evidence are inevitable. Adequately sifting through the layers of recontextualization to reconstruct the history of memory or seeing how new core spatial images impact older narratives is not always possible. Yet the western Serengeti case does show that in spite of the difficulty and paucity of sources, much can be learned about the past and about alternative ways of seeing the landscape that have had utility in particular historical contexts. Future research might test and augment this work by more extensive work in mapping that would follow more precisely the spatial patterns of oral tradition on the ground. In addition to paying attention to western Serengeti landscapes, this method also depends on the process of denaturalizing and historicizing the ways that those with power make decisions about the environment based on their ways of seeing the landscape.[198] Perhaps these concerns can reinvigorate the study of African history, where less and less is being written about the precolonial period as oral memory of that time seems to be gradually slipping away. This method, however, demonstrates that those memories are not gone but simply recontextualized and will survive if a connection to past landscapes is maintained or preserved in textual form. A spatial analysis of oral traditions introduces a whole new set of sources, including the distinctive Serengeti rocky outcroppings, acacia umbrella trees, and seasonal streams, making the Serengeti, even in its wilderness form, a book of memory. Imagining the Serengeti, or any other African landscape, is much more than aesthetic contemplation, it is a journey into human history marked on the landscape.

It is this sense of a humanized, historical landscape that might inspire rethinking of conservationist landscapes that separate nature and culture. Separation allows for the destruction of nature in productive spaces and the destruction of culture in leisure spaces like parks, where nature is absolutely protected and isolated. Although parks might now be necessary to preserve certain ecologies, it is also true that human spaces, even in industrialized settings, can respect and integrate "nature." In his article about the concept of wilderness, William Cronon concludes, "Wilderness gets us into trouble only if we imagine that this experience of wonder and otherness is limited to the remote corners of the planet, or that it somehow depends on pristine landscapes we ourselves do not inhabit."[199] He argues that instead we need to reimagine natural places as a home, where we honor its otherness and take responsibility for our actions in the ways that we use it.[200] None of us, and certainly not western Serengeti peoples, would like to see the end of the great wildebeest migration or the conversion of these incredibly beautiful landscapes into condominiums and golf courses or, worse yet, overgrazed pastures and denuded farmsteads. Yet more creative solutions than separation might be found. In the Serengeti this might mean finding ways to let people interact with the landscapes to which they have historical claim of calling home in ways that respect and preserve wild spaces.

Although this is a book about historical memory rather than a prescription for the future of conservation, one place to start in rethinking ways of being at home with nature, rather than being apart from it, might be to use the core spatial images from the past as the generative principles for reshaping new narratives about the Serengeti. Cronon claims that "narratives remain our chief moral compass in the world" and that "the stories we tell change the way we act in the world."[201] In the stories told about the Serengeti, it matters whether the landscape is imagined as a separate space of nature or a land inhabited by ancestors; its peoples as passive victims of state policy or as active agents in shaping the environment.[202] The generative principles inherent in the core spatial images of oral narratives have included human landscapes, interdependent ecologies, the strategies of diversification, inclusion, and distribution, sacred places of power and protection by encirclement. Imagine how those principles might be used as the basis for conservation programs that take human history in all its complexity and particularity seriously.[203] Likewise one would have to deal with the legacy of narratives from the last century, which included core spatial images of loss and dispersal, hiding and subterfuge, constriction and restriction, all of which present potentially negative attitudes of victimhood, passivity, and noncooperation. Yet western Serengeti peoples also use these images to create narratives of strength, self-reliance, and resilience in the face of severe challenges. Western Serengeti peoples have adapted their own stories to the reality of an interconnected globalized world on which the future of the Serengeti depends. A positive future also depends on creative new solutions to the problems of conservation emerging from the ground up, when western Serengeti peoples along with those involved in the park, take up the challenge and the responsibility of forging new ways from the old of imaging Serengeti.

Glossary

All terms are from Nata unless indicated otherwise.

abachama	see *omochama*
abakoro	see *omokoro*
abasimano	see *omosimano*
aghaso	ceremony to celebrate and purify a man who has killed a lion, a leopard, or a Maasai
ahaase	soil, earth; from the proto-Bantu root, *ase*
ahumbo	distant fields where people have temporary sleeping quarters during the farming season to guard the crops from wild animals
anchara	famine, hunger
anyumba	house; also a woman's household, family, the children of one mother
aring'a	blood brotherhood
ase	see *ahaase*
asega	see *saiga*
askari	colonial soldier(s) [Swahili]
asimoka	origin, emergence, the springing up; from the verb *sisimoka*; in Kuria, *semoka*
bao	board game played throughout the region [Swahili]
chama	council of elders; in Kuria, *injama*; in Maa, *kiama*
eghise	women's dance for mothers and to assist childbirth; black tail taken at initiation as an elder, at daughter's circumcision
eka	homestead; cattle corral with houses around the perimeter of its fence

ekehwe/ebehwe	spirits of the dead
ekimweso	purification ritual and, in Ishenyi, a fire ceremony for blessing; in Ikoma, *shishiga*; in Ngoreme, *ikoroso*
ekitana/ebitana	medicine bundle
ekyaro/ebyaro	territory, land; in Kuria, *ikiaro*
erisambwa/emisambwa	protective spirits of the land; ancestral spirits of rainmakers or prophets who protect the land
emigiro	totem, taboo, or avoidance, usually of a clan or descent group; prohibition that relates to specific animals or foods; emigiro often represent animals or food
hamate	clan or clan land; refers to a place
ikoroso	ritual fire for purification and blessing, for a single clan or homestead when confronted with the problems of death, sickness, or infertility, or the whole ethnic group; in Ngoreme, the ikoroso fire was kindled from particular trees (*esebe, omoreto, omorama*) that grew at an *erisambwa* site, also used to bless the new generation, or age-set
injama	see *chama*
kang'ati	leader of the age-set
kiama	see *chama*
kugaba	to distribute, give out
kuhemea	to search for food; to trade hunting products for grain
kukerera	ritual walk of the generation-set to heal the land
mbuga	black-cotton soil of the plains, marked by whistling thorn trees [Swahili]
mtemi	Ikizu chief
ntemi	scar given to Ikoma and Sonjo children for health; also a sign of belonging to the group and mark of brotherhood between Ikoma and Sonjo; in Sukuma, the name of the chiefship
nyangi	eldership titles and life-stage transitions with initiation rituals
nyika	wilderness; people who live in places for hunting or herding

obugabho/abagabho	prophet (from *kugaba*); rainmaker
oburwe	finger millet, food of first man and first woman in origin stories; the staple crop before colonialism
omochama/abachama	generation-set or age-set members
omokoro/abakoro	ancestral spirits, lit., big or old ones
omosimano/abasimano	strangers; people from other clans or ethnic groups who marry or are adopted in
omosimbe/abasimbe	independent woman (lit., female husband) who has inherited property and a homestead from her father
omotware/abatware	male wife; man married to independent woman with her own homestead and livestock inherited from her father
omunibi	cattle patron; wealthy man whose authority was based on his ability to distribute his wealth in livestock
omwame/abwame	patron; wealthy man whose authority was based on his ability to distribute his wealth and feed a large crowd at a feast; also known as *omunibi*
omwibororu/abibororu	person who is born in a place (native born), no matter what the bloodline
orokoba	medicines of protection that encircle the land
oruberi	settlement of a number of homesteads grouped together [Nata, Ikoma]
rikora/amakora	generation-set
risaga	work group that is given a beer party after work is completed
saiga	age-set [Nata, Ikoma]; in Ngoreme, *asega*
shishiga	see *ekimweso*
ugali	thick porridge; staple food throughout the region [Swahili]
ujamaa	family or community; philosophy of African Socialism, pursued by Pres. Julius K. Nyerere and the villagization scheme that followed his time in office [Swahili]
utemi	Ikizu chiefship, chiefly line

Notes

CO   Colonial Office
DSM   Dar es Salaam
EAF   East Africana Library, University of Dar es Salaam
FPS   Fauna Protection Society
FZS   Frankfurt Zoological Society
IKONA   Ikoma and Nata Wildlife Management Area, which includes
            Robanda, Nyichoka, Makundusi/Nyakitono, and Mbiso
LEAT   Lawyers Environmental Action Team
MDB   Musoma District Books, microfilm
PRO   Public Record Office, London
RHL   Rhodes House Library, Oxford
SNP   Serengeti National Park
SPFE   Society for the Preservation of the Fauna of the Empire
SRCP   Serengeti Regional Conservation Program
TNA   Tanzania National Archives, Dar es Salaam
TNP   Tanzania National Parks
TZ   Tanzania
WMA   Wildlife Management Area

Unless otherwise stated, all interviews were conducted by the author.

INTRODUCTION: LANDSCAPES OF MEMORY

1. For an analysis of this way of seeing in the founding of the park, see Roderick P. Neumann, "Ways of Seeing Africa: Colonial Recasting of African Society and Landscape in Serengeti National Park," *Ecumene* 2, no. 2 (1995): 149–69.

2. Bernhard Grzimek, *Serengeti darf nicht sterben* [Serengeti Shall Not Die] (1959; Berlin: Universal Family Entertainment, 2004), DVD9, 4:3 full frame.

3. Bernhard Grzimek and Michael Grzimek, *Kein Platz für wilde Tiere* [No Room for Wild Animals] (1956; Berlin Universal Family Entertainment, 2004), DVD9, 4:3 full frame.

4. Grzimek, *Serengeti darf nicht sterben*; Grzimek and Grzimek, *Kein Platz*.

5. For analysis of these images see Roderick P. Neumann, *Imposing Wilderness: Struggles over Livelihood and Nature Preservation in Africa* (Berkeley: University of California Press, 1998); Jonathan S. Adams and Thomas O. McShane, *The Myth of Wild Africa: Conservation without Illusion* (New York: Norton, 1992).

6. Interview with Wilson Shanyangi Machota, Wambura Edward Kora, Robanda, 29 August 2003. Thanks to Barbara Schachenmann, VIP/Grumeti Trust community officer, for taking us on this trip with her vehicle.

7. William Beinart and JoAnn McGregor, introduction to *Social History and African Environments*, ed. Beinart and McGregor (Athens: Ohio University Press: 2003), 4.

8. David William Cohen and E. S. Atieno Odhiambo, *Siaya: The Historical Anthropology of an African Landscape* (Athens: Ohio University Press, 1989), 9; Simon Schama, *Landscape and Memory* (New York: Knopf, 1995), 7.

9. Thomas R. Dunlap, *Nature and the English Diaspora: Environment and History in the United States, Canada, Australia, and New Zealand* (Cambridge: Cambridge University Press, 1999), 5.

10. Benedict Anderson, *Imagined Communities: Reflections on the Origin and Spread of Nationalism* (London: Verso, 1983).

11. W. J. T. Mitchell, introduction to *Landscape and Power*, ed. Mitchell (Chicago: University of Chicago Press, 1994), 2; Mitchell, "Imperial Landscape," ibid., 5, 16.

12. On theories of African heterarchy, see Susan Keech McIntosh, "Pathways to Complexity: An African Perspective," in *Beyond Chiefdoms: Pathways to Complexity in Africa*, ed. McIntosh (Cambridge: Cambridge University Press, 1999), 1–30.

13. I use Dadog to refer to the language and Tatoga to refer to the people, including both Rotigenga, Isimajek, and the larger Tatoga community in other places in Tanzania.

14. For recent analysis of literature on regions, see Allen M. Howard, "Nodes, Networks, Landscapes, and Regions: Reading the Social History of Tropical Africa, 1700s–1920," in *The Spatial Factor in African History: The Relationship of the Social, Material, and Perceptual*, ed. Howard and Richard M. Shain (Leiden: Brill, 2005), 21–140. Also see my contribution on region in that collection, Jan Bender Shetler, "'Region' as Historical Production: Narrative Maps from the Western Serengeti, Tanzania," 141–76. On some recent theoretical work still using the concept of hierarchical relations of exchange to define regions, see, for example, Claudio Lomnitz-Adler, "Concepts for the Study of Regional Culture," *American Ethnologist* 18, no. 2 (May 1991): 195–214; Eric Van Young, "Are Regions Good to Think?" introduction to *Mexico's Regions: Comparative History and Development* (San Diego: Center for U.S.-Mexican Studies, UCSD, 1992), 1–36. For a review of the literature, see Carol A. Smith, "Regional Economic Systems: Linking Geographical Models and Socioeconomic Problems," in *Economic Systems*, vol. 1 of *Regional Analysis* (New York: Academic Press, 1976), 3–63; Mary Beth Pudup, "Arguments within Regional Geography," *Progress in Human Geography* 12, no. 3 (September 1989): 369–91. As applied

to a historical study in Africa, see Charles H. Ambler, *Kenyan Communities in the Age of Imperialism: The Central Region in the Late Nineteenth Century* (New Haven, CT: Yale University Press, 1988); Allen M. Howard, "The Relevance of Spatial Analysis for African Economic History: The Sierra Leone–Guinea System," *Journal of African History* 17, no. 3 (1976): 365–88. My use of the concept of region is closer to that of Richard Waller for interior East Africa in "Ecology, Migration, and Expansion in East Africa," *African Affairs* 84 (July 1985): 356–57.

15. I had many interesting discussions with local people about what to call this group. Many voted for Rogoro, the people of the east, but just as many declared it was not an all-encompassing term. Some wanted the name of a mountain, to designate their unity, but could not agree on whether that should be Bangwesi (Mangwesi) or Chamuriho.

16. Some of the few include Gerald Hartwig, *The Art of Survival in East Africa: The Kerebe and Long-distance Trade, 1800–1895* (New York: Africana Publishing, 1976), about people on Lake Victoria to the east; A. O. Anacleti, "Pastoralism and Development: Economic Changes in Pastoral Industry in Serengeti, 1750–1961" (MA thesis, University of Dar es Salaam, 1975); Anacleti, "Serengeti: Its People and Their Environment," *Tanzania Notes and Records*, nos. 81/82 (1977): 23–34.

17. Published sources include Oskar Baumann, *Durch Massailand zur Nilquelle: Reisen und Forschungen der Massai-Expedition des deutschen Antisklaverei-Komite in den Jahren 1891–1893* (Berlin: Geographische Verlagshandlung Dietrich Reimer, 1894); Paul Kollmann, *The Victoria Nyanza: The Land, the Races and Their Customs, with Specimens of Some of the Dialects* (London: Swan Sonnenschein and Co., 1899); Max Weiss, *Die Völkerstämme im norden Deutsch-Ostafrikas* (Berlin: Carl Marschner, 1910). Early ethnographic accounts in the surrounding area include Edward Conway Baker, *Tanganyika Papers* (Oxford: Oxford University Press, 1935), microfilm; Otto Bischofberger, *The Generation Classes of the Zanaki (Tanzania)* (Fribourg, Switzerland: University Press, 1972); Hugo Huber, *Marriage and Family in Rural Bukwaya (Tanzania)* (Fribourg: University Press, 1973); Eva Tobisson, *Family Dynamics among the Kuria: Agro-Pastoralists in Northern Tanzania* (Göteborg: Acta Universitatis Gothoburgensis, 1986); M. J. Ruel, "Religion and Society among the Kuria of East Africa," *Africa* 35, no. 3 (1965): 295–306.

18. Those consulted include the White Father's Archives near Mwanza, the Sukuma Museum in Bujora, the Seventh-day Adventist and Mennonite Church archives (both in the Mara region and in the United States), the Public Record Office in London, and Rhodes House Library in Oxford, England. The Musoma District Books, the E. C. Baker Papers, and the Church Missionary Society papers are available on microfilm.

19. See, for example, J. E. G. Sutton, *The Archaeology of the Western Highlands of Kenya* (Nairobi: British Institute in Eastern Africa, 1973); for the earlier periods, see Stanley H. Ambrose, "The Introduction of Pastoral Adaptations to the Highlands of East Africa," in *From Hunters to Farmers: The Causes and Consequences of Food Production in Africa*, ed. J. Desmond Clark and Steven A. Brandt (Berkeley:

University of California Press, 1984), 212–39; Diane Gifford-Gonzalez, "Animal Disease Challenges to the Emergence of Pastoralism in Sub-Saharan Africa," *African Archaeological Review* 17, no. 3 (2000): 95–139; Gifford-Gonzalez, "Early Pastoralists in East Africa: Ecological and Social Dimensions," *Journal of Anthropological Archaeology* 17 (1998): 166–200.

20. Core vocabulary lists from David Lee Schoenbrun and Christopher Ehret and cultural vocabulary lists from the University of Dar es Salaam. Schoenbrun, *A Green Place, A Good Place: Agrarian Change, Gender, and Social Identity in the Great Lakes Region to the 15th Century* (Portsmouth, NH: Heinemann, 1998); Ehret, *Southern Nilotic History: Linguistic Approaches to the Study of the Past* (Evanston: Northwestern University Press, 1971); and Ehret, *An African Classical Age: Eastern and Southern Africa in World History, 1000 B.C. to A.D. 400* (Oxford: James Currey, 1998).

21. See, for example, the series from the Serengeti Research Institute, A. R. E. Sinclair and M. Norton-Griffiths, eds., *Serengeti: Dynamics of an Ecosystem* (Chicago: University of Chicago Press, 1979); Sinclair and Peter Arcese, eds., *Serengeti II: Dynamics, Management, and Conservation of an Ecosystem* (Chicago: University of Chicago Press, 1995).

22. Isabel Hofmeyr, *We Spend Our Years as a Tale That Is Told: Oral Historical Narrative in a South African Chiefdom* (Portsmouth, NH: Heinemann, 1993), 4; David W. Cohen, *Womunafu's Bunafu: A Study of Authority in a Nineteenth-Century African Community* (Princeton, NJ: Princeton University Press, 1977), 8–9; Isidore Okpewho, *African Oral Literature: Backgrounds, Character, and Continuity* (Bloomington: Indiana University Press, 1992). For more on the narrative forms of this region, see Jan Bender Shetler, "The Politics of Publishing Oral Sources from the Mara Region, Tanzania," *History in Africa* 29 (2002): 413–26.

23. See an account of oral styles in Jane I. Guyer and Samuel M. Eno Belinga, "Wealth in People as Wealth in Knowledge: Accumulation and Composition in Equatorial Africa," *Journal of African History* 36, no. 1 (1995): 110–11.

24. By contrast, among the Tatoga I was often taken to interviews with men in a younger generation.

25. In South Africa, Hofmeyr describes a similar gendered division of oral literature in which people said that men told "true" histories while women told "fictional narratives." She goes on to claim that the content and style of these stories are similar, the only difference being the spaces in which they are told. Hofmeyr, *We Spend Our Years*, 6.

26. On the analysis of spirit possession cults controlled by women as a counterhegemonic discourse to the dominant male, Islamic discourse, see Janice Boddy, *Wombs and Alien Spirits: Women, Men, and the Zar Cult in Northern Sudan* (Madison: University of Wisconsin Press, 1989), 5–7. For more on women's narratives, see Jan Bender Shetler, "The Gendered Spaces of Historical Knowledge: Women's Knowledge and Extraordinary Women in the Serengeti District, Tanzania," *International Journal of African Historical Studies* 36, no. 2 (2003): 283–307.

27. Jan Bender Shetler, "The Landscapes of Memory: A History of Social Identity in the Western Serengeti, Tanzania" (PhD diss., University of Florida, 1998); Shetler, *Telling Our Own Stories: Local Histories from South Mara, Tanzania* (Leiden: Brill, 2003).

28. Transcriptions and tapes deposited in the Mennonite Historical Library, Goshen College, Goshen, Indiana, and in the author's collection. Eventually tapes and transcriptions will be placed in a Tanzanian archive.

29. For an overview of African environmental history, see William Beinart, "African History and Environmental History," *African Affairs* 99 (April 2000): 269–302.

30. For the argument about the creation of wilderness, see William Cronon, "The Trouble with Wilderness, or, Getting Back to the Wrong Nature," *Uncommon Ground: Toward Reinventing Nature*, ed. William Cronon (New York: Norton, 1995), 1–55.

31. Schama, *Landscape and Memory*, 9, 13; Alfred W. Crosby, "The Past and Present of Environmental History," *American Historical Review* 100, no. 4 (October 1995): 1177–89; Dunlap, *Nature and the English Diaspora*, 6–8; Ute Luig and Achim von Oppen, "Landscape in Africa: Process and Vision, an Introductory Essay," *Paideuma* 43 (1997): 27; Neumann, *Imposing Wilderness*, 26–28.

32. Emmanuel Kwaku Akyeampong, *Between the Sea and the Lagoon: An Ecosocial History of the Anlo of Southeastern Ghana, c. 1850 to Recent Times* (Athens: Ohio University Press, 2001), 2.

33. For an argument about how frontier processes formed pan-African continuities, see Igor Kopytoff, "The Internal African Frontier: The Making of African Political Culture," in *African Frontier: The Reproduction of Traditional African Societies*, ed. Kopytoff (Bloomington: Indiana University Press, 1987), 3–84.

34. Beinart and McGregor, introduction, 4; see also Thomas Spear and Richard Waller, *Being Maasai: Ethnicity and Identity in East Africa* (Athens: Ohio University Press, 1993). For a West African example, see James F. Searing, *West African Slavery and Atlantic Commerce: The Senegal River Valley, 1700–1860* (Cambridge: Cambridge University Press, 1993).

35. On networks and reciprocity for social security, see Michael Watts, *Silent Violence: Food, Famine, and Peasantry in Northern Nigeria* (Berkeley: University of California Press, 1983); Steven Feierman, *Peasant Intellectuals: Anthropology and History in Tanzania* (Madison: University of Wisconsin Press, 1990); Gregory H. Maddox, "Leave Wagogo, You Have No Food: Famine and Survival in Ugogo, Tanzania, 1916–1961" (PhD diss., Northwestern University, 1988). On reciprocity, see Holly Hanson, *Landed Obligation: The Practice of Power in Buganda* (Portsmouth, NH: Heinemann, 2003); Jane I. Guyer, "Wealth in People, Wealth in Things — Introduction," *Journal of African History* 36, no. 1 (1995): 83–90.

36. James C. Scott, *Weapons of the Weak: Everyday Forms of Peasant Resistance* (New Haven, CT: Yale University Press, 1985); James C. Scott, *The Moral Economy of the Peasant: Rebellion and Subsistence in Southeast Asia* (New Haven, CT: Yale

University Press, 1977); Jonathon Glassman, *Feasts and Riots: Revelry, Rebellion, and Popular Consciousness on the Swahili Coast, 1856–1888* (Portsmouth, NH: Heinemann, 1995).

37. For accounts going back only to the nineteenth century in West Africa, see Ivor Wilks, *Asante in the Nineteenth Century* (Cambridge: Cambridge University Press, 1975); J. D. Y. Peel, *Ijeshas and Nigerians: The Incorporation of a Yoruba Kingdom, 1890s–1970s* (Cambridge: Cambridge University Press, 1983); R. S. Smith, *Kingdoms of the Yoruba*, 3rd ed. (Madison: University of Wisconsin Press,1988). For accounts in central southern Africa, some going back much earlier with written sources, see J. Matthew Schoffeleers, *River of Blood: The Genesis of a Martyr Cult in Southern Malawi, c. A.D. 1600* (Madison: University of Wisconsin Press, 1992); Jan Vansina, *The Children of Woot: A History of the Kuba Peoples* (Madison: University of Wisconsin Press, 1978); J. B. Peires, *The House of Phalo: A History of the Xhosa People in the Days of Their Independence* (Berkeley: University of California Press, 1981). For East Africa, see Steven Feierman, *The Shambaa Kingdom: A History* (Madison: University of Wisconsin Press, 1974); Christopher Wrigley, *Kingship and State: The Buganda Dynasty* (Cambridge: Cambridge University Press, 1996); David Newbury, *Kings and Clans: Ijiwi Island and the Lake Kivu Rift, 1780–1840* (Madison: University of Wisconsin Press, 1991). For East African historians returning to their home areas and reconstructing histories from oral tradition, see B. A. Ogot, *Migration and Settlement, 1500–1900*, vol. 1 of *History of the Southern Luo* (Nairobi: East African Publishing House, 1967); William R. Ochieng', *A Pre-colonial History of the Gusii of Western Kenya* (Nairobi: East African Literature Bureau, 1974); Isaria N. Kimambo, *A Political History of the Pare of Tanzania, c. 1500–1900* (Nairobi: East African Publishing House, 1969).

38. On blurring the boundary between the colonial and precolonial, see, for example, Jean Allman and Victoria Tashjian, *"I Will Not Eat Stone": A Women's History of Colonial Asante* (Portsmouth, NH: Heinemann, 2000).

39. James Leonard Giblin, *The Politics of Environmental Control in Northeastern Tanzania, 1840–1940* (Philadelphia: University of Pennsylvania Press, 1992); Feierman, *Peasant Intellectuals*; Gregory Maddox, James L. Giblin, Isaria N. Kimambo, *Custodians of the Land: Ecology and Culture in the History of Tanzania* (Athens: Ohio University Press, 1996). See also Yusuf Q. Lawi, "Where Physical and Ideological Landscapes Meet: Landscape Use and Ecological Knowledge In Iraqw, Northern Tanzania, 1920s–1950s," *International Journal of African Historical Studies* 32, nos. 2–3 (1999): 281–310.

40. William Cronon, "A Place for Stories: Nature, History, and Narrative," *Journal of American History* 78, no. 4 (March 1992): 1347–76.

41. Helge Kjekshus, *Ecology Control and Economic Development in East African History: The Case of Tanganyika, 1850–1950* (Heinemann Educational Books, 1977); Beinart and McGregor, introduction, 1–3; James C. McCann, *Green Land, Brown Land, Black Land: An Environmental History of Africa, 1800–1990* (Portsmouth, NH: Heinemann, 1999), 5; Melissa Leach and Robin Mearns, eds., *The Lie of the*

Land: Challenging Received Wisdom on the African Environment (Bloomington: Indiana University Press, 1996).

42. Beinart and McGregor, introduction, 17; Frederick Cooper, "Conflict and Connection: Rethinking Colonial African History," *American Historical Review* 99, no. 5 (1994): 1516–45; Helen Tilley, "African Environments and Environmental Sciences: The African Research Survey, Ecological Paradigms and British Colonial Development, 1920–1940," in *Social History and African Environments*, ed. William Beinart and JoAnn McGregor (Athens: Ohio University Press, 2003), 109–130; Richard Grove, *Green Imperialism: Colonial Expansion, Tropical Island Edens, and the Origins of Environmentalism, 1600–1860* (Cambridge: Cambridge University Press, 1995); Frederick Cooper and Ann Laura Stoler, "Tensions of Empire: Colonial Control and Visions of Rule," introduction, *American Ethnologist* 16, no. 4 (1989): 609–21.

43. Holly T. Dublin, "Dynamics of the Serengeti-Mara Woodlands: An Historical Perspective," *Forest and Conservation History* 35, no. 4 (October 1991): 169–78. See this story in McCann, *Green Land*, 71–74.

44. Neumann, *Imposing Wilderness*; Dan Brockington, *Fortress Conservation: The Preservation of the Mkomazi Game Reserve, Tanzania* (Bloomington: Indiana University Press, 2002); David Anderson and Richard Grove, *Conservation in Africa: People, Policies and Practice* (Cambridge: Cambridge University Press, 1987); Stuart A. Marks, *The Imperial Lion: Human Dimensions of Wildlife Management in Central Africa* (Boulder: Westview Press, 1984); Adams and McShane, *Myth of Wild Africa*; William Beinart and Peter A. Coates, *Environment and History: The Taming of Nature in the USA and South Africa, Historical Connections* (New York: Routledge, 1995).

45. Jane Carruthers, *The Kruger National Park: A Social and Political History* (Pietermaritzburg: University of Natal Press, 1995), 60–65; see also Carruthers, "Dissecting the Myth: Paul Kruger and the Kruger National Park," *Journal of Southern African Studies* 20, no. 2 (1994): 263–83; Carruthers, "Past and Future Landscape Ideology: The Kalahari Gemsbok National Park," in *Social History and African Environments*, ed. William Beinart and JoAnn McGregor (Athens: Ohio University Press, 2003), 255–66.

46. Neumann, *Imposing Wilderness*, 7–9, 11.

47. Grzimek, *Serengeti darf nicht sterben*.

48. Beinart and McGregor, introduction, 1–2; Neumann, *Imposing Wilderness*, 7–8; Agnes Kiss, *Living with Wildlife: Wildlife Resource Management with Local Participation in Africa* (Washington, DC: World Bank, 1990); Michael Wells, Katrina Brandon, and Lee Hannah, *People and Parks: Linking Protected Area Management with Local Communities* (Washington, DC: World Bank, 1992); Kevin A. Hill, "Zimbabwe's Wildlife Utilization Programs: Grassroots Democracy or an Extension of State Power?" *African Studies Review* 39, no. 1 (April 1996): 103–21; Elisabeth Croll and David Parkin, eds., *Bush Base: Forest Farm: Culture, Environment and Development* (New York: Routledge, 1992); David Hulme and Marshall Murphree,

eds., *African Wildlife and Livelihoods: The Promise and Performance of Community Conservation* (Oxford: James Currey, 2001); Alexander N. Songorwa, "Community-Based Wildlife Management (CWM) in Tanzania: Are the Communities Interested?" *World Development* 27, no. 12 (1999): 2061–79.

49. Luig and Oppen, "Landscape in Africa," 15. See also Eric Hirsch and Michael O'Hanlon, eds., *The Anthropology of Landscape: Perspectives on Place and Space* (Oxford: Clarendon Press, 1995); Alan R. H. Baker, "On Ideology and Landscape," introduction to *Ideology and Landscape in Historical Perspective: Essays on the Meanings of Some Places in the Past,* ed. Baker and Gideon Biger (Cambridge: Cambridge University Press, 1992),1–14.

50. Mitchell, "Imperial Landscape," 7–10; Luig and Oppen, "Landscape in Africa," 7. For a review of landscape in African environmental history see Beinart and McGregor, introduction, 4–5. In African history, see Karin Barber, *I Could Speak until Tomorrow: Oriki, Women and the Past in a Yoruba Town* (Edinburgh: Edinburgh University Press, 1991); Michelle Wagner, "Whose History Is History? A History of the Baragane People of Buragane, Southern Burundi, 1850–1932," 2 vols. (PhD diss., University of Wisconsin, Madison, 1991); Robert W. Harms, *Games against Nature: An Eco-cultural History of the Nunu of Equatorial Africa* (Cambridge: Cambridge University Press, 1987); Hofmeyer, *We Spend Our Years.*

51. Deborah Bird Rose, *Nourishing Terrains: Australian Aboriginal Views of Landscape and Wilderness* (Canberra: Australian Heritage Commission, 1996), 7–9.

52. Candace Slater, *Entangled Edens: Visions of the Amazon* (Berkeley: University of California Press, 2002).

53. Cohen and Odhiambo, *Siaya.*

54. Luig and Oppen, "Landscape in Africa," 20.

55. Tamara Giles-Vernick, *Cutting the Vines of the Past: Environmental Histories of the Central African Rain Forest* (Charlottesville: University of Virginia Press, 2002), 1–3.

56. Wagner, "Whose History?" 7.

57. Terence Ranger, "Women and Environment in African Religion: The Case of Zimbabwe," in *Social History and African Environments,* ed. William Beinart and JoAnn McGregor (Athens: Ohio University Press, 2003), 72–86; Carolyn Merchant, *The Death of Nature: Women, Ecology, and the Scientific Revolution* (San Francisco: Harper and Row, 1980); Henrietta L. Moore and Megan Vaughan, *Cutting Down Trees: Gender, Nutrition, and Agricultural Change in the Northern Province of Zambia, 1890–1990* (Portsmouth, NH: Heinemann, 1994).

58. The most prominent European example is Schama's study, *Landscape and Memory,* organized into sections dealing with wood, water, and rock. Terence Ranger studies one set of hills, while JoAnn McGregor is concerned with the Zambezi River. Ranger, *Voices from the Rocks: Nature, Culture, and History in the Matopos Hills of Zimbabwe* (Bloomington: Indiana University Press, 1999); McGregor, "Living with the River: Landscape and Memory in the Zambezi Valley, Northwest Zimbabwe," in *Social History and African Environments,* ed. William Beinart and

JoAnn McGregor (Athens: Ohio University Press, 2003). For literature on shrines and sacred places see, David Lan, *Guns and Rain: Guerillas and Spirit Mediums in Zimbabwe* (Oxford: James Currey, 1985); J. M. Schoffeleers, introduction to *Guardians of the Land: Essays on Central African Territorial Cults*, ed. Schoffeleers (Gwelo, Zimbabwe: Mambo Press, 1978), 1–45; Richard P. Werbner, *Regional Cults* (New York: Academic Press, 1977); Elizabeth Colson, "Places of Power and Shrines of the Land," *Paideuma* 43 (1997): 47–59; Michelle Wagner, "Environment, Community and History: 'Nature in the Mind' in Nineteenth- and Early Twentieth-Century Buha, Western Tanzania," in *Custodians of the Land: Ecology and Culture in the History of Tanzania*, ed. Gregory Maddox, James L. Giblin, and Isaria Kimambo (Athens: Ohio University Press, 1996), 175–99.

59. Sandra E. Greene, *Sacred Sites and the Colonial Encounter: A History of Meaning and Memory in Ghana* (Bloomington: Indiana University Press, 2002), 1–3.

60. McGregor, "Living with the River," 97.

61. Ranger, *Voices*; noted in Luig and Oppen, "Landscape in Africa," 26.

62. Neumann, *Imposing Wilderness*, 15–21; Gina Crandell, *Nature Pictorialized: "The View" in Landscape History* (Baltimore: Johns Hopkins University Press, 1993); W. G. Hoskins, *The Making of the English Landscape* (Harmondsworth: Penguin, 1970); Denis E. Cosgrove, *Social Formation and Symbolic Landscape* (Totowa, NJ: Barnes and Noble, 1985).

63. Dunlap, *Nature and the English Diaspora*, 21–41; Peder Anker, *Imperial Ecology: Environmental Order in the British Empire, 1895–1945* (Cambridge, MA: Harvard University Press, 2001); Grove, *Green Imperialism*; Tom Griffiths and Libby Robin, eds., *Ecology and Empire: Environmental History of Settler Societies* (Seattle: University of Washington Press, 1997); Michael L. Lewis, *Inventing Global Ecology: Tracking the Biodiversity Ideal in India, 1947–1997* (Athens: Ohio University Press, 2004).

64. John M. Mackenzie, *The Empire of Nature: Hunting, Conservation, and British Imperialism* (Manchester: Manchester University Press, 1988), 10–20, 26, 306.

65. Neumann, *Imposing Wilderness*, 21–25.

66. Luig and Oppen, "Landscape in Africa," 7.

67. Maurice Halbwachs, *Maurice Halbwachs on Collective Memory*, ed. and trans. Lewis A. Coser (Chicago, University of Chicago Press, 1992); Paul Connerton, *How Societies Remember* (Cambridge: Cambridge University Press, 1989); James Fentress and Chris Wickham, *Social Memory* (Oxford: Blackwell, 1992). On the proliferation of historical studies of memory, see Kerwin Lee Klein, "On the Emergence of Memory in Historical Discourse," *Representations* 69 (Winter 2000): 127–50.

68. Fentress and Wickham, *Social Memory*, xi, 7, 24–29, 47–49.

69. Elizabeth Tonkin, *Narrating Our Pasts: The Social Construction of Oral History* (Cambridge: Cambridge University Press, 1992; Joseph C. Miller, ed., *The African Past Speaks: Essays on Oral Tradition and History* (Folkestone, Eng.: Dawson, 1980); Jan Vansina, *Oral Tradition as History* (Madison: University of Wisconsin Press, 1985); Feierman, *Shambaa Kingdom*.

70. Fentress and Wickham, *Social Memory*, 32–40. For an example of changing traditions in changing context see Jan Bender Shetler, "Interpreting Rupture in Oral Memory: The Regional Context for Changes in Western Serengeti Age Organization (1850–1895)," *Journal of African History* 44, no. 3 (2004): 385–412.

71. Connerton, *How Societies Remember*, 2–3; Tonkin, *Narrating Our Pasts*. For an analysis of social memory outside of African history, see Patrick J. Geary, *Phantoms of Remembrance: Memory and Oblivion at the End of the First Millennium* (Princeton, NJ: Princeton University Press, 1994).

72. Part of this debate was reevaluated at the Words and Voices Conference in Bellagio, Italy, 24–28 February 1997, and in the follow-up conference. Papers from that conference published as Luise White, Stephan F. Miescher, and David William Cohen, eds., *African Words, African Voices: Critical Practices in Oral History* (Bloomington: Indiana University Press, 2001).

73. David Henige, "Oral Tradition and Chronology," *Journal of African History* 12, no. 3 (1971): 371–89; Miller, *African Past*; Paul Irwin, *Liptako Speaks: History from Oral Traditions in Africa* (Princeton, NJ: Princeton University Press, 1981); Fentress and Wickham, *Social Memory*, 77–86.

74. Fentress and Wickham, *Social Memory*, 11–14.

75. The theory is first argued in A. B. Lord, *The Singer of Tales* (Cambridge, MA: Harvard University Press, 1964); see also Joseph C. Miller, "Listening for the African Past," introduction to *The African Past Speaks: Essays on Oral Tradition and History*, ed. Miller (Folkestone, Eng.: Dawson, 1980), 5–9; Fentress and Wickham, *Social Memory*, 42–46.

76. Vansina, *Oral Tradition*, 144–46.

77. Miller, "Listening for the African Past," 8. Steven Feierman's structuralist interpretation of the core images in the Shambaa origin myth of Mbegha, in terms of the historical development of kingship, remains one of the best examples of this kind of interpretation. Feierman, *Shambaa Kingdom*, 40–69.

78. Fentress and Wickham, *Social Memory*, 59, 73.

79. Hofmeyr, *We Spend Our Years*, 106, 125, 132–33, 160.

80. In his paper for the 1997 Bellagio Conference, R. Newbury, "Contradictions at the Heart of the Canon," called this group the "fundamentalists," including historian Hartwig, who wrote about this region, for this region, G. Hartwig, "Oral Traditions Concerning the Early Iron Age in Northwestern Tanzania," *African Historical Studies* 4, no. 1 (1971): 93–114.

81. Gaston Bachelard, *The Poetics of Space* (New York: Orion Press, 1964); Francis S. Yates, *The Art of Memory* (Chicago: University of Chicago Press, 1966); Jonathan D. Spence, *The Memory Palace of Matteo Ricci* (New York: Viking, 1984); Mary J. Carruthers, *The Book of Memory: A Study of Memory in Medieval Culture* (Cambridge: Cambridge University Press, 1990); George Johnson, *In the Palaces of Memory: How We Build the Worlds inside Our Heads* (New York: Knopf, 1991). For a more detailed explanation of this spatial analysis of oral tradition, see Shetler, "Landscapes of Memory," ch. 1.

82. Bachelard, *Poetics of Space*, 9.

83. Ibid.

84. For other accounts that work with the "spatialization of time," see Renato Rosaldo, *Ilongot Headhunting, 1883–1974: A Study in Society and History* (Stanford: Stanford University Press, 1980), 42–58; for an Australian Aboriginal concept of history as paths called "songlines" or "dreaming tracks" that function as a map to remember the past, see Bruce Chatwin, *The Songlines* (New York: Viking, 1987), 2, 12.

85. Marc L. B. Bloch, *The Historian's Craft*, trans. from the French by Peter Putnam (Manchester: Manchester University Press, 1954), 61; Feierman, *Shambaa Kingdom*, 4; Miller, "Listening for the African Past," 6–8.

86. Allen M. Howard and Richard M. Shain present an analysis of much of the recent work on space in Africa. Howard and Shain, "African History and Social Space in Africa," introduction to *The Spatial Factor in African History: The Relationship of the Social, Material, and Perceptual*, ed. Howard and Shain (Leiden: Brill, 2005), 1–20. For other theoretical sources on the social construction of space, see Edward W. Soja, *Postmodern Geographies: The Reassertion of Space in Critical Social Theory* (London: Verso, 1989); Allan Pred, *Making Histories and Constructing Human Geographies: The Local Transformation of Practice, Power Relations, and Consciousness* (Boulder: Westview Press, 1990); David Harvey, *The Condition of Postmodernity: An Enquiry into the Origins of Cultural Change* (Oxford: Basil Blackwell, 1989); Derek Gregory, *Geographical Imaginations* (Cambridge, MA: Blackwell, 1994); Derek Gregory and John Urry, *Social Relations and Spatial Structures* (New York: St. Martin's, 1985); Cosgrove, *Social Formation*; Peter Gould and Rodney White, *Mental Maps* (Baltimore: Penguin, 1974); Shirley Ardener, *Women and Space: Ground Rules and Social Maps* (New York: St. Martin's, 1981); Henrietta L. Moore, *Space, Text, and Gender: An Anthropological Study of the Marakwet of Kenya* (Cambridge: Cambridge University Press, 1986).

87. See Henry Glassie's poetic treatment of the Irish landscape as a mnemonic artifact in which the past is entombed. Glassie, *Passing the Time in Ballymenone: Culture and History of an Ulster Community* (Philadelphia: University of Pennsylvania Press, 1982), 621–65. See also Barber, *I Could Speak*, 27, 34, on *oriki*, praise poems, as "fragments of the past."

88. Thanks to James Ellison for his archeological explanation of this concept: "Imagine a landscape inhabited [by people who] drop artifacts that reflect relations of exchange over a great distance and to the south. Then [these] people die, the site is covered by eolian deposits, and other people move in who drop artifacts that reflect trade in another set of directions and at a much closer radius. They die. Winds deflate the sediments leaving these quite different artifacts side by side." Ellison, pers. comm., 23 June 1997.

89. Connerton, *How Societies Remember*, 37.

90. Ibid., 20, 28.

91. For the social theory of space see note 86, above. For other African applications see Robert J. Thornton, *Space, Time, and Culture among the Iraqw of Tanzania* (New

York: Academic Press, 1980); David Parkin, *Sacred Void: Spatial Images of Work and Ritual among the Giriama of Kenya* (Cambridge: Cambridge University Press, 1991).

92. Emile Durkheim and Marcel Mauss, *Primitive Classification*, trans. from the French and ed. Rodney Needham (1903; Chicago: Chicago University Press, 1963), first published in French, 1903. Pierre Bourdieu enlarges this argument with his notion of habitus. Bourdieu, *Outline of a Theory of Practice*, trans. from the French Richard Nice (1972; Cambridge: Cambridge University Press, 1977), 1–71. For an application of spatial theory to African ethnography, see Anita Jacobson-Widding, ed., *Body and Space: Symbolic Models of Unity and Division in African Cosmology and Experience* (Uppsala: Almqvist and Wiksell International, 1991); Denise L. Lawrence and Setha M. Low, "The Built Environment and Spatial Form," *Annual Review of Anthropology* 19 (1990): 453–505.

93. Wrigley, *Kingship and State*, 49.

94. For a recent example of using ethnography and oral traditions along with linguistic data, see Kairn A. Klieman, *"The Pygmies Were Our Compass": Bantu and Batwa in the History of West Central Africa, Early Times to c. 1900 C.E.* (Portsmouth, NH: Heinemann, 2003). For an explanation of how sources can be used next to each other, see Jan Vansina, *Paths in the Rainforests: Toward a History of Political Tradition in Equatorial Africa* (Madison: University of Wisconsin Press, 1990).

95. Vansina, *Oral Tradition*, 23–24. See also Miller, introduction to *African Past Speaks*, 4; Feierman *Shambaa Kingdom*, 10–16, 40–69; Randall M. Packard, "The Study of Historical Process in African Traditions of Genesis: The Bashu Myth of Muhiyi," in Miller, *African Past Speaks*, 157–77; Thomas Spear, "Oral Traditions: Whose History?" *History in Africa* 8 (1981): 165–81; Spear, *Kenya's Past: An Introduction to Historical Method in Africa* (Harlow: Longman, 1981).

96. Fernand Braudel, *The Mediterranean and the Mediterranean World in the Age of Philip II*, trans. from the French by Siân Reynolds, ed. Richard Ollard (1949; Harper Collins, London, 1992).

97. Bourdieu, *Theory of Practice*, 1–71.

98. Ibid. Henrietta Moore suggests that spatial organization is like a "text" that can be "read." Moore, *Space, Text*, 79–86.

99. See Matthew Schoffeleers, "Oral History and the Retrieval of the Distant Past: On the Use of Legendary Chronicles as Sources of Historical Information," in *Theoretical Explorations in African Religion*, ed. Wim van Binsbergen and Matthew Schoffeleers (London: KPI, 1985), 164–88.

100. See Beinart and McGregor, introduction, 1–2.

101. Lucy Emerton and Iddi Mfunda, *Making Wildlife Economically Viable for Communities Living around the Western Serengeti, Tanzania* (London: International Institute for Environment and Development, 1999); Roderick P. Neumann, "Primitive Ideas: Protected Area Buffer Zones and the Politics of Land in Africa," *Development and Change* 28 (July 1997): 559–82;

102. White, Miescher, and Cohen, *African Words*.

103. Neumann, *Imposing Wilderness*.

1. Shetler, *Telling Our Own Stories*, 258.

2. M. Norton-Griffiths, D. Herlocker, and Linda Pennycuick, "The Patterns of Rainfall in the Serengeti Ecosystem, Tanzania," *East African Wildlife Journal* 13 (1975): 347.

3. A. R. E. Sinclair and M. Norton-Griffiths, eds., *Serengeti: Dynamics of an Ecosystem* (Chicago: University of Chicago Press, 1979). For an overview, see ch. 1, "Dynamics of the Serengeti Ecosystem: Process and Pattern," 1–30, and ch. 2, "The Serengeti Environment," 31–45, both by Sinclair. See also Sinclair, "Serengeti Past and Present," in *Serengeti II: Dynamics, Management, and Conservation of an Ecosystem*, ed. Sinclair and Peter Arcese (Chicago: University of Chicago Press, 1995), 23.

4. These dynamics are described in Sinclair and Norton-Griffiths, *Serengeti*, but Jonathan Scott writes for a popular audience in Scott, *The Great Migration* (London: Elm Tree Books, 1988). For an explanation of the nutritional reasons that wildebeests migrate to the plains, see Martyn G. Murray, "Specific Nutrient Requirements and Migration of Wildebeest," in *Serengeti II: Dynamics, Management, and Conservation of an Ecosystem*, ed. A. R. E. Sinclair and Peter Arcese (Chicago: University of Chicago Press, 1995), 231–56.

5. This genus includes the various species that share the common name crabgrass (esp. *D. ischaemum*, smooth crabgrass, a ubiquitous invasive weed in North America). Introduced spp. of *Cynodon* are collectively known as African Bermudagrass.

6. Sinclair, "Serengeti Environment," 36–42; Report on an Ecological Survey of the Serengeti National Park Tanganyika, November and December 1956, 1–2, Conservation Problems Overseas, Tanganyika, FT 3/599, PRO; Conservation Problems Overseas: Serengeti National Park, Tanganyika, FT 3/587, PRO.

7. Dennis Herlocker, *Woody Vegetation of the Serengeti National Park* (College Station: Caesar Kleberg Research Program in Wildlife Ecology/Texas A&M University, 1973), 9; Sinclair, "Serengeti Environment," 33.

8. Sinclair, "Serengeti Environment," 38–39; Report on an Ecological Survey, 26, PRO.

9. For the original accounts of this discovery, see L. S. B. Leakey, "The Newest Link in Human Evolution: The Discovery by L. S. B. Leakey of *Zinjanthropus boisei*," *Current Anthropology* 1, no. 1 (January 1960): 76–77. For the reevaluation of its classification, see L. S. B. Leakey, P. V. Tobias, and J. R. Napier, "A New Species of Genus Homo from Olduvai Gorge," *Current Anthropology* 6, no. 4 (October 1965): 424–27. For the context of contemporary evolutionary theory, see L. S. B. Leakey and Vanne Morris Goodall, *Unveiling Man's Origins: Ten Decades of Thought about Human Evolution* (Cambridge, MA: Schenkman Publishing, 1969), 157.

10. For a textbook account of hominid development, see Jerry H. Bentley and Herbert Ziegler, *Traditions and Encounters: A Global Perspective on the Past*, 2nd ed., 2 vols. (Boston: McGraw-Hill, 2003), 1:7–14.

11. Sinclair, "Serengeti Past and Present," 16.

12. Peter Atkins, Ian Simmons, and Brian Roberts, *People, Land and Time: An Historical Introduction to the Relations between Landscape, Culture and Environment* (London: Arnold, 1998), 4–5.

13. Sinclair, "Serengeti Past and Present," 3; Adams and McShane, *Myth of Wild Africa*, 55.

14. M. Norton-Griffiths, "The Influence of Grazing, Browsing, and Fire on the Vegetation Dynamics of the Serengeti," in *Serengeti: Dynamics of an Ecosystem*, ed. A. R. E. Sinclair and Norton-Griffiths (Chicago: University of Chicago Press, 1979), 327.

15. Sinclair, "Serengeti Environment," 39–40. This vegetation is found on the upper hill slopes specific to the red stony (skeletal) soils. Report on an Ecological Survey, 32.

16. Holly Dublin, "Vegetation Dynamics in the Serengeti-Mara Ecosystem: The Role of Elephants, Fire and other Factors," in *Serengeti II: Dynamics, Management, and Conservation of an Ecosystem*, ed. A. R. E. Sinclair and Peter Arcese (Chicago: University of Chicago Press, 1995), 81; Paul Mellars, "Fire Ecology, Animal Populations, and Man: A Study of Some Ecological Relationships in Prehistory," *Proceedings of the Prehistoric Society* 42 (1976): 15–45.

17. Norton-Griffiths, "Influence of Grazing," 332.

18. Dublin, "Vegetation Dynamics," 71–87; Dublin, "Dynamics of the Serengeti"; Tj Jager, *Soils of the Serengeti Woodlands, Tanzania*, Serengeti Research Institute Publications, no. 301 (Wageningen, Netherlands: Centre for Agricultural Publishing and Documentation, 1982), 41–43, 70.

19. Ken Campbell and Markus Borner, "Population Trends and Distribution of Serengeti Herbivores: Implications for Management," in *Serengeti II: Dynamics, Management, and Conservation of an Ecosystem*, ed. A. R. E. Sinclair and Peter Arcese (Chicago: University of Chicago Press, 1995), 117, 129, 141; Andy Dobson, "The Ecology and Epidemiology of Rinderpest Virus in Serengeti and Ngorongoro Conservation Area," in *Serengeti II: Dynamics, Management, and Conservation of an Ecosystem*, ed. A. R. E. Sinclair and Peter Arcese (Chicago: University of Chicago Press, 1995), 485–505.

20. Norton-Griffiths, "Influence of Grazing," 332–33, 341–48.

21. C. Winnington-Ingram, Administrative Officer, North Mara District, North Mara District (Tanzania) Survey of the African Farming and Land Utilization Problem in North Mara, Annual Report, 9 March 1950, 5, MSS. Afr. s. 1749, RHL.

22. A. R. E. Sinclair, "Equilibria in Plant-Herbivore Interactions," in *Serengeti II: Dynamics, Management, and Conservation of an Ecosystem*, ed. A. R. E. Sinclair and Peter Arcese (Chicago: University of Chicago Press, 1995), 108.

23. For a critique of the disequilibrium ecologists, who seem to downplay the effects of ecological degradation, see Beinart and McGregor, introduction, 1–7. For the romanticization of precolonial societies and the negative impact of this kind of thinking on conservation, see C. A. M. Atwell and F. P. D. Cotterill, "Postmodernism and African Conservation Science," *Biodiversity and Conservation* 9, no. 5

(2000): 559–77; A. W. Illius and T. G. O'Connor, "On the Relevance of Nonequilibrium Concepts to Arid and Semiarid Grazing Systems," *Ecological Applications* 9, no. 9 (1999): 798–813.

24. Sinclair, "Serengeti Environment," 37–40.

25. Stewart Edward White, *The Rediscovered Country* (Garden City, NY: Doubleday, Page and Co., 1915), 113; quoted also in Myles Turner, *My Serengeti Years: The Memoirs of an African Game Warden*, ed. Brian Jackman (New York: Norton, 1987), 29.

26. Paul L. Hoefler, *Africa Speaks: A Story of Adventure* (Chicago: John C. Winston Co., 1931), 292.

27. Bernhard Grzimek, *The Serengeti Shall Not Die*, trans. from the German by E. L. Rewald and D. Rewald, 1st American ed. (New York: E. P. Dutton, 1961), 304.

28. Interviews with Peter Mgosi Siwa, Morotonga, 23 August 2003; Mechara Masauta, Robanda, 22 August 2003; Nyamuko Soka, Morotonga, 28 August 2003. Grasses were identified in Nata by Nyawagamba Magoto and keyed to scientific name in D. M. Napper, *Grasses of Tanganyika*, Bulletin no. 18 (Dar es Salaam: Ministry of Agriculture, Forests and Wildlife, Tanzania, 1965), 132.

29. Testimonies differ on whether it was August or September, or a part of each. Interviews with Mang'oha Morigo, Bugerera, 24 June 1995; Wilson Shanyangi Machota, Morotonga, 12 July 1995.

30. Interviews with Peter Mgosi Siwa, Morotonga, 23 August 2003; Nyamuko Soka, Morotonga, 28 August 2003; Nyanchiwa Mesika, Morotonga, 27 August 2003.

31. Interview with Nyamuko Soka, Morotonga, 28 August 2003.

32. Memorandum on Masai History and Mode of Life, Prepared by National Game Parks, H. St. J. Grant, a District Officer in Masai District from August 1950 to December 1954, 8, 215/350/IV, TNA.

33. Interview with Nyamuko Soka, Morotonga, 28 August 2003. For scientific identification and problems today in rangeland infestation see Paul Mtoni, "Involve Them or Lose Both: Local Communities Surrounding Serengeti National Park in Bunda and Serengeti Districts in Relation to Wildlife Conservation" (MA thesis, Agricultural University of Norway, May 1999), 42, 91.

34. For more on trypanosomiasis, see John Ford, *The Role of the Trypanosomiases in African Ecology: A Study of the Tsetse Fly Problem* (Oxford: Clarendon Press, 1971); Richard Waller, "Tsetse Fly in Western Narok, Kenya," *Journal of African History* 31, no. 1 (1990): 81–101; James Giblin, "Trypanosomiasis Control in African History: An Evaded Issue?" *Journal of African History* 31, no. 1 (1990): 59–80. For a similar case in Sukuma and interaction with tick ecologies, see Martin H. Birley, "Resource Management in Sukumaland, Tanzania," *Africa* 52, no. 2 (1982): 1–29.

35. David F. Clyde, *History of the Medical Services of Tanganyika* (Dar es Salaam: Government Press, 1962), 28–29. Clyde cites traditions from Kerewe Island and Ikoma that describe a disease with the same symptoms as sleeping sickness. He

provides this as evidence that sleeping was sickness as an ancient disease. In Ikizu and Ikoma this disease was said to have almost depopulated the province over the last hundred years. Local informants said that the disease was contracted by the bite of the fly, beginning when the Ruwana and Mbalangeti rivers were in flood. There was a great deal of confusion as to whether this was sleeping sickness or severe hookworm disease in man, coincident with animal trypanosomiasis.

36. Juhani Koponen, *Development for Exploitation: German Colonial Policies in Mainland Tanzania, 1884–1914*, Studia Historica, no. 49 (Helsinki: Finnish Historical Society, 1994), 475–84; Ford, *Role of Trypanosomiases*; interview with Nyanchiwa Mesika, Morotonga, 27 August 2003.

37. Interviews with Peter Mgosi Siwa, Morotonga, 23 August 2003; Nyanchiwa Mesika, Morotonga, 27 August 2003.

38. Interviews with Wilson Shanyangi Machota, Morotonga, 12 July 1995; Surati Wambura, Morotonga, 13 July 1995; Rugayonga Nyamohega, Mugeta, 27 October 1995; C. F. M. Swynnerton to Honorable Chief Secretary, 26, 17 March 1924, Tsetse Fly, vol. 2, 2702, TNA.

39. H. G. Caldwell, Report on Sleeping Sickness in the Musoma District, July–August 1932, Sleeping Sickness, Musoma District, 215/463, TNA.

40. For another case of this in colonial Tanzania see Giblin, *Politics of Environmental Control.*

41. District Veterinary Officer, Musoma, to District Officer, Musoma, Annual Report 1927, 19 January 1928, 4, Provincial Administration, Mwanza Province, 1927–28, 246/PC/1/30, TNA.

42. J. F. Corson and Medical Officer, Ikoma, 15 April 1927, Third Note on Sleeping Sickness, Extracts of Report by District Veterinary Officer, Provincial Administration, Monthly Reports, Musoma District, 1926–29, 215/PC/1/7, TNA.

43. Heinrich Schnee, *Deutsches Kolonial-Lexikon*, 3 vols. (Leipzig: Quelle und Meyer, 1920), 1:89–90. Audrey Moore, *Serengeti* (London: Country Life, 1938), 207, 232–34.

44. Report on an Ecological Survey, 22, PRO.

45. Interview with Nyakerenge Nyamusaki, Morotonga, 25 August 2003.

46. Other dams or water sources constructed during the German era that were useful to wildlife were abandoned when they became part of the park and game control areas. Mtoni, "Involve Them," 40, 46, 114.

47. D. W. Malcolm, Report on Land Utilization in Sukuma, 1938, MSS. Afr. s. 1445/5, 59, 140, RHL.

48. Interviews with Mwalimu Nyamaganda Magoto, Mbiso, 3 August 2003; David Maganya Masama, Kemegesi, 12 August 2003; Wambura Tonte, Kemegesi, 12 August 2003; Mwalimu Nyamaganda Magoto, Mbiso, 3 August 2003.

49. Also mentioned in Mtoni, "Involve Them," 34.

50. Interview with Tetere Tumbo, Mbiso, 11 August 2003.

51. Mtoni, "Involve Them," 136.

52. Interview with Tetere Tumbo, Mbiso, 11 August 2003.

53. Shetler, *Telling Our Own Stories*, 251–52.

54. Interviews with Mzee Chengero, Wilson Machota, and Edward Kora, Robanda, 29 August 2003.

55. Augustine N. M. Kisigiro, *Kamusi ndogo ya Kinata-Kiswahili*, ed. Jan Shetler (Goshen, IN: Goshen College Printing Services, 2001).

56. S. M. Muniko, B. Muita oMagige, and Malcolm J. Ruel, eds., *Kuria-English Dictionary* (London: International African Institute, 1996), 32, 115.

57. Gilles Deleuze, "Rhizome," introduction to *A Thousand Plateaus*, ed. Felix Guattari (Minneapolis: University of Minnesota Press, 1987), 1–19. Thanks to Patrick Malloy for this reference and insight.

58. For the idea of emergence instead of origin, see Dennis Tedlock and Bruce Mannheim, *The Dialogic Emergence of Culture* (Chicago: University of Illinois Press, 1995), 8–15. On the study of origin mythology, see Charles H. Long, *Alpha: The Myths of Creation* (Chico, CA: Scholars Press, 1983).

59. For other emergence stories from this region, see Shetler, *Telling Our Own Stories*.

60. Interview with Jackson (Benedicto) Mang'oha Maginga, Mbiso, 18 March 1995. See also Shetler, *Telling Our Own Stories*, 257.

61. For a similar analysis of oral traditions see Packard, "Historical Process," 174. See also Packard's longer analysis in Packard, *Chiefship and Cosmology: An Historical Study of Political Competition* (Bloomington: Indiana University Press, 1981); Steven Feierman, "The Myth of Mbegha," in *The Shambaa Kingdom* (Madison: University of Wisconsin Press, 1974), 40–64.

62. Interview with Samweli M. Kiramanzera, Kurusanda, 3 August 1995. See also Shetler, *Telling Our Own Stories*, 251–52.

63. Interview with Machota Nyantitu, Morotonga, 28 May 1995. See also Shetler, *Telling Our Own Stories*, 266.

64. Shetler, *Telling Our Own Stories*, 155.

65. Interview with Silas King'are Magori, Kemegesi, 21 September 1995.

66. Shetler, *Telling Our Own Stories*, 221.

67. See ch. 4. Robert Harms dates his ethnographic material in this way: "The Nunu area can be subdivided into flooded forest, flooded grassland, riverbank, and dry land. These zones correspond with cultural subdivisions. By comparing the ethnography of the different microenvironments, one can distinguish the cultural traits shared by all of the Nunu from those that are distinct to a single environment. Traits shared by all are assumed to be old unless it can be demonstrated that they are recent innovations. Therefore, they define the more enduring and general features of the Nunu culture. In contrast, if an institution or practice is distinct to a certain micro-environment, we can assume that its existence or persistence has something to do with conditions unique to that area. It therefore represents innovation." Harms, *Games against Nature*, 6.

68. For the representation of hunter founders in African art, see Fritz W. Kramer, *The Red Fez: Art and Spirit Possession in Africa*, trans. from the German by Malcolm

Green (1987; London: Verso, 1993), 16. See also the Shambaa hunter myth in Feierman, *Shambaa Kingdom*.

69. John Bower, "The Pastoral Neolithic of East Africa," *Journal of World Prehistory* 5, no. 1 (1991): 74–76; Gifford-Gonzalez, "Early Pastoralists," 166, 169, 173, 189–90, 194–95; Gifford-Gonzalez, "Animal Disease Challenges," 95–139.

70. Ambrose, "Pastoral Adaptations," 222–33.

71. Ibid., 30, 238.

72. Michael G. Kenny, "Mirror in the Forest: The Dorobo Hunter-Gatherers as an Image of the Other," *Africa* 51, no. 1 (1981): 479; Ehret, *Southern Nilotic History*, 73. R. A. J. Maguire, a colonial officer in Maasailand during the 1920s, named eight different "Dorobo" groups, each speaking different languages and experiencing different levels of integration with other peoples. Maguire, "Il-Torobo," *Tanganyika Notes and Records* 25 (1948): 1–26.

73. Ehret, *African Classical Age*.

74. Ehret, *Southern Nilotic History*, 55–62, 130–32, tables D1, D2. It is not clear from historical linguistics when the Dadog speakers came to the western Serengeti. What we do know from the evidence of loanwords in present-day languages of the region is that Dadog speakers were in northern Tanzania, what is now Maasailand, perhaps as far west as the Mara, from the first millennium CE. They spread south into what is now the Maasai Steppe and southwest into parts of Kondoa, Mbulu, and Singida after 1000 CE. About the same time, incoming South Kalenjin–speaking peoples assimilated the northern Dadog-speaking settlements. Ehret, *Southern Nilotic History*, 40–43, 55–62. Loanwords from Dadog appear in Sonjo, Iraqw, and Aramanik. The impact of Dadog on the ancestors of the Sonjo was particularly significant. David L. Schoenbrun, "Early History in Eastern Africa's Great Lakes Region: Linguistic, Ecological, and Archaeological Approaches, ca. 500 B.C. to ca. A.D. 1000" (PhD diss., UCLA, 1990) 156–57, 182–204; Schoenbrun, "We Are What We Eat: Ancient Agriculture between the Great Lakes," *Journal of African History* 34, no. 3 (1993): 1–31. Eleusine millet: Also known as finger millet, *Eleusine corocana* was domesticated in East Africa and taken to India. Syngenta Foundation for Sustainable Agriculture, Crops, "Millet," http://www.syngentafoundation.org/millet.htm.

75. Evidence for these changes is found in the loanwords relating to herding, grain farming, and hunting adopted during this time by East Nyanza speakers. For linguistic evidence, see Schoenbrun, "Early History," 156–57, 182–204; Schoenbrun, "What We Eat," 1–31; Ehret, *Southern Nilotic History*, 40–43; Ehret, *African Classical Age*. For archaeological evidence, see Bower, "Pastoral Neolithic," 74–76; Ambrose, "Pastoral Adaptations," 222–33; Ambrose, "Hunter-Gatherer Adaptations to Non-marginal Environments: An Ecological and Archaeological Assessment," *Sprache und Geschichte in Afrika* 7, no. 2 (1986): 11–42.

76. "Shashi" is the Sukuma name given to all Mara peoples; it is still used as a derogatory term by the Sukuma today.

77. See Derek Nurse and Franz Rottland, "Sonjo: Description, Classification, History," *Sprache and Geschichte in Afrika* 12/13 (1991/92): 239–40; Derek Nurse and

Franz Rottland, "The History of Sonjo and Engaruka: A Linguist's View," *Azania* 28 (1993): 2. Guthrie's E50 group includes Gikuyu, Embu, Imenti Mero, Chuka, Tharaka, Tigania, Kitui Kamba, Machakos Kamba, and Daisu, while the East Nyanza languages are in the E40 group. See Malcolm Guthrie, *Comparative Bantu: An Introduction to the Comparative Linguistics and Prehistory of the Bantu Languages*, 4 vols. (Westmead: Gregg Press, 1967–71).

78. Ehret, *Southern Nilotic History*, 55. In distinction to the thesis of Nurse and Rottland, Ehret argues that Sonjo shows little influence from East Nyanza languages and much more from Southern Nilotic languages.

79. Other scholars have classified Sonjo as an Eastern Bantu language that stands on its own at the same level on the linguistic family tree with Swahili, Pokomo, Gikuyu, Kamba, Haya, and Luyia. Stanley H. Ambrose, "Archaeology and Linguistic Reconstructions of History in East Africa," in *The Archaeological and Historical Reconstruction of African History*, ed. Christopher Ehret and Merrick Posnansky (Berkeley: University of California Press, 1982), 110. Ambrose states, "The presence of Sonjo, Bantu cultivators, in the Lake Natron Basin of the northern Tanzania Rift Valley is an ecologically understandable exception to the distribution of Bantu speakers" (115). He also links the Later Iron Age Engaruka Complex with modern Sonjo irrigation agriculture (143). On Sonjo cultural characteristics, see Robert F. Gray, "Sonjo Lineage Structure and Property," in *The Family Estate in Africa: Studies in the Role of Property in Family Structure and Lineage Continuity*, ed. Robert F. Gray and Philip H. Gulliver (London: Routledge and Kegan Paul, 1964), 231–62; and Robert F. Gray, *The Sonjo of Tanganyika: An Anthropological Study of an Irrigation-Based Society* (London: Oxford University Press, 1963); Nurse and Rottland, "Sonjo: Description," 236.

80. See my various struggles with the question in Shetler, "Landscapes of Memory"; Shetler, "Interpreting Rupture."

81. Among the Ishenyi the Sagati, or Sageti, clan was reported by numerous informants, including Mang'ombe Morimi, Issenye/Iharara, 26 August 1995; Mikael Magessa Sarota, Issenye, 25 August 1995. The Sagari clan among the Ikoma was a hunting clan that has since disappeared as an independent clan. Interview with Mabenga Nyahega and Machaba Nyahega, Mbiso, 1 September 1995. On the origin story, interview with Marindaya Sanaya, Samonge, 5 December 1995.

82. Interviews with Peter Nabususa, Samonge, 5 December 1995; Emmanuel Ndenu, Sale, 6 December 1996.

83. Interview with Emmanuel Ndenu, Sale, 6 December 1995.

84. On the ntemi scar, see Robert F. Gray, *Sonjo of Tanganyika*, 15.

85. J. E. G. Sutton, "Engaruka and Its Waters," *Azania* 13 (1978): 37–38. For archaeological investigation of Engaruka, see L. S. B. Leakey, "Preliminary Report on Examination of Engaruka Ruins," *Tanganyika Notes and Records* 1 (1936): 57–60; John Sutton, "Engaruka etc.," *Tanzania zamani* 10 (1972): 7–10; Sutton, *A Thousand Years of East Africa* (Nairobi: British Institute in Eastern Africa, 1990), 33–40; Sutton, "The Irrigation and Manuring of the Engaruka Field System," *Azania* 21

(1986): 27–51; Sutton, "Towards a History of Cultivating the Fields," *Azania* 24 (1989): 99–112; Gray, *Sonjo of Tanganyika*, 53–56; Nurse and Rottland, "Sonjo and Engaruka."

86. Interview with Silas King'are Magori, Kemegesi, 21 September 1995.

87. Philip Curtin, Steven Feierman, Leonard Thompson, and Jan Vansina, *African History* (New York: Longman, 1978), 125. Feierman describes the pattern in the migrations of Bantu-speaking peoples who, when forced to move, chose those places where they could apply their environmental knowledge. David W. Cohen describes the same pattern in "The Face of Contact: A Model of a Cultural and Linguistic Frontier in Early Eastern Uganda," in *Nilotic Studies*, part 2, *Proceedings of the International Symposium on Languages and History of the Nilotic Peoples, Cologne, January 4–6, 1982*, ed. Rainer Vossen and Marianne Bechhaus-Gerst (Berlin: Dietrich Reimer Verlag, 1983), 339–56.

88. Interview with Emmanuel Ndenu, Sale, 6 December 1995; Ndenu stated that "Khambageu was a prophet and a god, he came from over toward Ikoma. His wife was Nankoni. They used to visit back and forth with Ikoma especially in the tenth through the twelfth month. They went to worship there and the ones that followed him went there to worship too." See also R. Gray, *Sonjo of Tanganyika*, 11–12, who relates the tradition of Khambageu coming from the Sonjo village of Tinaga, and then cursing the village, leading to its destruction. The name Khambageu comes from a Southern Nilotic Dadog word meaning great-grandfather. See also Nurse and Rottland, "Sonjo: Description," 217.

89. Nurse and Rottland, "Sonjo: Description," 236.

90. This thesis was first expounded by Anacleti, "Serengeti," 23–34; see also Anacleti, "Pastoralism and Development." The Maasai Loitai of the Loliondo highlands, whose territory extended north into Kenya and east to Lake Natron, reported to colonial anthropologist Henry Fosbrooke that the original inhabitants of this area were the "Ilmarau," the Maasai name for the Ikoma- or Bantu-speaking peoples of the western Serengeti in general. Fosbrooke, "Sections of the Masai in Loliondo Area," typescript, 1953, CORY 259, EAF.

91. Waller, "Ecology, Migration," 347–70.

92. Schoenbrun, *Green Place*, 160–62. See also Wagner, "Whose History?" 26–39.

93. J. M. Purseglove, *Tropical Crops: Monocotyledons* (London: Longman, 1972), 146–49.

94. Mtoni, "Involve Them," 45.

95. Interview with Pastor Wilson Shanyangi Machota, Morotonga, 12 July 1995.

96. White, *Rediscovered Country*, 241, 248.

97. Senior Agriculture Officer, Mwanza, to Director of Agriculture, DSM, 30 August 1941, Agriculture, General, 215/909/II, TNA.

98. Mtoni, "Involve Them," 89.

99. Kristen Alsaker Kjerland, "Cattle Breed; Shillings Don't: The Belated Incorporation of the abaKuria into Modern Kenya," (PhD thesis, University of Bergen, 1995), 37.

100. V. C. R. Ford, *The Trade of Lake Victoria* (Kampala: East African Institute of Social Research, 1955), 16.

101. Geographical Section of the Naval Intelligence Division, Naval Staff, Admiralty, *A Handbook of German East Africa* (1920; New York: Negro University Press, 1969), 159.

102. Interview with Mechara Masauta, Robanda, 22 August 2003.

103. H. St. J. Grant, District Officer, Report on Human Habitation of the Serengeti National Park, May 1954, and from the District Officer, Masai, Monduli, 28 May 1954 to the Provincial Commissioner, Northern Province, Arusha, Secretariat Files, AB 915 (3733) II, and National Game Parks, 215/350/III, TNA.

104. G. McL. Wilson, "The Tatoga of Tanganyika, Part One," *Tanganyika Notes and Records* 33 (1952): 40–41. Wilson describes the Tatoga Iseimajek and Rutageink, who live in the Ruwana valley of Mara, numbering thirteen hundred people in 1948.

105. Interview with Merekwa Masunga and Giruchani Masanja, Mariwanda, 6 July 1995.

106. Interview with Moremi Mwikicho, Sagochi Nyekipegete, Kenyatta Mosoka, Robanda, 12 July 1995. For the story of the bao game in Buganda tradition, see Wrigley, *Kingship and State*, 101.

107. Interview with Stephen Gojat Gishageta and Girimanda Mwarhisha Gishageta, Issenye, 27 July 1995. See also Shetler, *Telling Our Own Stories*, 282.

108. See ch. 3.

109. R. S. W. Malcolm, System of Government, extracts from a report, 1937, reel 24, 4, MDB.

110. Karte von Deutsch-Ostafrika, A.4, Ikoma (Berlin: D. Reimer [E. Vohsen], 1910), German Maps, GM 30/3, TNA; A. M. D. Turnbull, Senior Commissioner, Mwanza District, Report on Mwanza District for the Year 1925, 26 January 1926, Mwanza District Office Head: Provincial Administration, Sub-Head Annual Report,, 1925–26, 246/PC 1/17, TNA.

111. Ambrose, "Hunter-Gatherer Adaptations," 11–42; Schoenbrun, *Green Place*, 104–6; Curtis Marean, "Hunter to Herder: Large Mammal Remains from the Hunter-Gatherer Occupation at Enkapune Ya Muto Rock-Shelter, Central Rift Kenya," *African Archaeological Review* 10 (1992): 65–127.

112. This inherited knowledge was still evident in 1913, when a European hunter using Asi guides to cover this territory exclaimed that they "seem to know this country like a book," finding all the hidden waterholes and camps. White, *Rediscovered Country*, 112.

113. Ambrose, "Hunter-Gatherer Adaptations," 11–42; Schoenbrun, *Green Place*, 104–6.

114. Interview with Sira Masiyora, Nyerero, 17 November 1995.

115. Karte von Deutsch-Ostafrika, TNA.

116. One interpretation of this evidence could be that western Serengeti farming communities developed out of a preexisting hunter-gatherer society and that the emergence sites were their remembered hunting camps. Yet this kind of major shift

in identity from a hunter-gatherer society to one based on farming would presumably constitute a rupture in historical consciousness similar to that during the period of disasters described in chapter 4. Then places significant to an earlier way of life would be forgotten when divorced from their social context. The old sites of hunter-gatherer communities would not have figured in the historical imagination of these new communities. The evidence of historical linguistics and archaeology already presented points to two separate communities of hunters and farmers with different histories who met on the frontier.

117. See Edwin N. Wilmsen, *Land Filled with Flies: A Political Economy of the Kalahari* (Chicago: University of Chicago Press, 1989); Wilmsen and James R. Denbow, "Paradigmatic History of San-Speaking Peoples and Current Attempts at Revision," *Current Anthropology* 31, no. 5 (1990): 489–524; Andrew B. Smith, "The Kalahari Bushman Debate: Implications for Archeology in Southern Africa," *South African Historical Journal* 35 (1996): 1–15.

118. Spear and Waller, *Being Maasai.*

119. Interview with Wambura Edward Kora, Morotonga, 19 August 2003.

120. Interviews with Tetere Tumbo, Mbiso, 11 August 2003; Wambura Tonte, Kemegesi, 12 August 2003.

121. Interview with Jackson Mteba Mabura, Morotonga, 23 August 2003.

122. Ibid.

123. Interview with Paulo Machota Mongoreme, Kyandege, 8 August, 2003.

124. Interviews with Wilson Shanyangi Machota, Morotonga, 18 August 2003; George Wambura Gehamba and Samweli Muya Mongita, Morotonga, 21 August 2003; Stephano Makondo Karamanga, Robanda, 22 August 2003; Mechara Masauta, Robanda, 22 August 2003.

125. Interview with Wambura Edward Kora, Morotonga, 19 August 2003.

126. Geographical Section, *German East Africa*, 98–99.

127. See Kairn Klieman's analysis of hunter-gatherer history of the central African forest and how they assisted the Bantu-speaking settlers. Klieman, *"Pygmies Were Our Compass."*

128. Interview with Jackson (Benedicto) Mang'oha Maginga, Mbiso, 18 March 1995.

129. See Shetler, "Landscapes of Memory," 222; interview with Mang'oha Morigo, Bugerera, 24 June 1995; W. D. Raymond, "Tanganyika Arrow Poisons," *Tanganyika Notes and Records* 23 (1947): 49–65. On arrows, interview with Wambura Edward Kora, Morotonga, 19 August 2003.

130. Interviews with Jackson Mang'oha, Mbiso, 13 May 1995; Mahiti Gamba, Mayani Magoto, Bugerera, 3 March 1996; Nyamaganda Magoto, collection of culture vocabulary.

131. Interview with Mang'oha Morigo, Bugerera, 24 June 1995. In Ishenyi, *obutir* (September) is the month when the topi give birth and the time for preparing the fields. In October the first millet is planted. Interview with Nyambeho Marangini, Issenye, 7 September 1995.

132. Kenny, "Mirror in the Forest," 482. Unfortunately, I was unable to find any descendants of the Asi who could recount their oral traditions. The Asi have either totally assimilated into farming communities or have gone to live in Loliondo under Maasai patronage. Thus, until more research is done the view of Asi history presented here is entirely from the perspective of the farmers. But because western Serengeti farmers consider the Asi one of the original parents and first comers to the land, farmers respect their knowledge and history.

133. Interview with Mahiti Kwiro, Mchang'oro, 19 January 1996. A parallel process existed in the Kenyan highlands, where Kikuyu tradition recounts ritual adoptions of Kikuyu into Ndorobo kin groups to clear the land for farming. Greet Kershaw, *Mau Mau from Below* (Athens: Ohio University Press, 1997), 20–21. Kershaw dates the Kikuyu settlement of Kiambu to the era preceding the Kiraka famine of 1835.

134. See Spear and Waller, *Being Maasai*; interview with Paulo Machota Mongoreme, Kyandege, 8 August, 2003.

135. Interview with Wambura Edward Kora, Morotonga, 19 August 2003.

136. Geographical Section, *German East Africa*, 98–99.

137. Brad Weiss describes the Haya ceremony for blessing a new house, which involves lighting the fire for the first time by the father or a senior agnate. Weiss, *The Making and Unmaking of the Haya Lived World: Consumption, Commoditization, and Everyday Practice* (Durham: Duke University Press, 1996), 29–31, 51–52. For the Kuria, see Tobisson, *Family Dynamics*, 128–32.

138. For a discussion of the symbols of fire and water, see Anita Jacobson-Widding, "The Encounter in the Water Mirror," in *Body and Space: Symbolic Models of Unity and Division in African Cosmology and Experience*, ed. Anita Jacobson-Widding (Uppsala: Almqvist and Wiksell International, 1990), 177–216.

139. P. M. Mturi and S. Sasora, "Historia ya Ikizu na Sizaki"; reprinted in Shetler, *Telling Our Own Stories*, 59.

140. Shetler, *Telling Our Own Stories*, 61–63.

141. Robert A. LeVine, "The Gusii Family," in *The Family Estate in Africa: Studies in the Role of Property in Family Structure and Lineage Continuity*, ed. Robert F. Gray and P. H. Gulliver (London: Routledge and Kegan Paul, 1964), 70; Schoenbrun, *Green Place*, 98; Tobisson, *Family Dynamics*, 128–37.

142. These patterns are clearly observable in rural society today. During fieldwork the house in which my family and I lived was always referred to as "my" house and when men came to visit they always gathered in the courtyard shade to speak to my husband.

143. Interview with Sochoro Kabati, Nyichoka, 2 June 1995.

144. Interview with Megasa Mokiri, Motokeri, 4 March 1995. Megasa asserts, in contradiction to all other accounts, that first woman went to live in first man's hunting shelter.

145. For archeological evidence, see Sutton, *Archaeology of the Western Highlands*, 50–58.

146. Tobisson, *Family Dynamics*, 147–48.

147. Mtoni, "Involve Them," 35.

CHAPTER 2: SOCIAL LANDSCAPES

1. For an analysis of moral economy, see Scott, *Moral Economy*. For African famines, see Watts, *Silent Violence*.

2. Interview with Megasa Mokiri, Motokeri, 4 March 1995.

3. Interview with Mahewa Timanyi and Nyambureti Morumbe, Robanda, 27 May 1995.

4. For an example of this argument based on regional sources, see Michael G. Kenny, "The Stranger From the Lake: A Theme in the History of the Lake Victoria Shorelands," *Azania* 17 (1982): 9. See also Newbury, *Kings and Clans*, 200–226.

5. Shetler, *Telling Our Own Stories*, 255.

6. Vansina, *Paths in the Rainforests*, 104–10; Schoenbrun, *Green Place*, 94–101.

7. Kopytoff, "Internal African Frontier," 15, 37.

8. Watts, *Silent Violence*.

9. John Iliffe, *Africans: The History of a Continent* (Cambridge: Cambridge University Press, 1995), 1.

10. Jager, *Soils*, 120–21; Robert W. July, *A History of the African People*, 4th ed. (Prospect Heights, IL: Waveland Press, 1992), 4; Norton-Griffiths, Herlocker, and Pennycuick, "Patterns of Rainfall," 359.

11. Stephen Makacha, IKONA, "Extending Robanda Wildlife Management Area Now called IKONA to cover Four Villages, Mugumu District, Mara Region, Tanzania (Robanda, Nyichoka, Makundusi/Nyakitono, and Mbiso)," May 1999, Report to Pia Zimmerman c/o Frankfurt Zoological Society, SNP, TZ, author's collection.

12. A. R. E. Sinclair notes that the average rainfall for Musoma from 1921 to 1930 was 662.2 mm (26 in.). Sinclair, "Serengeti, Past and Present," 6; D. Thula, compiler, and H. C. Barlet, inspector, District Book Agriculture, District Agricultural Officer's Report, 6 May 1941, reel 24, 3–8, MDB.

13. Purseglove, *Tropical Monocotyledons*, 149. Most millet is raised with 1,000 to 1,500 mm of rainfall, while some types may be raised with 800 to 900 mm.

14. July, *African People*, 8.

15. Paul Richards, "Ecological Change and the Politics of African Land Use," *African Studies Review* 26, no. 2 (June 1983): 5, 25.

16. Makacha, IKONA, "Extending Robanda WMA."

17. East African Population Census, 1948, African Population of Musoma District, East Africa Statistical Department, Nairobi, 1 October 1948; African Population by Chiefdom, 1948, Secretariat Files, 40641, TNA. In the Native Affairs Census of 1926 the Ikoma Federation was listed with a total population of 14,799; 1,923 were Nata, 8,664 were Ikoma, and 4,212 were Issenye. Data from the Native Affairs Census 1926–29, 246/PC/3/21, TNA.

18. Yet the annual growth rate of 8.3 percent is the highest in the region, reflecting a large in-migration for open farmland and mining. Mara Regional Statistical Abstract, 1993, President's Office, Planning Commission, Bureau of Statistics, DSM, June 1995, 2, 12, EAF.

19. Iliffe, *Africans*, 1–4; Richards, "Ecological Change," 30; John Thornton, *Africa and Africans in the Making of the Atlantic World, 1400–1800*, 2nd ed. (Cambridge: Cambridge University Press, 1998), 105.

20. A number of important historical studies on famine in Africa have been based on the "loss of entitlement" theory of Amartya Kumer Sen, *Poverty and Famines: An Essay on Entitlement and Deprivation* (Oxford: Clarendon Press, 1981). For early applications of this theory in Africa, see Giblin, *Politics of Environmental Control*; Watts, *Silent Violence*; Megan Vaughan, *The Story of an African Famine: Gender and Famine in Twentieth-Century Malawi* (Cambridge: Cambridge University Press, 1987); James McCann, *From Poverty to Famine in Northeast Ethiopia: A Rural History, 1900–1935* (Philadelphia: University of Pennsylvania Press, 1987).

21. For the classical case, see E. E. Evans-Pritchard, *The Nuer: A Description of the Modes of Livelihood and Political Institutions of a Nilotic People* (Oxford: Clarendon Press, 1940); for segmentary theory's more recent manifestation, see R. Cohen, "Ethnicity: Problem and Focus in Anthropology," *Annual Review of Anthropology* 7 (1978): 379–403. See also Aidan Southall, *Alur Society: A Study in Processes and Types of Domination* (Cambridge: W. Heffer, 1956); Southall, "The Segmentary State: From the Imaginary to the Material Means of Production," in *Early State Economics*, ed. Henri J. M. Claessen and Pieter van de Velde (New Brunswick, NJ: Transaction, 1991), 75–96.

22. "Tribe" is a word no longer used by academics because of the assumptions behind it and the way it has been misused as a European construct imposed on African history. I use it only as it was used by colonial or anthropological writing of the time.

23. See Bourdieu, *Theory of Practice*; Moore, *Space, Text*, 79–86.

24. Schoenbrun, *Green Place*, 100–101.

25. Shetler, "Landscapes of Memory," 161–64, 184.

26. Karla O. Poewe, *Matrilineal Ideology: Male-Female Dynamics in Luapula, Zambia* (London: Academic Press, 1981), 3, 21, 25–26, 46–47. See also Christine Choi Ahmed, "Before Eve was Eve: 2200 Years of Gendered History in East-Central Africa" (PhD diss., UCLA, 1996), 143; Cynthia Brantley, "Through Ngoni Eyes: Margaret Read's Matrilineal Interpretations from Nyasaland," *Critique of Anthropology* 17, no. 2 (June 1997): 147–69.

27. Schoenbrun, *Green Place*, 178.

28. See Karen Sacks, *Sisters and Wives: The Past and Future of Sexual Equality* (Westport, CT: Greenwood Press, 1979).

29. Interview with Nyamaganda Magoto, Mbiso, 4 July 2002.

30. Richards, "Ecological Change," 26–27, 31.

31. Interviews with Wambura Tonte, Kemegesi, 12 August 2003; Peter Mgosi Siwa, Morotonga, 23 August 2003; Robi Chacha, Kemegesi, 13 August 2003.

32. *Tanganyika Territory Blue Book*, 1947, CO 726/2–30, PRO. Original figures: 207,800 total crop acres; 20,000 acres of legumes; 99,200 acres of sorghum and millet.

33. Director, Department of Agriculture, Dar es Salaam, report to the Honorable Chief Secretary, DSM, 8 December 1922, Agriculture: Tanganyika Territory, Report AB 1025 (7013), TNA.

34. Interview with Wambura Edward Kora, Morotonga, 19 August 2003.

35. Interview with Wambura Tonte, Kemegesi, 12 August 2003. This strategy is still practiced today, as observed by author while living in Bugerera, Nata, 1995–96.

36. D. W. Malcolm, Land Utilization in Sukuma, 17, 22, 101, 107, RHL.

37. Purseglove, *Tropical Monocotyledons*, 270.

38. Interviews with Joseph Sillery Magoto, Mbiso, 2 August 2003; Mwalimu Nyamaganda Magoto, Mbiso, 3 August 2003; George Wambura Gehamba and Samweli Muya Mongita, Morotonga, 21 August 2003; Mechara Masauta, Robanda, 22 August 2003; Nyawagamba Magoto and Mahiti Gamba, Bugerera, 1 November 1995; Simion Hunga Nason and Bhoke Mtoka, Kemegesi, 14 August 2003; Wambura Edward Kora, Morotonga, 26 August 2003.

39. Interview with Mashauri Ng'ana, Issenye, 2 November 1995.

40. Interview with Stephen Gojat Gishageta and Girimanda Mwarhisha Gishageta, Issenye, 27 July 1995.

41. Interview with Nyamuko Soka, Morotonga, 28 August 2003.

42. Malcolm, Land Utilization in Sukuma, 129.

43. Winnington-Ingram, Survey of African Farming, RHL.

44. See Catharine Newbury, *The Cohesion of Oppression: Clientship and Ethnicity in Rwanda, 1860–1960* (New York: Columbia University Press, 1988).

45. Richards, "Ecological Change," 37–39; F. G. McCall, Chief Veterinary Officer, Annual Report of the Department of Veterinary Science and Animal Husbandry, Tanganyika, 1921, 38, Veterinary Department, Annual Report, 1921, 3046/22, TNA.

46. Richards, "Ecological Change," 30; D. Thornton and N. V. Rounce, Ukara Island and the Agricultural Practices of the Wakara, Monthly Letter, October 1933, annexure 1, Tanganyika Territory Department of Agriculture; D. W. Malcolm, Report on Economic Conditions in the Island of Ukara with Special Reference to Soil Erosion, 1934, 22425, TNA; Sutton, "Cultivating the Fields," 99–112; Sutton, "Irrigation and Manuring," 27–51.

47. Interviews with Wambura Edward Kora, Morotonga, 19 August 2003; George Wambura Gehamba and Samweli Muya Mongita, Morotonga, 21 August 2003.

48. Interview with Bita Makuru, Bugerera, 11 February 1995. For Kuria settlement mobility, see Miroslava Prazak, "Cultural Expressions of Socioeconomic Differentiation among the Kuria of Kenya" (PhD diss., Yale University, 1992), 65–89. Prazak reports that an average homestead lasted for ten to twenty years (124).

49. Interview with Joseph Sillery Magoto, Mbiso, 2 August 2003.

50. Interviews with Wambura Edward Kora, Morotonga, 19 August 2003; Wambura Tonte, Kemegesi, 12 August 2003. These generalizations have been taken from many informal conversations, from the narratives of traditions about this period,

from the life stories of elders (including their narration of their parent's life histories), and especially from interviews with Surati Wambura, Morotonga, 13 July 1995, and Mariko Romara Kisigiro, Burunga, 31 March 1995.

51. Vansina, *Paths in the Rainforests*, 55. Vansina calculates that a spread of twenty-two kilometers in ten years is quite possible with this type of "drift." Ehret describes migration as a slow and gradual process of small groups moving onto the next pasture or field—as a process of assimilation rather than extermination. Ehret, *Southern Nilotic History*, 26–27. See also D. P. Collett, "Models in the Spread of the Early Iron Age," in *The Archaeological and Linguistic Reconstruction of African History*, ed. Christopher Ehret and Merrick Posnansky (Berkeley: University of California Press, 1982), 182–95.

52. In contrast to this is the system of land "ownership" by Kikuyu lineage groups in Kenya, who laboriously cleared farmland over a period of years out of the highland forest. Kershaw, *Mau Mau from Below*, 22–23.

53. Interview with Megasa Mokiri, Motokeri, 4 March 1995.

54. Yohana Kitena Nyitanga named 113 such places, in all directions from his present home. Nyitanga, Makundusi, 1 May 1995.

55. Interview with Nchota Chachamogohe, Kemegesi, 13 August 2003.

56. Interviews with Robi Chacha, Kemegesi, 13 August 2003; Stephano Makondo Karamanga, Robanda, 22 August 2003; Sarah Wanchota Noku, Morotonga, 25 August 2003.

57. Ouabain is used in Western medicine in ways similar to digitalis.

58. Interview with Mang'oha Morigo, Bugerera, 24 June 1995; Raymond, "Tanganyika Arrow Poisons," 49–65.

59. Interview with Makuru Nyang'aka, Nyichoka to Ryara, 7 March 1996.

60. Richards, "Ecological Change," 39.

61. Interviews with Wilson Shanyangi Machota, Morotonga, 18 August 2003; Robi Chacha, Kemegesi, 13 August 2003; George Wambura Gehamba and Samweli Muya Mongita, Morotonga, 21 August 2003; Sarah Wanchota Noku, Morotonga, 25 August 2003.

62. Interview with Joseph Sillery Magoto, Mbiso, 2 August 2003.

63. Interviews with Wambura Edward Kora, Morotonga, 28 August 2003. Nyanchiwa Mesika, Morotonga, 27 August 2003.

64. For theory of the gift economy see Marcel Mauss, *The Gift: Forms and Functions of Exchange in Archaic Societies* (New York: Norton, 1967), quoted at 3; Christopher C. Taylor, *Milk, Honey, and Money: Changing Concepts in Rwandan Healing* (Washington, DC: Smithsonian Institution Press, 1992).

65. Interview with Wambura Edward Kora, 19 August 2003, Morotonga.

66. Richards, "Ecological Change," 36–37.

67. Interviews with Weigoro Mincha, Kemegesi, 29 March 1996; Robi Chacha, Kemegesi, 13 August 2003; Nyakaho Magambo, Kemegesi, 13 August 2003.

68. Interviews with Sarah Wanchota Noku, Morotonga, 25 August 2003; Mashauri Ng'ana, Issenye, 2 November 1995; Nyanchiwa Mesika, Morotonga, 27 August 2003.

69. Information on women's exchanges comes from the author's participant-observation in rural western Serengeti communities for many years.

70. Interview with Robi Chacha, Kemegesi, 13 August 2003. She said that women did not go as far as men because they wore heavy iron anklets for beauty and dancing (*ebirangani*) that made it difficult to walk distances.

71. See Adam Kuper, "Lineage Theory: Critical Retrospect," *Annual Review for Anthropology* 11 (1982): 71–95; D. W. Hammond Tooke, "In Search of the Lineage: The Cape Nguni Case," *Man* 19 (March 1984): 77–93; Jane I. Guyer, "Household and Community in African Studies," *African Studies Review* 24, nos. 2–3 (June–September 1981): 87–137. See also Parker Shipton, "Strips and Patches: A Demographic Dimension in Some African Land-Holding and Political Systems," *Man*, n.s., 19, no. 4 (1984): 613–34.

72. Adam Kuper, "The 'House' and Zulu Political Structure in the Nineteenth Century," *Journal of African History* 34, no. 1 (1993): 469–87; David L. Schoenbrun, "Gendered Histories between the Great Lakes: Varieties and Limits," *International Journal of African Historical Studies* 29, no. 13 (1996): 461–92.

73. Schoenbrun, *Green Place*, 170–71, 159–60.

74. Sutton, *Archaeology of the Western Highlands*, 50–58; E. C. Baker, *Tanganyika Papers*; Tobisson, *Family Dynamics*, 129–33; H. S. Senior, "The Sukuma Homestead," *Tanganyika Notes and Records* 9 (1940): 42–44; Robert A. LeVine and Sarah E. LeVine, "House Design and the Self in an African Culture," in *Body and Space: Symbolic Models of Unity and Division in African Cosmology and Experience*, ed. Anita Jacobson-Widding (Uppsala: Almqvist and Wiksell International, 1991), 155–76; interview with Mayani Magoto, Bugerera, 18 February 1995.

75. LeVine, "Gusii Family," 70; Schoenbrun, *Green Place*, 98; Tobisson, *Family Dynamics*, 128–37.

76. See Regina Smith Oboler, "The House-Property Complex and African Social Organization," *Africa* 64, no. 3 (1994): 344, 351; Thomas Hakansson, "Family Structure, Bridewealth, and Environment in Eastern Africa: A Comparative Study of the House-Property Systems," *Ethnology* 28, no. 2 (1989): 117–35.

77. Interviews with Mahiti Gamba, Bugerera, 4 February 1996; George Wambura Gehamba and Samweli Muya Mongita, Morotonga, 21 August 2003; Peter Mgosi Siwa, Morotonga, 23 August 2003. Presumably low-land areas are connected with malaria or "fevers."

78. Noted by the European hunter Steward Edward White visiting Ngoreme in 1913. White, *Rediscovered Country*, 199, 242.

79. Interviews with Surati Wambura, Morotonga, 13 July 1995; Jackson (Benedicto) Mang'oha Maginga, Mbiso, 18 March 1995; Mariko Romara Kisigiro, Burunga, 31 March 1995. Much of this information is pieced together from conversations that contrasted the fort settlements of the disasters with this early pattern and with present village structure.

80. Malcolm, Land Utilization in Sukuma, 131, RHL.

81. Interviews with Wambura Tonte, Kemegesi, 12 August 2003; Nyanchiwa Mesika, Morotonga, 27 August 2003.

82. Information gained from living in the homestead of Nyawagamba Magoto in Bugerera (1995–96) and having my sons participate in watching his livestock herds. For a wonderful fictionalized version of these patterns in Ukerewe, see Aniceti Kitereza, *Mr. Myombekere and His Wife Bugonoka, Their Son Ntulanalwo and Daughter Bulihwali: The Story of an Ancient African Community,* trans. from the Kikerewe by Gabriel Ruhumbika (Dar es Salaam: Mkuki na Nyota Publishers, 2002).

83. Mtoni, "Involve Them," 41; interviews with Wambura Edward Kora, Morotonga, 19 August 2003; George Wambura Gehamba and Samweli Muya Mongita, Morotonga, 21 August 2003; Peter Mgosi Siwa, Morotonga, 23 August 2003; Nyanchiwa Mesika, Morotonga, 27 August 2003.

84. This early German report also states of the "Washashi and Wangorimi" that "the fields in some cases are several hours' journey from the houses." Geographical Section, *German East Africa,* 97.

85. Acting Provincial Commissioner, Mwanza Province, to Honorable Chief Secretary, 14 July 1930; Agricultural Development, Mwanza Province, Secretariat Files, 19080, TNA.

86. Extract from Mr. C. G. Clay's Broadcast Talk in the Third Programme, 9 July 1949, Prospects for Colonial Agriculture, 513, Agriculture: General, 1934–48, 215/909/II, TNA.

87. Iona Mayer demonstrates that four generations is "the usual maximum limit for exact genealogical tracing in many African kinship systems" because this is the "natural limit of historical record in a preliterate culture." Mayer, "From Kinship to Common Descent: Four Generation Genealogies among the Gusii," *Africa* 35, no. 4 (October 1965): 366. In the naming system today a person uses his personal name, followed by his father's name, followed by that grandfather's name.

88. Evans-Pritchard, *Nuer,* 366. Based on similar observations in Gusii, Iona Mayer argues that lineage and clan, or what she calls "kin lineage" and "ancestral lineage," should not be merged in one model of kinship. Instead, the anthropologist can properly interpret the kin lineage only as a kinship group. Mayer, "Kinship to Common Descent," 383.

89. Lan, *Guns and Rain,* 23.

90. For the classic anthropological arguments about territory and kinship, see Sir Henry Sumner Maine, *Ancient Law: Its Connection with the Early History of Society and Its Relation to Modern Ideas* (London: James Murray, 1861); Lewis Henry Morgan, *Ancient Society* (New York: World Publishing, 1877). These ideas were picked up by later anthropologists as the foundation of much ethnography in Africa; see A. R. Radcliffe-Brown, *The Study of Kinship Systems: Structure and Function in Primitive Society* (New York: Free Press, 1965), 49–51; Meyer Fortes, *Kinship and the Social Order* (London: Routledge and Kegan Paul, 1969).

91. M. J. Ruel, "Kuria Generation Classes," *Africa* 32, no. 1 (1962): 14–36. On provinces, see Bischofberger, *Generation Classes,* 15–16. In Zanaki the clan "provinces" are called ekyaro while the "subclans," which are not territorial and move between provinces, are called hamati. Clans are not exogamous.

92. E. B. Dobson, "Comparative Land Tenure of Ten Tanganyika Tribes," *Tanganyika Notes and Records* 38 (1955): 31–39.

93. Ruel, "Kuria Generation Classes," 14–36; Hans Cory, "Report on the Pre-European Tribal Organization in Musoma (South Mara District) and . . . Proposals for Adaptation of the Clan System to Modern Circumstances," 1945, 1–14, CORY 173, EAF.

94. Tobisson, *Family Dynamics*, 97.

95. Interview with Nyamaganda Magoto, Bugerera, 4 October 1995.

96. Interview with Jackson Mteba Mabura, Morotonga, 23 August 2003. Other information supplied by interviews with Mayani Magoto, Bugerera, 4 April 1995, and Nyawagamba Magoto, Bugerera, 27 February 1995, and 6 October 1995; Mahewa Timanyi and Nyambureti Morumbe, Robanda, 27 May 1995; Nyambeho Marangini, Issenye, 7 September 1995; Morigo (Mchombocho) Nyarobi, Issenye, 28 October 1995.

97. Interview with Nyamaganda Magoto, Bugerera, 4 October 1995.

98. There are various instances of this—for example, interviews with Mwamedi Hassan, Bugerera, 3 May 1995; Jackson (Benedicto) Mang'oha Maginga, Mbiso, 18 March 1995; Tetere Tumbo, Mbiso, 5 April 1995; Mwenge Elizabeth Magoto, Mbiso, 6 May 1995; Isaya Charo Wambura, Buchanchari, 22 September 1995.

99. Interview with Jackson Mteba Mabura, Morotonga, 23 August 2003.

100. Interview with Nyawagamba Magoto, Bugerera, 6 October 1995.

101. I asked each informant I interviewed to tell their own life history. This information generalizes from many of those histories and from specific conversations particularly with the Magoto family on strangers. Informal discussions in Ikoma on stranger wives confirmed these ideas.

102. Interviews with Mzee Mswaga, Bugerera, 29 March 1995; Nyawagamba Magoto, Bugerera, 6 October 1995.

103. Interview with Nyawagamba Magoto, Bugerera, 27 February 1995.

104. Also discussed by Huber, *Marriage and Family*, 95–96; Machota Sabuni, Issenye, 14 March 1996.

105. Interviews with Nyawagamba Magoto, Bugerera, 3 September, 2 October, 4 October 1995; Machota Sabuni, Issenye, 14 March 1996.

106. Interview with Jackson Mteba Mabura, Morotonga, 23 August 2003.

107. Many told stories of selling children during famines, including Mahewa Timanyi and Nyambureti Morumbe, Robanda, 27 May 1995; Elfaresti Wambura Nyetonge, Kemegesi, 20 September 1995.

108. Interviews with Mohere Mogoye, Bugerera, 25 March 1995; Yohana Kitena Nyitanga, Makundusi, 1 May 1995.

109. Interview with Nyanchiwa Mesika, Morotonga, 27 August 2003.

110. Interview with Bhoke Wambura, Maburi, 7 October 1995.

111. Maryknoll Fathers, Iramba Parish, "Ngoreme-English Dictionary," n.d., author's collection; interviews with Zabron Kisubundo Nyamamera and Makang'a Magigi, Bisarye, 9 November 1995; Sarya Nyamuhandi and Makanda Magige, Bu-

mangi, 10 November 1995; Daudi Katama Maseme and Samueli Buguna Katama, Bwai, 11 November 1995; Elfaresti Wambura Nyetonge, Kemegesi, 20 September 1995; Bhoke Wambura and Atanasi Kebure Wambura, Maburi, 7 October 1995. See ch. 4 for more on slavery.

112. Interview with Mahiti Kwiro, Mchang'oro, 19 January 1996.

113. Chief Secretary to the Government, DSM, Native Land Tenure and Land Rights, 17 July 1934, Compilation of Economic Maps, 1934–36, Secretariat Files 23275, TNA; Shetler, *Telling Our Own Stories*, 254–56.

114. This is not a reference to the present-day town of the same name in North Mara.

115. Samweli Kiramanzera and Kihenda Manyorio, Kurasanda, 3 August 1995.

116. See a similar analysis in Jan Bender Shetler, "'A Gift for Generations to Come': A Kiroba Popular History from Tanzania and Identity as Social Capital in the 1980s," *International Journal of African Historical Studies* 28, no. 1 (1995): 69–112.

117. David William Cohen, "Doing Social History from Pim's Doorway," in *The African Past Speaks: Essays on Oral Tradition and History*, ed. Joseph C. Miller (Folkestone, Eng.: Dawson, 1980), 196; Cohen, "Reconstructing a Conflict in Binafu: Seeking Evidence outside the Narrative Tradition," in Cohen, *Womunafu's Bunafu*, 48–67.

118. Carole A. Buchanan, "Perceptions of Ethnic Interaction in the East African Interior: The Kitara Complex," *International Journal of African Historical Studies* 11, no. 3 (1978): 425–26.

119. Interviews with Nyamaganda Magoto, Mbiso, 3 August 2003; Nyamuko Soka, Morotonga, 28 August 2003; Nyanchiwa Mesika, Morotonga, 27 August 2003.

120. Kopytoff, "Internal African Frontier," 71–75.

121. Mtemi Seni Ngokolo, "Historia ya Utawala wa Nchi ya Kanadi ilivyo andikwa na Marahemu Mtemi Seni Ngokolo mnamo tarehe 10/6/1928," provided by his son, Mtemi Mgema Seni, to Buluda Itandala, 20 May 1971, author's collection. Thanks to Dr. Itandala for his help on Sukuma traditions about the Mara region.

122. Guyer and Belinga, "Wealth in People," 91–120.

123. David Schoenbrun, *The Historical Reconstruction of Great Lakes Bantu Cultural Vocabulary: Etymologies and Distributions* (Cologne: Rüdiger Köppe Verlag, 1996), entries 287, 288.

124. Interviews with Megasa Mokiri, Motokeri, 4 March 1995; Mayani Magoto, Bugerera, 5 April 1996.

125. Paul Asaka Abuso, *A Traditional History of the Abakuria, c. A.D. 1400–1914* (Nairobi: Kenya Literature Bureau, 1980), 83–86, 143.

126. E. C. Baker, "Notes on Tribes," *Tanganyika Papers*, 13–14.

127. Zedekia Oloo Siso, Oral Traditions of North Mara, Buturi, Tanzania, 1965–90, unpublished accounts, author's collection.

128. David William Cohen, ed., *Towards a Reconstructed Past: Historical Texts from Busoga* (Oxford: Oxford University Press, 1986); Wrigley, *Kingship and State;*

Apolo Kaggwa, *The Kings of Buganda* [Basekabaka be Buganda], trans. M. S. M. Ki-wanuka (Nairobi: East African Publishing House, 1971); Hartwig, *Art of Survival*, 40. Hartwig shows that many of the wild animal avoidances in Ukerewe traced their origins to Bunyoro. For the influence of these kingdoms on the other side of the lake in the nineteenth century, see David W. Cohen, "Food Production and Food Exchange in the Precolonial Lakes Plateau Region," in *Imperialism, Colonialism and Hunger: East and Central Africa*, ed. Robert I. Rotberg (Lexington, MA: D. C. Heath and Co., 1983), 1–18.

129. Otto Bischofberger reports that blacksmiths from Zanaki went to Uzinza to get iron heart-shaped hoes. Bischofberger, *Generation Classes*, 51. Iron smelting tra-dition is also established in Luo areas to the north. The Turi ethnic group, with its own territory called Buturi, now exists in North Mara; it adopted Luo speech and customs within the last two generations. Zedekia Siso reports that the people of Bu-turi in North Mara used to smelt iron. Siso, Oral Traditions of North Mara.

130. The word *mwiro/bwiro* comes from the Great Lakes Bantu root *mwiru*, meaning farmer—in distinction to the *batúá*, or original hunter-gatherers. Schoen-brun, *Green Place*, 157; Schoenbrun, *Historical Reconstruction*, 196, 331.

131. Interviews with Sarya Nyamuhandi and Makanda Magige, Bumangi, 10 No-vember 1995; Kinanda Sigara, Bugerera, 27 May 1995; Isaya Charo Wambura, Buchanchari, 22 September 1995; Apolinari Maro Makore, Megasa, 29 September 1996; Bhoke Wambura and Atanasi Kebure Wamburi, Mburi, 7 October 1995; Bischofberger, *Generation Classes*, 51, describes the avoidance of Turi by Bwiro in Zanaki; Gray, *Sonjo of Tanganyika*, on Sonjo Turi blacksmiths, 78.

132. Interviews with Riyang'ang'ara Nyang'urara, Sarawe, 20 July 1995; Silas King'are Magori, Kemegesi, 21 September 1995; Makuru Moturi, Maji Moto, 29 September, 1995; Bhoke Wambura, Maburi, 7 October 1995; Sarya Nyamuhandi and Makanda Magige, Bumangi, 10 November 1995.

133. Interviews with Nyamaganda Magoto, Mbiso, 3 August 2003; Wambura Tonte, Kemegesi, 12 August 2003; George Wambura Gehamba and Samweli Muya Mongita, Morotonga, 21 August 2003; Jackson Mteba Mabura, Morotonga, 23 Au-gust 2003; Nyanchiwa Mesika, Morotonga, 27 August 2003; Nyamuko Soka, Moro-tonga, 28 August 2003.

134. Interview with Nyamaganda Magoto, Mbiso, 3 August 2003.

135. Interview with Peter Mgosi Siwa, Morotonga, 23 August 2003.

136. Interviews with Mariko Romara Kisigiro, Burunga, 31 March 1995; Mahewa Timanyi and Nyambureti Morumbe, Robanda, 27 May 1995. For more detail on these exchanges, see ch. 4.

137. For a similar trading system described for western Kenya, see Margaret Jean Hay, "Local Trade and Ethnicity In Western Kenya," *Economic History Review* 2, no. 1 (1975): 7.

138. Nyamaganda Magoto, cultural vocabulary list, Bugerera, 1995.

139. Interviews with Maro Mchari Maricha, Maji Moto, 28 September 1995; Mashauri Ng'ana, Issenye, 2 November 1995.

140. In distinction to the oathing ceremonies of friendship described for the Ishenyi and the Nata, this ritual (*kubarisi aring'a, gutarana*) seals the relationship between individuals and affects only their immediate families. In the ritual, an elder cuts the fingers of each party, puts the blood of each onto a bit of porridge, and feeds it to each party. The two are then lifelong friends (*omwisani wa sarago, omwisani bo maguta*). A friend cannot kill, harm, or steal from his friend; he cannot take his friend's wife or betray his interests in any arrangements he makes; and he must always look out for his friend's interests. Nyamaganda Magoto, Bugerera, 4 October 1995. These friendships were also often used as the basis for cattle trusteeship. Efaristi Bosongo Gikaro, Masinki, 30 September 1995. If this trust was broken, the blood brothers must perform a ritual of purification such as that described above for murder.

141. Schoenbrun, *Historical Reconstruction*, 162, 164.

142. Schoenbrun, *Green Place*, 187.

143. E. C. Baker, "Rain" (typescript), *Tanganyika Papers*, 4. Baker found this to be the case in Ikizu, a number of Jita clans, Zanaki, and Ukerewe.

144. Provincial Administration, Monthly Reports, Musoma District, 3 May 1929, 1926–29, 215/PC/1/7, TNA. "Two of the age grades of the Wanata indulged in a fight, the origin of which is at present obscure, but had something to do with rain and rainmaking. Only sticks were used and so the damage done was slight."

145. Interviews with Maro Mugendi and Maria Maseke, Busawe, 22 September 1995; Njaga Nyasama, Kemegesi, 14 September 1995; E. C. Baker, Monthly report for February 1928, 10 March 1928, District Officer to Provincial Commissioner, 1926–29, Provincial Administration, Monthly Reports, Musoma District, 215/PC/1/7, TNA.

146. E. C. Baker, "Rain," 4.

147. Interviews with Mashauri Ng'ana, Issenye, 2 November 1995, and Sarya Nyamuhandi and Makanda Magige, Bumangi, 10 November 1995, dealt with these topics in particular.

148. Interviews with Stephano Makondo Karamanga, Robanda, 22 August 2003; Jackson Mteba Mabura, Morotonga, 23 August 2003; Mabenga Nyahega and Machaba Nyahega, Mbiso, 1 September 1995.

CHAPTER 3: SACRED LANDSCAPES

1. For discussion on sacred landscapes in oral tradition, see Greene, *Sacred Sites*; Parkin, *Sacred Void*; Thornton, *Space, Time*; Wagner, "Environment, Community." For an economic analysis of nodes as places of power, see Howard, "Nodes, Networks."

2. Heike Schmidt argues that constructing the forest as a spiritual landscape in Zimbabwe served to protect environmental resources. Schmidt, "'Penetrating' Foreign Lands: Contestations over African Landscapes: A Case Study from Eastern Zimbabwe," *Environment and History* 1, no. 3 (1995): 360–61.

3. Interview with Megasa Mokiri, Motokeri, 4 March 1995.

4. In much the same way, Terence Ranger describes pilgrimage places in Zimbabwe. Ranger, "Taking Hold of the Land: Holy Places and Pilgrimages in Twentieth-Century Zimbabwe," *Past and Present* 117 (November 1987): 158–94. See also Hofmeyer, *We Spend Our Years.*

5. This shift in the reckoning of time from generation-set to age-set is argued more fully in ch. 4. For telescoping and the middle time period, see Vansina, *Oral Tradition*, 23–24; Spear, *Kenya's Past*, 46–47.

6. Schoenbrun demonstrates the connection between the power of speech and healing or divination. Schoenbrun, *Green Place*, 199–200.

7. Interview with Matias Mahiti Kebumbeko, Torogoro, 2 April 1996.

8. This information about spirits comes from innumerable discussions but in particular interviews with Kinanda Sigara, Nyawagamba Magoto (who were my assistants), Chamuriho, and Mahiti Kwiro, Mchang'oro, 19 January 1996.

9. For outline of different kinds of spirits among the Kuria, see Ruel, "Religion and Society," 295–306.

10. Schoenbrun, *Historical Reconstruction*, 347. See the discussion of spirits in Schoenbrun, *Green Place*, chs. 5, 6. This substantive meaning was itself derived from a verb meaning to judge or, possibly, to grant blessing, found more widely in Savannah Bantu. See Schoenbrun, *Green Place*, 199–201.

11. Michael Kirwen, *The Missionary and the Diviner: Contending Theologies of Christian and African Religions* (Maryknoll, NY: Orbis Books, 1987), 6; Ruel, "Religion and Society," 296.

12. Schoenbrun, *Historical Reconstruction*, 347.

13. Interview with Maro Mchari Maricha, Maji Moto, 28 September 1995.

14. Interviews with Stephen Gojat Gishageta and Girimanda Mwarhisha Gishageta, Issenye, 28 March 1996; Merekwa Masunga and Giruchani Masanja, Mariwanda, 6 July 1995; Gilumughera Gwiyeya, Girihoida Masaona, and Gorobani Gesura, Issenye, 28 July 1995; Wambura Nyikisokoro, Sang'anga Buchanchari, 23 September 1995; Marunde Godi, Juana Masanja, Mayera Magondora, Manawa, 24 February 1996.

15. The Isimajega elders requested that I ask the park for permission for them to return there to propitiate the erisambwa. Tatoga ethnographies report large funeral mounds built for important elders in their cattle kraals. Poles that were planted in the mound grew into trees and a "sacred grove" was established, which would be visited by the man's ancestors for propitiation of his spirit. George J. Klima, *The Barabaig: East African Cattle-Herders* (New York: Holt, Rinehart and Winston, 1970), 102–7. Klima reports that only rarely are these mounds built for women (105). See also Wilson, "Tatoga of Tanganyika, Part One," 34–47.

16. Interview with Yohana Kitena Nyitanga, Makundusi, 1 May 1995.

17. The word *nyichoka* means "the place of a snake."

18. A similar prophetic tradition is described for the Cwezi spirit possession cults in Uganda by Renee Louis Tantala, "The Early History of Kitara in Western Uganda: Process Models of Religion and Political Change" (PhD diss., University of Wis-

consin, Madison, 1989); John Beattie, "Group Aspects of the Nyoro Spirit Mediumship Cult," *Rhodes-Livingston Institute Journal* (1961): 11–35; and more generally for ngoma in John Janzen, *Ngoma: Discourses of Healing in Central and Southern Africa* (Berkeley: University of California Press, 1992).

19. While most informants would not date the Machaba story, one elder said that the Ishenyi were at Nyigoti, which would put it during the period of late nineteenth-century disasters. Mang'ombe Morimi, Issenye/Iharara, 26 August 1995. Others dated it to the time when the Ishenyi were still at Nyeberekera, just before the disasters. Interviews with Morigo (Mchombocho) Nyarobi, Issenye, 28 October 1995; Machota Sabuni, Issenye, 14 March 1996. Tatoga informants dated it to the time of the prophet Saigilo's father, which would also date it to the mid-nineteenth century, just before the disasters (ca. 1850–1870). The fact that the Ikoma went to the Tatoga prophet because of infertility problems would suggest that the disasters had already begun.

20. Interview with Machota Sabuni, Issenye, 14 March 1996.

21. Interview with Bokima Giringayi, Mbiso, 26 October 1995.

22. Game Preservation Department, Annual Report, 1936, Tanganyika Territory (Dar es Salaam: Government Printer, 1937), 9–10. The report was made because the Game Department confiscated one of the tusks, weighing twenty-seven kilograms, which was later recovered by the Ikoma after they came and made their case that the Ikoma would "cease to exist" if the Machaba were taken away. The delegation sent to free the Machaba had a piece of the second tusk and part of a buffalo horn in hand to make their case. The Ikoma elders collected twenty-two rhinoceros horns and fifty-six tusks in return for getting the Machaba back, and the game officer was allowed to see the Machaba ceremony in which the people were anointed with butter smeared on the tusks.

23. Interview with Mabenga Nyahega and Machaba Nyahega, Mbiso, 1 September 1995.

24. Interview with Machota Sabuni, Issenye, 14 March 1996. Sabuni was an important informant on all aspects of the Machaba.

25. Interviews with Mabenga Nyahega, Bugerera, 5 September 1995; Moremi Mwikicho, Sagochi Nyekipegete, Kenyatta Mosoka, Robanda, 12 July 1995.

26. For this process in other parts of the Great Lakes region, see Schoenbrun, *Green Place*, ch. 5.

27. They finally agreed that I could take a picture, but, on a roll of otherwise good pictures, that photo did not turn out. It was the last picture on the roll, and when I went to change film I could not find the roll of unexposed film that I was sure I put in my bag that morning.

28. Interview with Keneti Mahembora, Gitaraga, 9 February 1996.

29. Interview with Makuru Magambo, Geteku, 8 March 1996.

30. Interview with Sochoro Kabati, Nyichoka, 2 June 1995.

31. Feierman, *Peasant Intellectuals*; Packard, *Chiefship and Cosmology*.

32. G. Chouin, "Sacred Groves in History: Pathways to the Social Shaping of Forest Landscapes in Coastal Ghana," Institute for Development Studies, *IDS*

Bulletin 33, no. 1 (1 January 2002): 39–46; Lewis, *Inventing Global Ecology*, 138–40, 224–28.

33. The most important informants on the Nata emisambwa rituals were Gabuso Shoka, Mbiso, 30 May 1995; Makuru Nyang'aka and Keneti Mahembora, Gitaraga and Nyichoka, 9 February 1996; Makuru Magambo, Geteku, 8 March 1996.

34. Interview with Reterenge Nyigena, Maji Moto, 23 September 1995.

35. Interview with Sochoro Kabati, Nyichoka, 2 June 1995.

36. For an example elsewhere in Tanzania of ritual precedence given to "first comers" in ceremonies concerning the land, see H. A. Fosbrooke, "A Rangi Circumcision Ceremony: Blessing a New Grove," *Tanganyika Notes and Records* 50 (1958): 30–38.

37. Mturi, "Historia ya Ikizu"; reprinted in Shetler, *Telling Our Own Stories*, 25–27.

38. See Peter Schmidt's analysis of the "tree of iron" in *Historical Archaeology: A Structural Approach in an African Culture* (Westport, CT: Greenwood Press, 1978).

39. Abuso, *Traditional History*, 35; Ogot, *Migration and Settlement*, 38–39.

40. For a discussion of "firstness" as a principle of legitimacy, see Schoenbrun, *Green Place*, chs. 4, 5; Kopytoff, "Internal African Frontier."

41. Mturi, "Historia ya Ikizu"; in Shetler, *Telling Our Own Stories*, 45–49.

42. The literature on ritual is vast and sophisticated. For example, see Maurice Bloch, *From Blessing to Violence: History and Ideology in the Circumcision Ritual of the Merina of Madagascar* (Cambridge: Cambridge University Press, 1986); Newbury, *Kings and Clans*; Emile Durkheim, *Elementary Forms of Religious Life* (London: Allen and Unwin, 1954); Victor Turner, *The Forest of Symbols: Aspects of Ndembu Ritual* (Ithaca, NY: Cornell University Press, 1967); Max Gluckman, *Rituals of Rebellion in South-East Africa* (Manchester: Manchester University Press, 1954).

43. Connerton, *How Societies Remember*, 71.

44. Ibid., 41–71, esp. 70. See also Renee Tantala, "Verbal and Visual Imagery in Kitara (Western Uganda): Interpreting 'The Story of Isimbwa and Nyinamwiru,'" in *Paths toward the Past: African Historical Essays in Honor of Jan Vansina*, ed. Robert Harms (Atlanta: African Studies Association Press, 1994), 223–43.

45. Yet informants often seemed confused about how each of these groups functioned and in what contexts. Both seemed to be described in similar ways with similar functions. In addition, the exact way that age and generation organization interacted in each of the small ethnic groups of Nata, Ikoma, Ngoreme, and Ishenyi varied. Often the great variety seemed random and inconsistent, indicating that perhaps because these structures had ceased to be functional and that knowledge about them was no longer based on experience.

46. Some anthropologists use the term *class* interchangeably with *set*.

47. P. T. W. Baxter and Uri Almagor, introduction to *Age, Generation, and Time: Some Features of East African Age Organization* (New York: St. Martin's, 1978), 1–2.

48. R. G. Abrahams, "Aspects of Labwor age and generation grouping and related systems," in *Age, Generation, and Time: Some Features of East African Age Organizations*, ed. P. T. Baxter and Uri Almagor (New York: St. Martin's, 1978), 37–38.

49. Interview with Mwenge Elizabeth Magoto, Mbiso, 6 May 1995.

50. Ehret, *Southern Nilotic History*, 45.

51. The Zanaki and Ikizu practice a somewhat hybrid form. See Bischofberger, *Generation Classes*; Malcolm Ruel, "Kuria Generation Sets," Makerere Institute of Social Research Conference, Kampala, 1954–58; Ruel, "Kuria Generation Classes," 14–36.

52. "Mara, Luhyia, Forest and Rwenzori branches of Great Lake Bantu have maintained these institutions." Schoenbrun, *Good Place*, 176–77.

53. In the Suguti language of Kwaya, *-kukura*, but not in Kuria. Kisigiro, "Kamusi ndogo."

54. However, it is conceivable that some form of saiga may have developed any time after 1000 CE in relationship with Tatoga pastoralists.

55. Ehret, *Southern NiloticHistory*, 55–62; Shetler, "Landscapes of Memory," ch. 4.

56. LeVine and LeVine, "House Design," 162.

57. Ruel makes this point in "Kuria Generation Classes," 32.

58. Eisei Kurimoto and Simon Simonse discuss that this aspect of age systems, the importance of ritual over military functions, goes back to Evans-Pritchard's assertions about the Nuer age system in 1936. Introduction to *Conflict, Age and Power in North East Africa: Age Systems in Transition*, ed. Eisei Kurimoto and Simon Simonse (Athens: Ohio University Press, 1998): 1–29. The authors hope to renew interest in the political aspects of age-systems.

59. Baxter and Almagor, introduction, 1–3. Tanganyika government anthropologist Hans Cory wrote that the "supreme power" in Ikoma was the "warrior age-grade," which was "divided into three military units." After they retired from warriorhood, the elders constituted a "chama" under the leadership of a "mukina," who "managed the civil affairs of the respective military units only." Although Cory admitted to his lack of information on the political functioning of the "age-grade system in pre-European times," he assumed that it was "best adapted to meet military emergencies" and was thus now "obsolete." Cory, "Pre-European Tribal Organization," EAF. See also Bischofberger, *Generation Classes*, 11–16, 100–101; Kemal Mustafa, "The Concept of Authority and the Study of African Colonial History," *Kenya Historical Review Journal* 3, no. 1 (1975): 55–83. Ruel saw generation-sets among the Kuria as "embodying a systematic pattern of relationships which serves to determine the status of any individual person *vis-à-vis* others." He preferred to see generation-sets as "social categories" rather than "corporate groups." Ruel, "Kuria Generation Classes," 17, 33. Here he refers mainly to the relational norms between generations: those in adjacent classes have a relationship of respect and deference; those in alternate classes or the same class, one of joking and intimacy. The neglect

of this topic is due in large part to the reluctance of Mara peoples to talk about the nature of these secret rituals.

60. Bischofberger, *Generation Classes*, 65–67.

61. Connerton, *How Societies Remember*, 66.

62. Interview with Kirigiti Ng'orogita, Mbiso, 8 June 1995.

63. Thornton, *Space, Time*, 19.

64. David L. Schoenbrun, pers. comm. For an example of the importance of walking the land to heal it in Australia, see Elizabeth A. Povinelli, *Labor's Lot: The Power, History, and Culture of Aboriginal Action* (Chicago: University of Chicago Press, 1993).

65. Interview with Ikota Mwisagija and Kiyarata Mzumari, Kihumbo, 5 July 1995; Mturi and Sasora, "Historia ya Ikizu"; in Shetler, *Telling Our Own Stories*, 82–83.

66. Interview with Mashauri Ng'ana, Issenye, 2 November 1995. Each particular kind of tree used in rituals has a symbolic significance.

67. The main interviews on the Ishenyi *kerera* (*kukerera*) were with Mang'ombe Morimi, Issenye/Iharara, 26 August 1995; Rugayonga Nyamohega, Mugeta, 27 October 1995; Morigo (Mchombocho) Nyarobi, Issenye, 28 September 1995; Mashauri Ng'ana, Issenye, 2 November 1995.

68. Interview with Rugayonga Nyamohega, Mugeta, 27 October 1995. Two children who wet their beds accompanied the two leaders and actually did the work of spreading the medicine.

69. The most important interviews for information on Nata saiga ritual were: Mang'oha Machunchuriani (who had the title of *mwekundu*, elder who makes preparations for the saiga ceremony), Mbiso, 24 March 1995; Sochoro Kabati (who had the title of *kang'ati* of the Gikwe for the Bongirate Saiga), Nyichoka, 2 June 1995; Kirigiti Ng'orogita (who had the title of *rikora mchama*), Mbiso, 8 June 1995; Mang'oha Morigo (who had the title of *kang'ati* of the Moriho for the Bongirate Saiga), Bugerera, 24 June 1995.

70. Interviews with Sochoro Kabati, Nyichoka, 2 June 1995; Mang'oha Machunchuriani, Mbiso, 24 March 1995.

71. Interview with Machota Sabuni, Issenye, 14 March 1996. The last time the Machaba appeared in Ikoma ritual was in 1994.

72. Interview with Stephen Gojat Gishageta and Girimanda Mwarhisha Gishageta, Issenye, 28 March 1996.

73. Interview with Machota Sabuni, Issenye, 14 March 1996. The Ikoma were some of the most reluctant to talk about the rikora; they kept secret from outsiders much of this information.

74. Interview with Mwita Maro, Maji Moto, 29 September 1995.

75. The oryx has a thin, stright horn; that of the kudu is thicker and twisted.

76. The most important Ngoreme interviews on these rituals were, Nsaho Maro, Kenyana, 14 September 1995; Mwita Maro, Maji Moto, 29 September 1995 who kept the horn for the Iregi of Bumare and was the Maina *mtangi* (firstborn), having obtained the horn in 1957 he was well overdue to pass it on to the next generation-

set, the Saai. Ritual horns were common symbols of generational authority throughout the region. Each Ishenyi clan also had a horn (*enchobe*) that was passed on as
the older generation retired. Interview with Rugayonga Nyamohega, Mugeta, 27
October 1995.

77. Interview with Paulo Maitari Nyigana and Ibrahim Mutatiro Kemuhe, Maji
Moto, 29 September 1995.

78. Ruel, "Kuria Generation Classes," 21–28.

79. Benjamin Mkirya, *Historia, mila na desturi ya Wazanaki* (Peramiho, Tanzania: Benedictine Publications, Ndanda Mission Press, 1991) 39–43, 59–62; interview
with Zabron Kisubundo Nyamamera, Bisarye, 9 November 1995.

80. Interviews with Stephen Gojat Gishageta and Girimanda Mwarhisha Gishageta, Issenye, 27 July 1995; Marunde Godi, Juana Masanja, Mayera Magondora,
Manawa, 24 February 1996.

81. Thornton, *Space, Time*, 19.

82. Ibid.

83. Christopher Gray, *Modernization and Cultural Identity: The Creation of
National Space in Rural France and Colonial Space in Rural Gabon*, Occasional
Paper no. 21 (Bloomington: Indiana Center on Global Change and World Peace,
February 1994), 37–38.

84. This debate is discussed in Thornton, *Space, Time*, 8–16.

85. Interviews with Jackson (Benedicto) Mang'oha Maginga, Mbiso, 18 March
1995; Mahiti Kwiro, Mchang'oro, 19 January 1996; Pastor Wilson Shanyangi Machota, Morotonga, 12 July 1995.

86. See Hans Cory, "Land Tenure in Bukuria," *Tanganyika Notes and Records* 23
(1947): 70–79; Prazak, "Cultural Expressions," 97.

87. Characterized by John Ford as the *Grenzewilderness*. Ford, *Role of Trypanosomiases*.

88. The original dimensions of the tsetse fly belt were thirty to thirty-five miles
wide. H. J. Lowe, Senior Veterinary Officer, Northern Province, to Provincial Commissioner in Arusha, 6 March 1931, Land and Mines, Chiefdoms' Boundary Dispute Files, North Mara District, 83, 3/127, TNA; J. F. Corson, Medical Officer,
Ikoma, 15 April 1927, Third Note on Sleeping Sickness, Extracts of Report by District Veterinary Officer, 1926–29, Provincial Administration, Monthly Reports, Musoma District, 215/PC/1/7, TNA.

89. Schoenbrun, *Green Place*, 126–28.

90. Grant, District Officer, Report on Human Habitation of the SNP, TNA.
Maasai testifying for Serengeti National Park Committee of Enquiry in 1957 claimed
that "Moru was occupied by the Maasai of the Serenget section at least from the
time of the circumcision of Il Merishari (1806) until the great rinderpest epidemic."
Memorandum on the Serengeti National Park, signed on behalf of the Masai of the
National Park by Oltimbau ole Masiaya, 1957, MSS. Afr. s. 1237, RHL. Colonial officers seeking to reestablish the boundary between Lake and Eastern provinces suggested the ecological or "natural dividing line on the Serengeti between the open

grass steppe and the Savanna bush and line of hills bordering the plain to the west." German maps B4 Eyasi and A4 Ikoma. The explanation could also simply be that the Germans had Maasai or Ndorobo guides. H. St. J. Grant, District Commissioner, Masai Monduli, to Provincial Commissioner, Northern Province, 29 September 1954, Western Serengeti Plains, National Game Parks, 215/350/III, TNA.

91. Klaus Gerresheim, *The Serengeti Landscape Classification*, Serengeti Research Institute Publication no. 165, Serengeti Monitoring Programme, African Wildlife Leadership Foundation, 1974, 2.

92. Interviews with Joseph Sillery Magoto, Mbiso, 2 August 2003; Wambura Edward Kora, Morotonga, 19 August 2003.

93. For the literature on identity formation in opposition to others, see Fredrik Barth, *Ethnic Groups and Boundaries: The Social Organization of Culture Difference* (Bergen: Universitetsforlaget; London: George Allen and Unwin, 1969); Edward Said, *Orientalism* (New York: Vintage Books, 1994); Anderson, *Imagined Communities.*

94. Mang'oha Morigo says the Ndorobo walk around like wild animals following the herds. Interview, Bugerera, 24 June 1995. Kiyarata Mzumari says that the plains were a fearsome place because of the danger of Maasai, lions, and buffalo, interview, Mariwanda, 8 July 1995.

95. Interviews with Zamberi Masambwe, Gisuge Chabwasi, Sanzate, 22 June 1995; Zabron Kisubundo Nyamamera and Makang'a Magigi, Bisarye, 9 November 1995; Criminal Case no. 92 of 1931, Villagers of Watende living in Bunjari, 9 September 1931, Native Affairs, Collective Punishments and Prosecution of Chiefs, 1926–31, 246/PC/3/4, TNA.

96. R. E. S. Tanner, "Cattle Theft in Musoma, 1958–59," *Tanzania Notes and Records* 65 (1966): 31–32.

97. Robert David Sack, *Human Territoriality: Its Theory and History* (Cambridge: Cambridge University Press, 1986), 19.

98. Maurice Bloch, *Blessing to Violence.* See also Feierman's analysis of the concepts of "healing the land" and "harming the land," reinterpreted in changing historical contexts of Shambaa in Tanzania. Feierman, *Peasant Intellectuals.*

99. Ruel, "Kuria Generation Classes," 15. Ruel estimated that the Kuria ebyaro of the past consisted of between three thousand and ten thousand people, which is about the size of all of Nata or Ikoma or Ishenyi today.

100. Interviews with Sochoro Kabati, Nyichoka, 2 June 1995; Kirigiti Ng'orogita, Mbiso, 8 June 1995.

101. Ruel, "Kuria Generation Classes," 33.

102. Interview with Sochoro Kabati, Nyichoka, 2 June 1995.

103. Interview with Nsaho Maro, Kenyana, 14 September 1995.

104. Thornton, *Space, Time,* 88–97.

105. Maryknoll Fathers, "Ngoreme-English Dictionary," author's collection.

106. For the importance and interpretation of folk models of procreation, see Poewe, *Matrilineal Ideology,* 4–7; Anita Jacobson-Widding, ed., *The Creative Com-*

munion: African Folk Models of Fertility and the Regeneration of Life (Uppsala: Almqvist and Wiksell International, 1990).

107. Interview with Nyamaganda Magoto, cultural vocabulary list, Mbiso and Bugerera, 1995–96.

108. Brad Weiss discusses the Haya use of protection medicines in wedding ceremonies, new houses, as well as death itself, involving "binding rites" to ensure protection and peace. Weiss, *Making and Unmaking*, 39.

109. Schoenbrun, *Historical Reconstruction*, no. 335, pers. comm.

110. B. Weiss, *Making and Unmaking*, 47.

111. Malcolm Ruel, "Piercing: Ritual Actions of the Kuria, East Africa" June 1958, Makerere Institute of Social Research Conference Papers (1954–58). For an in-depth analysis of "blockage and flow" in Rwanda, see Taylor, *Milk, Honey*, 11–12.

112. Brad Weiss describes the Haya ceremony for blessing a new house, which involves lighting the fire for the first time by the father or a senior agnate. Weiss, *Making and Unmaking*, 29–31, 51–52. These fires also have medicines to protect the house made from specific trees.

113. Interview with Makuru Nyang'aka, Sochoro Kabati, Barichera Machage Barichera, Riyara, 7 March 1996.

114. Interviews with Mang'ombe Morimi, Issenye/Iharara, 26 August 1995; Mashauri Ng'ana, Issenye, 2 November 1995; Paulo Maitari Nyigana and Ibrahim Mutatiro Kemuhe, Maji Moto, 29 September 1995.

115. Interview with Rugayonga Nyamohega, Mugeta, 27 October 1995. Each homestead extinguishes its fire and a new one is ceremonially lit with medicines and distributed. Elders ritually slaughter a goat and sprinkle its stomach contents over the people (komusa), then they light the new fire and distribute it to each homestead.

116. Interview with Mashauri Ng'ana, Issenye, 2 November 1995.

117. Interview with Nyambeho Marangini, Issenye, 7 September 1995.

118. B. Weiss, *Making and Unmaking*, 48–50.

119. Interviews with Mashauri Ng'ana, Issenye, 2 November 1995; Jackson (Benedicto) Mang'oha Maginga, Mbiso, 18 March 1995; Mariko Romara Kisigiro, Burunga, 31 March 1995; Sagochi Nyekipegete, Robanda, 12 June 1995; Mabenga Nyahega and Machaba Nyahega, Mbiso, 1 September 1995; Kiyarata Mzumari, Mariwanda, 8 July 1995.

120. Interview with Rugayonga Nyamohega, Mugeta, 27 October 1995. In Kiswahili, *kuchunga nchi*. Among the abachama there were specialists with other names. The Ishenyi *ekereri*, whose job it was to "guard the land," led both the age- and generation-set and had eight abachama from each clan to support him." The Nata chose eight abachama for the generation-set in power and eight abachama for each age-set, four from each clan moiety and a leader (kang'ati, or *omotiro*) from each. Another Ishenyi rikora official was the omusamu who prepared the meat on sticks over the fire for the ceremony. Interviews with Mang'ombe Morimi, Issenye/Iharara, 26 August 1995; Nyambeho Marangini, Issenye, 7 September 1995.

121. Each ritual leader of the Maasai age-set also had to have a "pure" body and come from a "pure" family; he was a "man of peace" and stayed in the rear during a raid. John Lawrence Berntsen, "Pastoralism, Raiding, and Prophets: Maasailand in the Nineteenth Century" (PhD diss., University of Wisconsin, Madison, 1979), 79–81.

122. Although this description comes from interviews with Nata, the qualifications for rikora leadership were the same in all the groups I interviewed, and the long stick a universal symbol of leadership, called the *ekinara* in Ishenyi. Interviews with Mang'oha Machunchuriani, Mbiso, 24 March 1995; Sochoro Kabati, Nyichoka, 2 June 1995; Kirigiti Ng'orogita, Mbiso, 8 June 1995; Mang'ombe Morimi, Issenye Iharara, 26 August 1995; Rugayonga Nyamohega, Mugeta, 27 October 1995. The Maasai delegations to see the prophets went unarmed with a long black stick as the sign of peace. Berntsen, "Pastoralism, Raiding," 202.

123. Interview with Mang'oha Morigo, Bugerera, 24 June 1995.

124. Shetler, "Landscapes of Memory," 9–10, 485.

125. Koponen, *Development for Exploitation*, 649–53; Cory, "Land Tenure in Bukuria," 71; Huber, *Marriage and Family*, 16; Shetler, "Landscapes of Memory," 532.

126. Baxter and Almagor, introduction, 10–19.

127. Interviews with Paulo Maitari Nyigana, Maji Moto, 29 September 1995; Isaya Charo Wambura, Buchanchari, 22 September 1995; Nsaho Maro, Kenyana, 14 September 1995. The abachama of the orokoba passed the protection medicine on behalf of the prophet who provided the medicine (*omogitana*), the abachama of the rain passed the rain medicine (*omusano*) on behalf of the rainmaker (*omogemba*).

128. Kopytoff, "Internal African Frontier," 22. Christopher Gray also argues that district identity was "almost always overridden by clan and lineage identity" and that "territory was defined socially as there was no real ownership tie to the land per se and territoriality was merely one strategy among others to enhance the wealth of one's lineage or clan." Gray, "The Disappearing District? Territorial Transformation in Southern Gabon, 1850–1950," in *The Spatial Factor in African History: The Relationship of the Social, Material, and Perceptual*, ed. Allen M. Howard and Richard M. Shain (Leiden: Brill, 2005), 221–44.

129. Biographical accounts of Nyerere include, Judith Listowel, *The Making of Tanganyika* (London: Chatto and Windus, 1965); Kemal Mustafa, "The Development of Ujamaa in Musoma: A Case Study of Butiama Ujamaa Village" (MA thesis, University of Dar es Salaam, 1975); Mustafa, "Concept of Authority"; William Edgett Smith, *Nyerere of Tanzania* (London: Victor Gollancz, 1973).

130. Interviews with Wilson Shanyangi Machota and Wambura Edward Kora, Robanda, 29 August 2003; Machota Sabuni, Issenye, 14 March 1996; Wambura Edward Kora, Morotonga, 19 August 2003; Nyamuko Soka, Morotonga, 28 August 2003.

CHAPTER 4: THE TIME OF DISASTERS

This chapter uses much of the same material that appeared in Shetler, "Rupture in Oral Memory." For other accounts of social transformation during the period of disasters in Tanzania, see Giblin, *Politics of Environmental Control*; Maddox, "Leave Wagogo."

1. Spear and Waller, *Being Maasai*, 1–2.

2. For kingdoms in Eastern African, see Feierman, *Shambaa Kingdom*; Newbury, *Kings and Clans*; Packard, *Chiefship and Cosmology*; Wrigley, *Kingship and State*.

3. Berntsen, "Pastoralism, Raiding," 112.

4. Rosaldo describes a similar phenomenon in his study of Illongot society in the Philippines. He found that history was divided into two major periods—before and after the Japanese invasion of 1945: "The stories of 1945 were so numerous, so vivid, so detailed, so often told that it took me over a year to realize that they represented but a narrow strip in time." This amplified moment in the historical imagination became "the great divide that separated a bygone past from one that merged into the present." Rosaldo's interpretation of various oral narratives had to take into consideration that this brief period had been generalized to represent the whole period before 1945. Rosaldo, *Ilongot Headhunting*, 38–54. For similar analysis in relation to the Maasai, see Richard Lamprey and Richard Waller, "The Loita-Mara Region in Historical Times: Patterns of Subsistence, Settlement and Ecological Change" in *Early Pastoralists of South-western Kenya*, ed. Peter Robertshaw (Nairobi: British Institute in East Africa, 1990), 19.

5. Interview with Mashauri Ng'ana, Issenye, 2 November 1995. He says this is where people from the clan of Abang'ohe from Bene Okinyonyi went to fish and got swallowed up or disappeared; but all Ishenyi go there to propitiate the spirit. This is not the site where Shang'angi is propitiated.

6. Interview with Mikael Magessa Sarota, Issenye, 25 August 1995. See also Shetler, *Telling Our Own Stories*, 275–76.

7. Interview with Machota Sabuni, Issenye, 14 March 1996.

8. Interviews with Rugayonga Nyamohega, Mugeta, 27 October 1995; Mashauri Ng'ana, Issenye, 2 November 1995; Morigo (Mchombocho) Nyarobi, Issenye, 28 October 1995.

9. Interview with Kirigiti Ng'orogita, Mbiso, 8 June 1995. Kirigiti is the last surviving Nata generation-set leader of his section.

10. Interview with Morigo (Mchombocho) Nyarobi, Issenye, 28 September 1995.

11. For an account of the environmental disasters in Tanzania, see John Iliffe, *A Modern History of Tanganyika* (Cambridge: Cambridge University Press, 1979), ch. 5; Kjekshus, *Ecology Control*. For a critique of the "degradation narrative," see McCann, *Green Land*, introduction; Leach and Mearns, *Lie of the Land*.

12. Interview with Mang'ombe Morimi, Issenye/Iharara, 26 August 1995.

13. Mike Davis, *Late Victorian Holocausts: El Niño Famines and the Making of the Third World* (London: Verso, 2001), 6–7.

14. Sharon Elaine Nicholson, "A Climatic Chronology for Africa: Synthesis of Geological, Historical, and Meteorological Information and Data" (PhD diss., University of Wisconsin, Madison, 1976)," 119–120, 156–57, 227–28.

15. Société des Missionnaires d'Afrique (Pères-Blancs), Visitations Book, Nyegina, Mwanza 1, 1931–32, 67–69, White Fathers Regionals' House, Nyegezi, Mwanza.

16. Société des Missionnaires d'Afrique, "Ukerewe," *Chronique trimestrielle* 27 (1905): 133.

17. Société des Missionnaires d'Afrique, *Rapport annuel*, 1910–11, 383; L. Bourget, Trip Diary, 1904, "Report of a Trip in 1904 from Bukumbi to Mwanza, Kome? Ukerewe, Kibara, Ikoma—Mara Region, together with some stories" (n.p., n.d.), M-SRC54, Sukuma Archives.

18. Société des Missionnaires d'Afrique, "Nyegina," *Rapport annuel*, no. 11 (1915–16): 328–30; "Nyegina," no. 13 (1919–20): 354; "Nyegina," no. 17 (1921–22): 520. For background on the White Fathers Mission, see J. Bouniol, *The White Fathers and Their Missions* (London: Sands and Co., 1929).

19. For more on the Maasai at the lake, see Baumann, *Durch Massailand*, 44–46; the section on the Ngorongoro crater is available in English translation: Baumann, *Ngorongoro's First Visitor, Being an Annotated and Illustrated Translation from Dr. O. Baumann's* Durch Mas[s]ailand zur Nilquelle, trans. G. E. Organ, annotated H. A. Fosbrooke (Kampala: East African Literature Bureau, 1963), 12–14.

20. Interview with Tirani Wankunyi, Issenye, 7 April 1995. Kjerland cites evidence from the Kenya District Books that rihaha was a different cattle disease, one that preceded rinderpest by a considerable period. Kjerland, "Cattle Breed," 134–35. Among my informants *rihaha* was also used to describe the cattle diseases of the colonial period, including rinderpest and East Coast fever. Incidence of these diseases was confirmed in a report by District Veterinary Officer, Musoma, to District Officer, Musoma, Annual Report 1927, January 1928, Mwanza Province 1927–28, Provincial Administration, Annual Report, 246 PC/1/30, TNA. Cattle lung disease said to have been introduced only after 1916 by Maasai stock crossing the Kenya border. F. G. (?) McCall, Chief Veterinary Officer, Annual Report of the Department of Veterinary Science and Animal Husbandry, Tanganyika, 1921, Veterinary Department Annual Report, 1921, 3046/22, TNA; Richard Waller, "The Maasai and the British, 1895–1905: The Origins of an Alliance," *Journal of African History* 17, no. 4 (1976): 530–32; Berntsen, "Pastoralism, Raiding," 83, 276–79; A. H. Jacobs, "The Traditional Political Organization of the Pastoral Maasai" (PhD diss., Oxford University, 1965), 95–99. All show that the rinderpest followed nearly three decades of livestock diseases and was immediately preceded by an epidemic of livestock pleuropneumonia.

21. The epidemiology of syphilis is not well understood. By the 1870s syphilis is assumed to be rapidly rising along trade routes. The problem is that this disease may have been yaws, which presents similar symptoms. The vast majority of childhood complaints were about yaws, not syphilis, while lesions developing in adults after the turn of the century were probably syphilis. Yet with the eight hundred years or more of precolonial contact on the coast with Arabs and three hundred years of contact with Portuguese, it is difficult to say when and where syphilis was introduced. It is improbable that yaws mutated into syphilis, so we can assume that

syphilis was introduced. Anne Stacie Canning Colwell, MD, pers. comm., 5 February 1998.

22. Interview with Morigo (Mchombocho) Nyarobi, Issenye, 28 September 1995. Dating of the Tatoga leaving the crater due to Maasai pressure around midcentury according to Maasai age-set chronology. It is difficult to distinguish stories of infertility from the early colonial years resulting from migrant labor and those resulting from the caravan trade.

23. Interviews with Masosota Igonga, Ring'wani, 6 October 1995; Sira Masiyora, Nyerero, 17 November 1995.

24. Interviews with Mariko Romara Kisigiro, Burunga, 31 March 1995; Marimo Nyamakena and Katani Magori Nyabunga, Sanzate, 10 June 1995. The disease is known locally as *kyamunda* (Nata) or *nyamugwa* (Ikizu); other names are *oborondo*, *egesaho*, etc.

25. Interview with Machota Sabuni, Issenye, 14 March 1996.

26. Iliffe, *Modern History*; Ruth Rempel, "Trade and Transformation: Participation in the Ivory Trade in Late 19th Century East and Central Africa," *Canadian Journal of Development Studies* 19, no. 3 (1998): 529–52; Stephen J. Rockel, "A Nation of Porters: The Nyamwezi and Their Labor Market in Nineteenth-Century Tanzania," *Journal of African History* 41, no. 2 (2000): 173–95.

27. T. Wakefield, "Wakefield's Notes on the Geography of Eastern Africa, Routes of Native Caravans from the Coast . . . ," *Journal of the Royal Geographical Society* 40 (1870): 303–39. Wakefield, "Native Routes through the Masai Country from information obtained by the Rev. T. Wakefield," *Proceedings of the Royal Geographical Society*, n.s., 4 (1882): 742–47; Hartwig, *Art of Survival*, 78.

28. Baumann, *Durch Massailand*, 38–41; Early History section entered by E. C. Baker, 9 December 1929, 1–20; "Musoma Neighbourhood in 1892, A translation of Dr. Oskar Baumann's account of his safari through South Mara and Ukara Island with an introduction and notes by G. W. Y. Hucks," in original Baumann, *Durch Massailand*, Musoma District Books, reel 24, 36–38; Carl Uhlig, *Die ostafrikanische Bruchstufe und die angrenzenden Gebiete zwischen den Seen Magad und Lawa ja Mweri sowie dem Westfluss des Meru*, part 1: *Die Karte* (Berlin: Ernst Siegfried Mittler und Sohn, 1909), 4–5.

29. Joseph Thomson, *Through Masai Land: A Journey of Exploration among the Snowclad Volcanic Mountains and Strange Tribes of Eastern Equatorial Africa . . .* (London: Sampson Low, Marston, Searle, and Rivington, 1887; reprint, London: Frank Cass, 1968). Thomson states that whatever is known about the land beyond Kilimanjaro is from the Wakefield accounts, either because the risks were too great or the cost too high. Thomson was commissioned specifically to find a way through Maasailand to the lake by the Royal Geographical Society.

30. Hartwig, *Art of Survival*, 541; Kollmann, *Victoria Nyanza*, 177. See also Von Hauptmann Schlobach, "Die Volksstämme der deutschen Ostkuste des Victoria-Nyansa," *Mitteilungen aus den deutschen Schutzgebieten* 14 (1901): 183. On "Shashi," see ch. 1, note 76.

31. Alan H. Jacobs, "A Chronology of the Pastoral Maasai," *Hadith: Proceedings of the Annual Conference of the Historical Association of Kenya* (Nairobi: East African Publishing House, 1968), 1:28.

32. Ralph A. Austen, *Northwest Tanzania under German and British Rule: Colonial Policy and Tribal Politics, 1889–1939* (New Haven, CT: Yale University Press, 1968), 6–30.

33. For slavery in East Africa, see R. W. Beachy, *The Slave Trade of Eastern Africa* (New York: Barnes and Noble, 1976); Abdul Sheriff, *Slaves, Spices, and Ivory in Zanzibar: Integration of an East African Commercial Empire into the World Economy, 1770–1873* (Athens: Ohio University Press, 1987); Edward A. Alpers, *Ivory and Slaves in East Central Africa: Changing Patterns of International Trade to the Later Nineteenth Century* (London: Heinemann, 1975).

34. On abarondo hunting associations, see ch. 5. For elephant associations related to Ukerewe, see Hartwig, *Art of Survival*, 66–67.

35. For an overview, see Iliffe, *Modern History*, 40–77.

36. Tobisson, *Family Dynamics*. See also, R. Malcolm, System of Government, MDB.

37. Hartwig, *Art of Survival*, 66–71, 80–81, 116; A. Buluda Itandala, "A History of the Babinza of Usukuma, Tanzania, to 1890" (PhD diss., Dalhousie University, 1986), 213–18; E. A. Chacker, "Early Arab and European Contacts with Ukerewe," *Tanzania Notes and Records* 68 (1968): 75–86; C. F. Holmes, "Zanzibar Influence at the Southern End of Lake Victoria: The Lake Route," *African Historical Studies* 4, no. 3 (1971): 479–503.

38. Austen, *Northwest Tanzania*, 6–30.

39. Valdemar E. Toppenberg, *Africa Has My Heart* (Mountain View, CA: Pacific Press Publishing Association, 1958), 45.

40. In the age of Merishari (ca. 1806–26) they took the Lake Manyara area from the Tatoga, and in subsequent ages, perhaps as late as the 1840s, the Maasai forced the Tatoga to withdraw from the Ngorongoro Crater and the Engaruka area. John G. Galaty, "Maasai Expansion and the New East African Pastoralism" in *Being Maasai: Ethnicity and Identity in East Africa*, ed. Thomas Spear and Richard Waller (Athens: Ohio University Press, 1993), 74. See also, Berntsen, "Pastoralism, Raiding," 31, and, on the process of migrational drift, 40; Memorandum on Masai History and Mode of Life, Prepared by H. St. J. Grant, A District Officer in the Masai District from August 1950 to December 1954, National Game Parks, vol. 4, 215/350/IV, TNA.

41. Paul Spencer, "Age Systems and Modes of Predatory Expansion," in *Conflict, Age and Power in North East Africa: Age Systems in Transition*, ed. Eisei Kurimoto and Simon Simonse (Athens: Ohio University Press, 1998), 172.

42. Memorandum on Masai History and Mode of Life.

43. Fosbrooke, "Sections of the Masai," EAF; Galaty, "Maasai Expansion," 72. R. Waller, "Economic and Social Relations in the Central Rift Valley: The Maa-Speakers and Their Neighbors in the Nineteenth Century," in *Kenya in the 19th*

Century, ed. Bethwell A. Ogot (Kisumu, Kenya: Anyange Press, 1985), 116, 120. A colonial officer from Monduli claimed that the Salei, Endulen Laitaiyuk, and Serenget Maasai sections were the first to enter the area that is now the park, while the Kisongo only came in the twentieth century. Grant, District Officer, Report on Human Habitation of the SNP, TNA.

44. In the intense debates surrounding establishment of Serengeti National Park's boundaries in the 1930s, Fosbrooke and others claimed that the Serenget Maasai had for the past century used the western Serengeti as fallback grazing in times of drought. Fosbrooke, "Sections of the Masai in Loliondo Area." For the entire debate see National Game Parks files, vols. 1–4, 215/350/I-IV, TNA.

45. Galaty, "Maasai Expansion," 72. The Loitai confederation included the Siria, Laitayok, Salei, Serenget, and Loitai.

46. Spear and Waller, *Being Maasai*, intro. For twentieth-century Maasai relations in Ngoreme, interview with Wambura Tonte, Kemegesi, 12 August 2003.

47. Richard Waller uses the expression, "the lords of East Africa" to describe the regional dominance of the Maasai. Waller, "The Lords of East Africa: The Maasai in the Mid-nineteenth Century (c. 1840–c.1885)" (PhD diss., Cambridge University, 1979). See also Berntsen, "Pastoralism, Raiding," 32. On economic symbiosis, see Spear and Waller, *Being Maasai*, 2–4.

48. J. E. G. Sutton, "Becoming Maasailand," in *Being Maasai: Ethnicity and Identity in East Africa*, ed. Thomas Spear and Richard Waller (Athens: Ohio University Press, 1993), 42.

49. Bourget, Trip Diary, Sukuma Museum; See also Société des Missionnaires d'Afrique, *Chronique trimestrielle* 24, no. 94 (April 1902): 94.

50. They administered the area with one European officer and thirty-five African *askari* of the fourteenth company of protectorate troops. Schnee, *Deutsches Kolonial-Lexikon*, 89–90; R. Malcolm, System of Government, MDB; M. Turner, *My Serengeti Years*, 32–33, reported from Lt. Paul Deisner. Lemi G. George says that the Germans built the fort to provision the trade route to the coffee and sisal plantations to the east and to exploit the gold and ivory resources in the area. George, *Report Paper on Ft. Ikoma Monument Research Project* (Dar es Salaam: TANAPA, 2001), 7–8.

51. J. E. G. Sutton, "Some Reflections on the Early History of Western Kenya," in *Hadith* 2, ed. Bethwell A. Ogot (Nairobi: East African Publishing House, 1970): 24; Berntsen, "Pastoralism, Raiding," 112–43, 172, 224; Waller, "Economic and Social Relations," 89, 107, 133; Thomas P. Ofcansky, "The 1889–97 Rinderpest Epidemic and the Rise of British and German Colonialism in Eastern and Southern Africa," *Journal of African Studies* 8, no. 1 (1981): 31–38.

52. Société des Missionnaires d'Afrique, "Nyegina (Notre-Dame de Consolation)," *Rapport annuel*, 1911–12, 392; Société, *Chronique trimestrielle* 24, no. 94 (April 1902): 94.

53. Toppenberg, *Africa Has My Heart*, 63.

54. This insight thanks to ongoing conversations with Richard Waller.

55. For a description of this regional economy, see Spear and Waller, *Being Maasai*, 2; Waller, "Ecology, Migration," 347–70.

56. See Iliffe's analysis for all of Tanganyika, in *Modern History*, 75–77, 163–67.

57. Koponen, *Development for Exploitation*, 475–77.

58. White, *Rediscovered Country*, 109, 167, 217, 226.

59. M. Turner, *My Serengeti Years*, 25.

60. Hoefler, *Africa Speaks*, 57–58, 136.

61. White, *Rediscovered Country*, 194.

62. The Geographical Section of the Naval Intelligence Division stated that "during the great migration of the Masai, the Wandorobo were either driven out or forced to submit." Geographical Section, *German East Africa*, 98–99. They differentiate between Ndorobo who speak a Maasai language and those in the Serengeti who speak a different language for which the "Washashi in Ikoma" act as interpreters. This implies a long-term relationship between Ikoma and these hunter-gatherers, who had only recently become clients of the Maasai.

63. Interviews with Stephen Gojat Gishageta and Girimanda Mwarhisha Gishageta, Issenye, 27 July 1995; Gilumughera Gwiyeya and Girihoida Masaona, Issenye, 28 July 1995; Marunde Godi, Manawa, 24 February 1996. For a popular written account recommended to me by Tatoga elders, see Institute for Swahili Research, *Zamani mpaka siku hizi* (Dar es Salaam: East African Literature Bureau, 1930), 44–46.

64. H. A. Fosbrooke, Senior Sociologist, Tanganyika, "Masai History in Relation to Tsetse Encroachment," Arusha, 1954, CORY 254, EAF.

65. Baumann, *Durch Massailand*, 38–42; Kollmann, *Victoria Nyanza*, 176.

66. This way of categorizing regional relationships does not agree with the way that Maasai scholars have interpreted mutually exclusive identities in reference to "differential access to resources" and economic specialization. Spear and Waller, *Being Maasai*, 6.

67. Interview with Gilumughera Gwiyeya and Girihoida Masaona, Issenye, 28 July 1995.

68. Sutton cites linguistic and oral evidence that the Tatog-speaking peoples once occupied the Loita-Mara plains and across Serengeti to the crater highlands, being pushed out or absorbed by Maasai expansion. Sutton, "Becoming Maasailand," 48.

69. Interview with Marindaya Sanaya, Samonge, 5 December 1995. Wilson reports an extraordinarily similar practice by the Tatog Barabaig where any killing of cattle thieves or lions may be used to collect lots of cattle. He speculates that the group most likely to engage in this activity are youngest sons without other outlets for status. Among the Barabaig the anointing of the killer with butter is a propitiation of the ancestors and the cattle given to him equivalent to blood compensation offered to a kinsmen. Wilson, "Tatoga of Tanganyika, Part Two," 35–56.

70. Among the Barabaig the killer "adorns himself with women's ornaments, which symbolize that he is like a woman who has given birth. Killing an enemy of

the people and giving birth is symbolically equated. The killer of an enemy must observe a convalescence period (one month) for having given 'birth,' and is restricted from touching food or doing any work." Klima, *Barabaig*, 58–60. Western Serengeti peoples also allow women who are courageous in birth to dance the aghaso with the men. Interviews with Zamberi Masambwe and Gisuge Chabwasi, Mariwanda, 22 June 1995; Mang'oha Morigo, Bugerera, 24 June 1995; Merekwa Masunga and Giruchani Masanja, Mariwanda, 6 July 1995; Elfaresti Wambura Nyetonge, Kemegesi, 20 September 1995; Zabron Kisubundo Nyamamera and Makang'a Magigi, Bisarye, 9 November 1995; Marindaya Sanaya, Samonge, 5 December 1995.

71. Kramer shows how African representation of the European other in sculpture was used to define self. Kramer, *Red Fez*, 2. Boddy demonstrates that the zar possession cult in Sudan fosters an "alien world at the heart of culture." Boddy, *Wombs and Alien Spirits*, 342.

72. Interview with Mang'oha Morigo, Bugerera, 24 June 1995.

73. Spear and Waller, *Being Maasai*, 2.

74. For Kuria Maasai clan ancestors, see also Cory, "Land Tenure in Bukuria," 71–72. See an account of a scare of a Maasai raid because a woman was carrying out one of these rituals on behalf of her Maasai ancestor in Sukuma, and a young herd boy saw the proceedings and ran to alert everyone that a raid was in progress in Raids by Masai, 1936, vol. 1, Secretariat Files, 23384, TNA. See also interviews with Gabuso Shoka, Mbiso, 30 May 1995; Bokima Giringayi, Mbiso, 26 October 1995; Tetere Tumbo, Mbiso, 5 April 1995; Mwita Magige, Mosongo, 9 September 1995.

75. Interview with Elia Masiyana Mchanake and Robi Nykisokoro, Borenga, 21 September, 1995. There are variations of this story, including that they met on a hunt and that Saroti ate porridge (ugali). In some versions of this story Matiti is said to be an Iregi, the clan that left the Nyeberekera dispersal center in the Ishenyi story. Saroti is sometimes said to be Maasai only in that he was a "vagabond, traitor or outcast"; claiming that his origin was actually Gosi, from the Shirati area. All accounts confirm that the spring at Kiru is a powerful erisambwa place. And all are both proud of and embarrassed by this important Maasai ancestor. Interviews with Isaya Charo Wambura, Buchanchari, 22 September 1995; Charwe Matiti, Nyeboko, 22 September 1995.

76. Grant, Report on Human Habitation of the Serengeti National Park, 10–11, TNA.

77. Schnee, *Deutsches Kolonial-Lexikon*, 121, 679–81; M. Weiss, *Völkerstämme*, 244–45; Baumann, *Durch Massailand*, 57, 196–99, 246.

78. Interview with Mang'oha Morigo, Bugerera, 24 June 1995.

79. Bourget, Trip Diary, Sukuma Museum; interview with Machota Sabuni, Issenye, 14 March 1996.

80. E. C. Baker asserts that the Kuria abaNgibabe age set initiated in 1858–62 in Nyabasi first began piercing the tops of their ears for the insertion of small sticks as ornaments. E. C. Baker, "Age-Grades in Musoma District," *Man* 27 (December

1927): 223. It would be interesting to know if western Serengeti use of Maasai orna-mentation differed slightly enough, as Donna Klumpp and Corinne Kratz show for the Okiek, that it is a visual display of both submission and resistance to Maasai dominance. Klumpp and Kratz, "Aesthetics, Expertise, and Ethnicity: Okiek and Maasai Perspectives on Personal Ornament," in *Being Maasai: Ethnicity and Identity in East Africa*, ed. Richard Waller and Thomas Spear (London: James Currey, 1993), 195–221. Unfortunately, there are few of these ornaments left as they are not currently in use.

81. Kjerland, "Cattle Breed," 123.

82. Musoma District, notes from Musoma district books on local tribes and chieftains, in German (ca. 1912?), manuscript, CORY 348, EAF.

83. Wakefield, "Wakefield's Notes," 303–39; Wakefield, "Native Routes," 742–47.

84. Interviews with Mahewa Timanyi and Nyambureti Morumbe, Robanda, 27 May 1995; Machota Sabuni, Issenye, 14 March 1996.

85. Annual Report 1933, Musoma District, Annual Reports, Native Affairs Section, Lake Province, 215/924/2, TNA.

86. Interview with Mariko Romara Kisigiro, Burunga, 31 March 1995. See also Shetler, *Telling Our Own Stories*, 263–64.

87. Others say that skulls and bones can still be found on the battlefield.

88. Philipo claims that these battles took place between 1700 and 1970. P. Haimati and P. Houle, "Mila na matendo ya Wangoreme," unpublished mimeo, Iramba Mission, 1969, author's collection. But in an interview he mentions the "wars of the Mairabe" in connection with the Maasai falling into the hole at Kimeri. Interview with Philipo Haimati, Iramba, 15 September 1995.

89. Haimati and Houle, "Mila na matendo."

90. Ibid.; interview with Philipo Haimati, Iramba, 15 September 1995.

91. Interview with Mashauri Ng'ana, Issenye, 2 November 1995.

92. Visitations Book, Nyegina, Mwanza 1, 1931–32, White Fathers Regionals' House, Nyegezi, 67–69.

93. Both Kjerland and Hartwig cite the Mwanza District Books. Kjerland "Cattle Breed," 135; Hartwig, *Art of Survival*, 127–28. Hartwig also states that there were a lot of Luo, "Gaya" slaves on Kerewe (125–26); for an assessment of Ukerewe slavery, see 114–28. These reports are confirmed by Kuria informants in Tobisson, *Family Dynamics*, 12–13.

94. See Glassman, *Feasts and Riots*; Frederick Cooper, *Plantation Slavery on the East Coast of Africa in the Nineteenth Century* (New Haven, CT: Yale University Press, 1977).

95. For a description of a similar process in central Kenya, see Ambler, *Kenyan Communities*, 134–35.

96. Mtemi Seni Ngokolo, "Historia ya Utawala wa Nchi ya Kanadi," author's collection.

97. Interview with Ikota Mwisagija, Kihumbo, 5 July 1995. Ikota is from Nya-kinywa's clan. For more on the Sukuma chiefship, see Per Brandstrom, "Seeds and

Soil: The Quest for Life and the Domestication of Fertility in Sukuma-Nyamwezi Thought and Reality," in *The Creative Communion: African Folk Models of Fertility and the Regeneration of Life*, ed. Anita Jacobson-Widding and Walter van Beek (Uppsala: Almqvist and Wiksell International, 1990), 167–86; Hans Cory, *The Indigenous Political System of the Sukuma and Proposals for Political Reform* (Nairobi: East African Institute of Social Research/Eagle Press, 1954).

98. Other traditions collected in the 1930s showed that Muesa was "the first remembered chief" (ca. 1895). Richard C. Thurnwald, *Black and White in East Africa: The Fabric of a New Civilization: A Study in Social Contact and Adaptation of Life in East Africa* (London: George Rutledge and Sons, 1935), 46–47.

99. For a recent critique of king lists, see Wrigley, *Kingship and State*, 27–41.

100. Sonjo is the name given to the ethnic group by the colonial officers, after the sonjo bean. The Sonjo call themselves Bantemi, after the ntemi scar that they share with the Ikoma.

101. For one of the few ethnographies of Sonjo, see Gray, *Sonjo of Tanganyika*, 88. However, a connection with the western Serengeti is suggested by the fact that the name of the first age-set that elders remember is Olnyamburete, about eleven age-sets ago (like the western Serengeti generation-set name Nyambureti).

102. White, *Rediscovered Country*, 55.

103. Alan H. Jacobs, "The Irrigation Agricultural Maasai of Pagasi: A Case of Maasai-Sonjo Acculturation," paper presented at Dar es Salaam Social Science Conference, 2–5 January 1968, 1–12; John L. Berntsen, "The Maasai and Their Neighbors: Variables of Interaction," *African Economic History* 2 (Fall 1976): 1–11.

104. Interviews with Peter Nabususa, Samonge, 5 December 1995;Marindaya Sanaya, Samonge, 5 December 1995.

105. Interview with Samweli Ginduri, Samonge, 6 December 1995.

106. Interviews with Peter Nabususa, Samonge, 5 December 1995; Samweli Ginduri, Samonge, 6 December 1995. The Tinaga site was visited by Gray, *Sonjo of Tanganyika*, 13.

107. Interview with Nsaho Maro, Kenyana, 14 September 1995. Philipo Haimati says that the Ngoreme came from Sonjo "Nyahaba." Haimati, handwritten notebook on Ngoreme history, which I saw on 14 September 1995.

108. Interview with Mzee Taranka, Bugerera, 10 May 1995.

109. Interview with Samweli Ginduri, Samonge, 6 December 1995. Maasai history would also confirm this date. Berntsen, "Pastoralism, Raiding," 112–43, 172, 224.

110. Interview with Emmanuel Ndenu, Sale, 6 December 1995 (Sonjo); Gray, *Sonjo of Tanganyika*, 11–12; interviews with Peter Nabususa, Samonge, 5 December 1995; Marindaya Sanaya, Samonge, 5 December 1995; Samweli Ginduri, Samonge, 6 December 1995. F. G. Finch, "Hambageu: Some Additional Notes on the God of the Wasonjo," *Tanganyika Notes and Records* 47/48 (1957): 203–8; H. A. Fosbrooke, "Hambageu, the God of the Wasonjo," *Tanganyika Notes and Records* 35 (1953): 38–43; E. Simenauer, "The Miraculous Birth of Hambageu, Hero-God of the Sonjo," *Tanganyika Notes and Records* 38 (1955): 23–30.

111. Interview with Emmanuel Ndenu, Sale, 6 December 1995.

112. For early colonial recording of these traditions, see E. C. Baker, comp., Tribal History and Legends, 29 December, reel 24, 8–9, MDB; H. C. Barlet, Population Distribution, 11 February 1941, reel 24, 1–2, MDB.

113. Schnee, *Deutsches Kolonial-Lexikon*, 680–81.

114. Baumann, *Durch Massailand*, 56.

115. Kollmann, *Victoria Nyanza*, 177–78. Fortified villages had walls 5 feet high, 3 feet wide, and 800 feet long in original.

116. Elfaresti Wambura Nyetonge, Kemegesi, 20 September 1995.

117. Société des Missionnaires d'Afrique, "Ukerewe," *Chronique trimestrielle* 24, no. 95 (July 1902): 281; Bourget, Trip Diary, Sukuma Museum. The White Fathers attributed this to a period of famine to intertribal war and the raids of the Maasai and Luo.

118. Kollmann, *Victoria Nyanza*, 177.

119. For a description of Sonjo fortifications, see White, *Rediscovered Country*, 76. For western Kenya, see R. T. K. Skully, "Fort Sites of East Bukusu, Kenya," *Azania* 4 (1969): 105–14; Skully, "Nineteenth Century Settlement Sites and Related Oral Traditions from the Bungoma Area, Western Kenya," *Azania* 14 (1979): 81–96. For Nyamwezi, see Richard F. Burton, *The Lake Regions of Central Africa: A Picture of Exploration*, 2 vols. (London: Longman, Green, Longman, and Roberts, 1860). For Kuria, see Cory, "Land Tenure in Bukuria," 70–79. For Sonjo, see Gray, *Sonjo of Tanganyika*, 33–34.

120. Interviews with Stephen Gojat Gishageta and Girimanda Mwarhisha Gishageta, Issenye, 27 July 1995; Gilumughera Gwiyeya and Girihoida Masaona, Issenye, 28 July 1995.

121. Interview with Mohere Mogoye, Bugerera, 25 March 1995. Mohere is from the chief's descent group.

122. Iliffe, *Modern History*, 75. In southern Tanzania the unrest caused by Ngoni incursions from the south were responsible for concentrated settlements. For a nationalist biography of Mirambo, see Norman Robert Bennett, *Mirambo of Tanzania, 1840?–1884* (New York: Oxford University Press, 1971).

123. A. M. D. Turnbull, Senior Commissioner, Mwanza, report to Game Warden, Kilossa, 28 March 1924, Game Regulations, 1923–29, 215/PC/14/1, TNA.

124. Koponon, *Development for Exploitation*, 349–50.

125. Information on age-set chronologies and about the new age-sets comes from many interviews and informal conversations.

126. Interview with Kirigiti Ng'orogita, Mbiso, 8 June 1995.

127. Interview with Morigo (Mchombocho) Nyarobi, Issenye, 28 October 1995.

128. Kikao cha mila, desturi na asili ya kabila la Waishenyi kilichokutana tarehe 6/6/1990, Nyiberekera, Ishenyi, author's collection.

129. Haimati and Houle, "Mila na matendo," author's collection.

130. Spear and Waller, *Being Maasai*, 6; Galaty, "Maasai Expansion," 81–82.

131. Galaty, "Maasai Expansion," 80. The Maasai word for a local circumcision group, *isirit*, is also used by western Serengeti peoples, as *siriti*.

132. Interview with Morigo (Mchombocho) Nyarobi, Issenye, 28 October 1995.

133. Explicitly stated in interview with Rugayonga Nyamohega, Mugeta, 27 October 1995.

134. John Lamphear reported a similar change from generation to age-sets among the Turkana, to create a more efficient system of military mobilization. Among the Turkana too, the Maasai system was not accepted in its entirety, and many generational aspects of the original system survived. Lamphear, "Aspects of 'Becoming Turkana': Interactions and Assimilation between Maa- and Ateker-Speakers," in *Being Maasai: Ethnicity and Identity in East Africa*, ed. Thomas Spear and Richard Waller (Athens: Ohio University Press, 1993), 94–95.

135. Ibid., 95–97; Bob J. Walter, *Territorial Expansion of the Nandi of Kenya, 1500–1905*, Papers in International Studies, Africa Series, no. 9 (Athens: Ohio University Center for International Studies, 1970).

136. Interviews with Megasa Mokiri, Motokeri, 4 March 1995; Mang'oha Morigo, Bugerera, 24 June 1995.

137. Paul Spencer, "Becoming Maasai, Being in Time," in *Being Maasai: Ethnicity and Identity in East Africa*, ed. Thomas Spear and Richard Waller (Athens: Ohio University Press, 1993), 141.

138. Lamphear, "Becoming Turkana," 94.

139. Interviews with Morigo (Mchombocho) Nyarobi, Issenye, 28 October 1995; Machota Sabuni, Issenye, 14 March 1996; Mang'oha Morigo, Bugerera, 24 June 1995; Ali Maro Wambura, Masinki, 30 September 1995.

140. For the lakeshore trade and prosperity, see D. Cohen, *Reconstructed Past*, 2–3; D. Cohen, "Food Production," 1–18; Hartwig, *Art of Survival*, 106, 116, 127.

141. See Marcia Wright, *Strategies of Slaves and Women: Life-Stories from East and Central Africa* (New York: L. Barber Press, 1993).

142. Interview with Baginyi Mutani and Mayenye Nyabunga, Sanzate, 8 September 1995. Women who have been courageous in birth are allowed to dance in the aghaso for lion killers; this is also noted for Tatoga.

143. See Boddy, *Wombs and Alien Spirits*, 39, 109. The village and interior space was represented as female.

144. Tobisson, *Family Dynamics*, 89; Hay, "Local Trade," 7–12; Cora Ann Presley, "The Mau Mau Rebellion, Kikuyu Women and Social Change," *Canadian Journal of African Studies* 22, no. 3 (1988): 502–27; Claire C. Robertson, *Trouble Showed the Way: Women, Men, and Trade in the Nairobi Area, 1890–1990* (Bloomington: Indiana University Press, 1997).

145. Interviews with Nyambeho Marangini, Issenye, 7 September 1995; Philipo Haimati, Iramba, 15 September 1995.

146. Interviews with Morigo (Mchombocho) Nyarobi, Issenye, 28 October 1995; Mwita Magige, Mosongo, 9 September 1995.

147. For examples of this practice in other places, see Peter Rigby, *Cattle and Kinship among the Gogo: A Semi-pastoral Society of Central Tanzania* (Ithaca, NY: Cornell University Press, 1969), 148; Cohen and Odhiambo, *Siaya*, 12–15.

148. Interview with Nyamaganda Magoto, Bugerera, 3 March 1995.

149. Many elders in the eastern areas refer to the saiga cycle as the ekyaro. In particular, the explicit testimony of Mang'ombe Morimi, Issenye/Iharara, 26 August 1995.

150. Interview with Morigo (Mchombocho) Nyarobi, Issenye, 28 October 1995.

151. Interview with Mashauri Ng'ana, Issenye, 2 November 1995 says this most explicitly as the "rule" (*utawala*, in Kiswahili) of the saiga being their responsibility for the *uganga*, or healing, needed for the land.

152. Kollmann, *Victoria Nyanza*, 175.

153. White, *Rediscovered Country*, 16.

154. M. Turner, *My Serengeti Years*, 26.

155. Grzimek, *Serengeti Shall Not Die*, 70.

156. Schmidt, "'Penetrating' Foreign Lands," 359.

157. M. Turner, *My Serengeti Years*, 25.

158. White, *Rediscovered Country*, 4–9.

CHAPTER 5: RESISTANCE TO COLONIAL INCORPORATION

1. David Anderson, "Depression, Dust Bowl, Demography, and Drought: The Colonial State and Soil Conservation in East Africa during the 1930s," *African Affairs* 83 (July 1984): 324, 336; Dunlap, *Nature and the English Diaspora*, 26–31, 45–48; MacKenzie, *Empire of Nature*, 36.

2. Bernhard Gissibl," The Nature of Colonialism: Being and Elephant in East Africa," Paper presented at the Animals in History Conference, Cologne, May 2005, 1, 9, 11; MacKenzie, *Empire of Nature*; Neumann, "Ways of Seeing Africa," 149–51; Koponen, *Development for Exploitation*, 24–28.

3. Among the large bibliography of sources on labor history in Africa for Tanzania, see Thaddeus Sunseri, *Vilimani: Labor Migration and Rural Change in Early Colonial Tanzania* (Portsmouth, NH: Heinemann, 2002); Fred J. Kaijage, *Labor Conditions in the Tanzanian Mining Industry, 1930–1960*, Working Papers, no. 83 (Boston: African Studies Center, Boston University, 1983); Sidney J. Lemelle, "Capital, State, and Labor: A History of the Gold Mining Industry in Colonial Tanganyika, 1890–1942" (PhD diss., UCLA, 1986). On labor dynamics, see Elias C. Mandala, *Work and Control in a Peasant Economy: A History of the Lower Tchiri Valley in Malawi, 1859–1960* (Madison: University of Wisconsin Press, 1990).

4. Nicola Swainson, *The Development of Corporate Capitalism in Kenya, 1918–77* (Berkeley: University of California Press, 1980); Sunseri, *Vilimani*.

5. Anderson, "Depression, Dust Bowl," 336–37; Dunlap, *Nature and the English Diaspora*, 45–50.

6. On "Shashi," see ch. 1, note 76.

7. Schlobach, *Volksstämme der deutschen Ostkuste*, 183; Austen, *Northwest Tanzania*, 53–58; R. Malcolm, System of Government, MDB. For a history of the development of Mwanza as a colonial base, see Laird Revis Jones, "The District Town

and the Articulation of Colonial Rule: The Case of Mwanza, Tanzania, 1890–1945," 2 vols. (PhD diss., Michigan State University, 1992). For the German presence in this region, see Ralph Austen, "Memoirs of a German D.C. in Mwanza, 1907–16: Extracts from the Memoirs of Theodor Gunzert," *Tanzania Notes and Records* 66 (1966): 171–79. For accounts of some of the few expatriates living in the area during these years, see Carl Jungblut, *Vierzig Jahre Afrika, 1900–1940* (Berlin-Friedenau: Spiegel Verlag Paul Lippa, 1941); Jungblut owned a plantation in Majita; see also Anne Luck, *Charles Stokes in Africa* (Nairobi: East African Publishing House, 1972). Stokes was a CMS missionary (arrived 1879) who quit (1886) and became a caravan operator, living in Ukerewe and Mwanza.

8. See ch. 4, note 50.

9. George, *Report Paper on Ft. Ikoma*, 7–9.

10. Ibid., 12–13; Austen, *Northwest Tanzania*, 111–12; M. Turner, *My Serengeti Years*, 33.

11. Scott, *Weapons of the Weak*, 31.

12. Interview with Mahewa Timanyi and Nyambureti Morumbe, Robanda, 27 May 1995. Wilson Shanyangi Machota was my guide for this interview, and Samuel, the son of Mahewa, was also at the interview.

13. This is a tactic described by James Scott as a "weapon of the weak." People who are not strong enough to resist openly do so covertly. Scott, *Weapons of the Weak*.

14. Interview with Mohere Mogoye, Bugerera, 25 March 1995.

15. Iliffe, *History of Tanzania*; Leroy Vail, ed., *The Creation of Tribalism in Southern Africa* (Berkeley: University of California Press, 1991); Kopytoff, "Internal African Frontier," 3–16.

16. For some of the literature on resistance in Africa, see Cooper, "Conflict and Connection;" Michael F. Brown, "On Resisting Resistance," *American Anthropologist* 98, no. 4 (1996): 729–35; Allen Isaacman, "Peasants and Rural Social Protest in Africa," *African Studies Review* 33, no. 2 (September 1990): 1–120; Sherry B. Ortner, "Resistance and the Problem of Ethnographic Refusal," *Comparative Studies in Society and History* 37, no. 1 (1995): 173–93. For East Africa, see Isaria N. Kimambo, *Penetration and Protest in Tanzania: The Impact of the World Economy on the Pare, 1860–1960* (Athens: Ohio University Press, 1991); Edward I. Steinhart, *Conflict and Collaboration: The Kingdoms of Western Uganda, 1890–1907* (Princeton, NJ: Princeton University Press, 1977); Glassman, *Feasts and Riots*.

17. Austen, *Northwest Tanzania*, 31–61.

18. See Iliffe, *History of Tanganyika*; Austin, *Northwest Tanzania*.

19. Geographical Section, *German East Africa*, 96.

20. Sebastiani Muraza Marwa, *Mashujaa wa Tanzania: Mtemi Makongoro wa Ikizu: Historia ya Mtemi Makongoro na kabila lake la Waikizu, mwaka 1894 hadi 1958* (Peramiho, Tanzania: Benedictine Publications Ndanda, 1988), 20–21.

21. Geographical Section, *German East Africa*, 97.

22. Kollmann, *Victoria Nyanza*, 175–76.

23. R. Malcolm, System of Government, MDB.

24. Marwa, *Mashujaa wa Tanzania*, 17–18.

25. E. C. Baker, Tribal History and Legends, MDB; R. Malcolm, System of Government, MDB.

26. E. C. Baker, Administrative Officer in Charge, Musoma Sub-District, to Senior Commissioner, Mwanza, 17 February 1926, Native Administration Successions in Sultanates, 1926–31, 246, PC/3/3, TNA.

27. White, *Rediscovered Country*, 173, 204, 223.

28. Siso, Oral Traditions of North Mara, author's collection; Mustafa, *Concept of Authority*, 72. In Zanaki, Fundi Kenyeka became chief because he could make nails. He was sent as a messenger to greet the Germans on behalf of the Busegwe rainmaker but instead told the Germans that his son, Kitara, was the chief. E. C. Baker, Tribal History and Legends, MDB; R. Malcolm, System of Government, MDB.

29. Interview with Nyamaganda Magoto, Bugerera, 3 March 1995.

30. Cory, Pre-European Tribal Organization," 1, EAF.

31. E. C. Baker, Tribal History and Legends, MDB.

32. Musoma District Book, Musoma Sub-District, 1916–27, 62, TNA.

33. R. Malcolm, System of Government, MDB.

34. Austen, *Northwest Tanzania*, 112–13; R. Malcolm, System of Government, 55, 6, MDB.

35. Austen, *Northwest Tanzania*, 58.

36. Schnee says that a customs post was opened in Nyabange in 1912 and moved to Musoma in 1913. Schnee, *Deutsches Kolonial-Lexikon*, 290. See also P. J. C. Marchant, letter home "to Ma and Smallest," 13 July 1951, P. J. C. Marchant, Personal Letters, 1950–52, MSS. Afr. s. 1701 (1), RHL. The Musoma District Book puts the date at 1913. Musoma System of Government, Extracts from a report by R. S. W. Malcolm, 1937, reel 24, 4, MDB. See also translation of a letter from Shirati, 16 December 1911, from Schultz to Royal District Officer, Mwanza (Kaiserliche Bezirksnebenstelle) Ref. 612 of 26–9–1911, reel 24, MDB; White, *Rediscovered Country*, 220–21, 250.

37. Musoma District, notes on local tribes, EAF.

38. Sultanates of Musoma District, 3, Native Courts, Mwanza, 1924, 2860/32, TNA.

39. Musoma District, notes on local tribes.

40. District Book Natural History (Game), H. C. Barlet, 16 December 1940, 1–2; H. C. Barlet, Illicit Hunting, 14 August 1941, reel 24, 3–4, MDB.

41. Interview with Nyamaganda Magoto, Mbiso, 3 August 2003.

42. H. C. Barlet, Distribution of Population in the Ikoma Federation, 16 August 1941, reel 24, 3–4, MDB.

43. Monthly report for August 1927, 13 September 1927, 1926–29 Provincial Administration, Monthly Reports, Musoma District, 215/PC/1/7, TNA.

44. MacKenzie, *Empire of Nature*, 81, 141, 149, 164, 206–7; Koponon, *Development for Exploitation*, 536. On resistance, see Neumann, *Imposing Wilderness*, 13.

45. Koponen, *Development for Exploitation*, 481–84; Game Warden, Kilossa, to Honorable Chief Secretary to the Governor DSM, 26 February 1924, 62, Game Regulations, 1923–29, 215/PC/14/I, TNA; Third Note on Sleeping sickness, signed by J. F. Corson, Medical Officer, Ikoma, 15 April 1927, 1926–29 Provincial Administration Monthly Reports Musoma District, 215/PC/1/7, TNA; interview with David Maganya Masama, Kemegesi, 12 August 2003; Marchant, letters "to Ma and Smallest," 10 September, 23 September, 11 November 1950, 13 July 1951.

46. A. M. D. Turnbull, Senior Commissioner Mwanza, to Director of Game Preservation, Kilossa, 2 July 1924, Game Regulations, 1923–29, 215/PC/14/I, TNA.

47. Swynnerton to Honorable Chief Secretary, 17 March 1924, Proposals for Tsetse Fly Control in Usukuma and in the Territory Generally, 26, Tsetse Fly, vol. 2, 2702, TNA.

48. A. M. D. Turnbull, Senior Commissioner, Mwanza, to Director of Game Preservation, Kilossa. 2 July 1924; Senior Commissioner, Mwanza, to Administrative Officer, Maswa, 9 July 1925, Game Regulations, 1923–29, 215/PC/14/I, TNA; interviews with Joseph Sillery Magoto, Mbiso, 2 August 2003; Nyamaganda Magoto, Mbiso, 3 August 2003.

49. Tsetse Fly, vol. 1, 2702, TNA; Swynnerton to the Honorable Chief Secretary, Dar es Salaam, 23 July 1921, 24.

50. E. C. Baker, District Officer, to Provincial Commissioner, Mwanza, 29 June 1926, Game Regulations, 1923–29, 215/PC/14/I, TNA.

51. Koponen, *Development for Exploitation*, 479–81.

52. For the Month of November 1928, 5 December 1928, 1926–29 Provincial Administration Monthly Reports, Musoma District, 215/PC/1/7, TNA.

53. Interviews with Joseph Sillery Magoto, Mbiso, 2 August 2003; Nyamaganda Magoto, Mbiso, 3 August 2003.

54. Koponen, *Development for Exploitation*, 480.

55. Outbreak of Sleeping Sickness in Mwanza Province, Secretariat 20909, TNA.

56. District Veterinary Officer, Musoma, to District Officer, Musoma, 19 January 1928, Annual Report 1927, 19 January 1928, Mwanza Province 1927–28, Provincial Administration, 246 PC/1/30, TNA.

57. Annual Report on the Territory for 1926 to the Council of the League of Nations on Administration of Tanganyika Territory 1926, Secretariat Files, 1733/4–1733/12, reel 2, TNA, microfilm; Third Note on Sleeping Sickness, signed by J. F. Corson, Medical Officer, Ikoma, 15 April 1927, 1926–29 Provincial Administration, Monthly Reports, Musoma District, 215/PC/1/7, TNA.

58. District Commissioner, Musoma, to Provincial Commissioner, Lake Province, 21 April 1932, Sleeping Sickness, Outbreak of Sleeping Sickness, TNA.

59. Interview with Mang'oha Morigo and Paulo Machota Mongoreme, Kyandege, 8 August 2003.

60. District Commissioner, Medical—General Establishment 1940–1941, reel 24, MDB.

61. MacKenzie, *Empire of Nature*, 164, 209, 299–306; Dunlap, *Nature and the English Diaspora*, 59–61.

62. White, *Rediscovered Country*, 91.

63. A. M. J. Turnbull, Senior Commissioner, Mwanza, to Game Warden, Shinyanga, 25 July 1924, in reference to the report of Mr. Klein, a game hunter, Game Regulations, 1923–29, 215/PC/14/I, TNA.

64. Acting Provincial Commissioner, Mwanza, to Director of Game Preservation, 11 March 1929, Game Regulations, 1923–29, 215/PC/14/I, TNA.

65. Hoefler, *Africa Speaks*, 139–44.

66. MacKenzie, *Empire of Nature*, 89–121.

67. This case is parallel to that of the criminalization of Kuria cattle raiding. Michael L. Fleisher, *Kuria Cattle Raiders: Violence and Vigilantism on the Tanzania/Kenya Frontier* (Ann Arbor: University of Michigan Press, 2000), 52. See also Neumann, *Imposing Wilderness*, 5. For a history of hunting in East Africa, see Edward I. Steinhart, "Hunters, Poachers and Gamekeepers: Towards a Social History of Hunting in Colonial Kenya," *Journal of African History* 30, no. 2 (1989): 247–64; Steinhart, *Black Poachers, White Hunters: A Social History of Hunting in Colonial Kenya* (Athens: Ohio University Press, 2006).

68. Ordinance no. 41 of 1921, Tanganyika Territory Gazette, Dar es Salaam, vol. 2, no. 41, 16 December 1921, 567–93, Game Regulations, 1923–29, 215/PC/14/I, TNA.

69. Game Regulations, 1923–29, 567–93, 215/PC/14/I, TNA.

70. In 1941 the district officer reported that there were only thirty muzzle-loading guns in all of Musoma District. H. C. Barlet, District Book Natural History (Game), 16 December 1940, 1–2; Barlet, Illicit Hunting, 14 August 1941, reel 24, 3–4, MDB.

71. Mackenzie, *Empire of Nature*, 10–20, 26.

72. Gissibl, "Nature of Colonialism,"13, 1; Koponen, *Development for Exploitation*, 540.

73. Government Memorandum on the Legal History of the Area Now Forming the Serengeti National Park, Government, 10 June 1957, distributed by the Secretariat in reference to Serengeti Committee of Enquiry, National Game Parks, 215/350/IV, TNA.

74. Government Notice 177; H. C. Barlet, District Book Natural History (Game), 16 December 1940, 1–2. Barlet also reports that from 1930–32 all of Musoma District was considered a game reserve. Barlet, Illicit Hunting, 14 August 1941, reel 24, 3–4, MDB. See also Grzimek, *Serengeti Shall Not Die*, 71; Report of the Serengeti Committee of Enquiry, 1957, Government Printer, DSM, 1957, Conservation Problems Overseas, Tanganyika, FT 3/599, PRO.

75. Interviews with Joseph Sillery Magoto, Mbiso, 2 August 2003; Mang'oha Morigo and Paulo Machota Mongoreme, Kyandege, 8 August 2003.

76. Chief Secretary, DSM, to all Provincial Commissioners, 27 March 1926, Minute by Governor, Circular Letter 36 of 1926, Game, 215/13/I, TNA; Chief Secretary to Provincial Commissioner, Mwanza, 8 October 1926; Assistant District Of-

ficer, Musoma, to Provincial Commissioner, Mwanza, 11 November 1926, Game Regulations, 1923–29, 215/PC/14/I, TNA.

77. F. J. Durman, for Chief Secretary, to Game Warden, Arusha, 27 Mary 1932; Chief Secretary to Provincial Commissioner Northern, 6 January 1932, Killing of Game by Natives, vol. 1, Secretariat Files, 13371, TNA; Provincial Commissioner, Lake, to Honorable Chief Secretary, May 1941, Killing of Game by Natives, vol. 2, TNA. In 1941 the district officer suggested that the only way to stop poaching in the western Serengeti was to outlaw and abolish the peoples' main weapon, poisoned arrows. H. C. Barlet, District Book Natural History (Game), 16 December 1940, 1–2; Barlet, Illicit Hunting, 14 August 1941, reel 24, 3–4, MDB.

78. Interviews with Nyamaganda Magoto, Mbiso, 3 August 2003; Mang'oha Morigo and Paulo Machota Mongoreme, Kyandege, 8 August 2003; Tetere Tumbo, Mbiso, 11 August 2003.

79. Response of Chief Secretary to Provincial Commissioner, Replies, 22 October 1934, Killing of Game by Natives, vol. 1, TNA.

80. R. A. J. Maguire for Chief Secretary to Government, to all Provincial Commissioners, 11 January 1947, Game Licenses, 215/1153/I, TNA; P. E. Mitchell, Senior Commissioner, to Provincial Commissioner, Lake, 26 January 1933, 115, Game Licenses, 215/1153/I, TNA.

81. R. H. Robertson to all Provincial Commissioners, 21 July 1951, Member of Local Government's Circular letter no. 19 of 1951, Game, Including National Parks, 41 G 1/1, TNA; Game Ranger, Banagi, to Provincial Commissioner, Lake, 5 September 1953, Game, Including National Parks, Vermin Destruction, 41 G1/2, TNA.

82. Acting Chief Secretary A. E. Stack to all Senior Commissioners, Administrative Officers, District and Sub-Districts, Game Warden, 15 September 1923, Game Regulations, 1923–29, 215/PC/14/I, TNA.

83. Swynnerton, Game Warden, Kilosa, to Chief Secretary, DSM, 7 August 1923, p. 1, Native Settlement in Game Reserves 7227, TNA.

84. A. M. D. Turnbull, Senior Commissioner, Mwanza, to Director of Game Preservation, Kilossa, 2 July 1924, Game Regulations, 1923–29, 215/PC/14/I, TNA.

85. Provincial Commissioner to District Officer, Musoma, 13 July 1928, Game Regulations, 1923–29, 215/PC/14/I, TNA; F. D. Arundell, Game Ranger, Monthly Report, Banagi hill, 30 June 1929, Game: Indiscriminate Slaughter, Newspaper Protest, Secretariat Files, 13582, TNA. In same file, see statistics for game convictions in the 1920s in Provincial Commissioner, Mwanza to Chief Secretary, 30 September 1929. See also reports from Banagi in Game Regulations, 1923–29, 215/PC/14/I, TNA; Annual Report of Game Department 1959, CO 822/2662, PRO.

86. A. M. D. Turnbull, Senior Commissioner, Mwanza, to Administrative Officer in Charge, Musoma, 19 January 1924, Game Regulations, 1923–29, 215/PC/14/I, TNA.

87. Acting Provincial Commissioner to District Officers, 1 March 1929, 253, Game Regulations, 1923–29, 215/PC/14/I, TNA.

88. Government Notice no. 368 of 1942, TNA.

89. A. M. J. Turnbull, Senior Commissioner, Mwanza, to Game Warden, Shinyanga, 25 July 1924, Game Regulations, 1923–29, 215/PC/14/I, TNA.

90. Administrative Officer, Musoma, to Senior Commissioner, Mwanza, 3 March 1924, Game Regulations, 1923–29, 215/PC/14/I, TNA.

91. C. F. M. Swynnerton, Director, to Honorable Chief Secretary, Kilosa, 22 November 1924, Game Circular on Native Scouts, 20 April 1922, Secretariat Files, AB 914 (3733) I, TNA; Game, Including National Parks, Vermin Destruction, 41 G1/2, TNA.

92. Extract from Honorable Secretary for Native Authority's Minute of 26 November 1926, 6, Secretariat Files, AB 915 (3733) II, TNA; Provincial Commissioner, Lake, to Honorable Chief Secretary, 8 May 1941, Killing of Game by Natives, Secretariat Files, 13371, II, TNA.

93. Game Warden, Kilossa, to Honorable Chief Secretary, DSM, 26 February 1924, 62, Game Regulations, 1923–29, 215/PC/14/I, TNA.

94. Game Warden to Honorable Chief Secretary, DSM, 6 August 1923, Native Settlement in Game Reserves, 7227, TNA.

95. A. M. D. Turnbull, Senior Commissioner, to Game Warden, Kilossa, 28 March 1924, Game Regulations, 1923–29, 215/PC/14/I, Dar es Salaam, TNA.

96. A. M. D. Turnbull, Senior Commissioner, Mwanza, to Director of Game Preservation, Kilossa, 2 July 1924, Game Regulations, 1923–29, 215/PC/14/I, TNA.

97. For Kuria, see Prazak, "Cultural Expressions," 51.

98. White, *Rediscovered Country*, 177.

99. Native Court of Ikoma, 57, Musoma Sub-District, 1916–27, MDB, TNA.

100. Interview with David Maganya Masama, Kemegesi, 12 August 2003.

101. Koponen, *Development for Exploitation*, 649–53; Cory, "Land Tenure," 71; Huber, *Marriage and Family*, 16.

102. The same report gives another reason for this tendency: "to evade the demands of the combatants during the war, [native people] moved into Game Reserves, near elephant watering holes." C. F. M. Swynnerton, Game Warden, to Honorable Chief Secretary, DSM, 6 August 1923, Native Settlement in Game Reserves, 7227, TNA.

103. Senior Commissioner, Mwanza, to Director of Game Preservation, Kilossa, 28 August 1925, Game Regulations, 1923–29, 215/PC/14/I, TNA.

104. Résumé of a speech by his excellency Sir Harold MacMichael to Waikoma at Ikoma on Friday, 24 July 1936, Killing of Game by Natives, vol. 1, TNA.

105. M. S. Moore, Game Ranger, to Provincial Commissioner, Lake, 25 July 1938; Acting Game Warden to Honorable Chief Secretary, 10 August 1938; Game Warden Arusha to the Honorable Chief Secretary, 17 August 1938, Killing of Game by Natives, vol. 1, TNA.

106. Provincial Commissioner, Lake, to Honorable Chief Secretary, 18 August 1938; E. C. Baker, Acting Provincial Commissioner, Lake Province, 25 August 1938, Killing of Game by Natives, vol. 1, TNA. See also, Provincial Commissioner, Lake, to Honorable Chief Secretary, 4 May 1939, Killing of Game by Natives, vol. 2, TNA.

107. Interview with Mang'ombe Morimi, Issenye/Iharara, 26 August 1995.

108. Nyamaganda Magoto describes lion hairballs, lion oil, and elephant or rhino dung as ingredients for medicines in Sukuma. Interview with Nyamaganda Magoto, Mbiso, 3 August 2003. Also, interviews with Mang'oha Morigo and Paulo Machota Mongoreme, Kyandege, 8 August 2003; Yohana Kitena Nyitanga, Makundusi, 1 May 1995.

109. Interviews with Mashauri Ng'ana, Issenye, 2 November 1995; Zamberi Masambwe and Gisuge Chabwasi, Mariwanda, 22 June 1995; Makongoro Nyamwitweka and Nyawagamba Magoto, Rubana, 4 April 1996. See also Odhiambo Anacleti, Kijiji cha Butiama: Mila na Desturi Zinazohusiana na Mahari na Ndoa: Tarafa ya Makongoro, Wilaya ya Musoma, Mara, 26 March 1979, Utafiti na Mipango/Utamaduni, Dar es Salaam, UTV/M 13/22, Mara Region Office of Culture, Musoma.

110. Baumann, Durch Massailand, 67. When Baumann reached the area of Magu, just south of the Mara region in Sukuma, he observed, "The natives are great travelers, almost all of them were young people who had been to the coast."

111. Interview with Zamberi Masambwe and Gisuge Chabwasi, Mariwanda, 22 June 1995.

112. Baumann, Durch Massailand, 59.

113. Interview with Nyamaganda Magoto, Mbiso, 3 August 2003.

114. Interviews with Tetere Tumbo, Mbiso, 11 August 2003; Wambura Tonte, Kemegesi, 12 August 2003.

115. Interview with Wambura Tonte, Kemegesi, 12 August 2003.

116. Interview with Mang'oha Morigo and Paulo Machota Mongoreme, Kyandege, 8 August 2003.

117. Kollmann, Victoria Nyanza, 199.

118. A. M. J. Turnbull, Senior Commissioner, Mwanza, to Game Warden, Shinyanga, 25 July 1924, Game Regulations, 1923–29, 215/PC/14/I, TNA.

119. White, Rediscovered Country, 168, 229, 235.

120. Kollmann, Victoria Nyanza, 181–83.

121. Interview with Mabenga Nyahega, Bugerera, 5 September 1995.

122. See Iliffe, Modern History, 40–87. In Sukuma one tusk had to be given to the ntemi. Itandala, "History of the Babinza," 218. See also Edward I. Steinhart, "Elephant Hunting in 19th Century Kenya: Kamba Society and Ecology in Transformation, International Journal of African Historical Studies 33, no. 2 (2001): 335–49; Gissibl, "Nature of Colonialism"; Edward A. Alpers, "The Ivory Trade in Africa: An Historical Overview," in Elephant: The Animal and Its Ivory in African Culture, ed. Doran H. Ross (Los Angeles: Fowler Museum of Cultural History, UCLA, 1992), 349–63.

123. MacKenzie, Empire of Nature, 148; Gissibl, "Nature of Colonialism," 3.

124. Interview with Mabenga Nyahega and Machaba Nyahega, Mbiso, 1 September 1995.

125. Interviews with Joseph Sillery Magoto, Mbiso, 2 August 2003; Tetere Tumbo, Mbiso, 11 August 2003.

126. Interview with Wambura Tonte, Kemegesi, 12 August 2003.

127. Interview with Mang'oha Morigo and Paulo Machota Mongoreme, Kyandege, 8 August 2003.

128. Interviews with David Maganya Masama, Wambura Tonte, Kemegesi, 12 August 2003. On elephant hunting and political resistance, see Thaddeus Sunseri, "Reinterpreting a Colonial Rebellion: Forestry and Social Control in German East Africa, 1874–1915," *Environmental History* 8, no. 3 (2003): 430–51.

129. Gissibl, "Nature of Colonialism," 12.

130. A. M. J. Turnbull, Senior Commissioner, Mwanza, to the Game Warden, Shinyanga, 25 July 1924, in reference to the report of Mr. Klein, a game hunter, Game Regulations, 1923–29, 215/PC/14/I, TNA; Acting Provincial Commissioner to District Officers, 1 March 1929, 253, Game Regulations, 1923–29, 215/PC/14/I, TNA.

131. Game Warden, Kilossa, to Honorable Chief Secretary to the Government, DSM, 26 February 1924, 62, Game Regulations, 1923–29, 215/PC/14/I, TNA.

132. Ibid.

133. Department of Veterinary Science and Animal Husbandry, Mpwapwa, to Honorable Chief Secretary, 12 August 1935, Ghee Industry, vol. 1, Secretariat Files, 18638, TNA.

134. Acting Provincial Commissioner, Mwanza, to Honorable Chief Secretary, 14 July 1930; Agricultural Development, Mwanza Province, TNA.

135. White, *Rediscovered Country*, 167, 227.

136. F. Longland, Provincial Commissioner, Report on Native Labour in Gold Mining Areas, Interim Report no. 1, Musoma District, 20 November 1935, Recruitment, Supply and Administration of Labour for the Mines, Secretariat Files, 23047, TNA.

137. *Blue Books*, 1929–48.

138. Interview with Zamberi Manyeni and Guti Manyeni Nyabwango, Sanzate, 15 June 1995.

139. Fleisher, *Kuria Cattle Raiders*, 58.

140. Interviews with Surati Wambura, Morotonga, 13 July 1995; Mang'ombe Morimi, Issenye/Iharara, 26 August 1995; Mang'oha Morigo and Paulo Machota Mongoreme, Kyandege, 8 August 2003; Philipo Haimati, Iramba, 15 September 1995; Efaristi Bosongo Gikaro, Masinki, 30 September 1995.

141. Interviews with Mwita Magige, Mosongo, 9 September 1995; Jackson (Benedicto) Mang'oha Maginga, Mbiso, 13 May 1995.

142. Andrew Roberts, "The Nyamwezi," in *Tanzania before 1900*, ed. Andrew Roberts (Nairobi: East African Publishing House, 1968), xv, 129.

143. E. P. Thompson defines the process of class formation in this way: "Class happens when some men, as a result of common experiences (inherited or shared), feel and articulate the identity of their interests as between themselves, and as against other men whose interests are different from (and usually opposed to) theirs." Thompson, *The Making of the English Working Class* (New York: Vintage Books, 1966), 9.

144. Following are partial nyangi lists for each ethnic group. Ishenyi: Msanga (birth of first child), Ebinyenyi (first teeth), Richawa, Asaro, and Moroko (circumcision), Titinyo, Egishe or Ngaruki (black-tail eldership). Interviews with Nyambeho Marangini, Issenye, 7 September 1995; Morigo (Mchombocho) Nyarobi, Issenye, 28 October 1995. Ikoma: Rigamba Rina (naming), Gotusi (for women when child weaned), Asaro, Atitinyo, Ekiriri Atato or Ekiriratero, Egesubero (for women), Ekirara Nyumba (men), Aguho, Egishe (black-tail eldership with circumcision of first daughter; the white tail is automatic when your son takes the black tail). Interviews with Surati Wambura, Morotonga, 13 July 1995; Bokima Giringayi, Mbiso, 26 October 1995. Nata: Asaro, Etitinyo, Amaka Nyangi, Aguho, Egise, Morokingi, Ekirang'ani, Egisikero, Omoroseke, Omongibo, Omurara. Interviews with Gabuso Shoka, Mbiso, 30 May 1995; Yohana Kitena Nyitanga, Makundusi, 1 May 1995. Ngoreme: Esaro, Borano, Kukerera, Isuba, Risancho–Ekise. Interviews with Mwita Magige, Masongo, 9 September 1995; Francis Sabayi Maro, Masinki, 6 October 1995; Njaga Nyasama and Nyabori Marwa, Kemegesi, 14 September 1995. Ikizu: Amatwe (piercing of ears), Esaro (circumcision, Rosarangi), Titinyo or Borano, Isubo (woman's first child), Ekise or Ehimbo (men get black tail and women a walking stick), Amarungweta, or Magiha (guards for top elders), Esega (knows ritual for women's first child), Kibage (knows ritual for circumcision), Kegoro (for girls ritual), Murungweta, Kirang'ani (wear iron bracelets), Kirundu (advisor to the generation-set), Nebwe (white-tail advisor), Mhimaye (ivory armbands, white tail), Mchiero (two white tails). Interviews with Marimo Nyamakena and Katani Magori Nyabunga, Sanzate, 10 June 1995; Samweli M. Kiramanzera, Kurasanda, 3 August 1995; Ikota Mwisagija, Kiyarata Mzumari, Kihumbo, 5 July 1995.

145. Interview with Gabuso Shoka, Mbiso, 30 May 1995. This man holds one of the highest nyangi titles in Nata: Omongibo.

146. Ibid.

147. Interview with Jackson (Benedicto) Mang'oha Maginga, Mbiso, 13 May 1995.

148. Interview with Zamberi Masambwe, Mariwanda, 22 June 1995.

149. For descriptions of Sukuma secret societies that may have been one model for this elaboration, see Hans Cory, "Sukuma Secret Societies," 1938, CORY 191, EAF.

150. Hans Koritschoner, Laws, Manners and Customs, Society of the Bayeye, 14–3/7, n.d., reel 24, MDB.

151. H. C. Barlet, Laws Manners and Customs; Hints on a Few Common Customs, General, 11 July 1941, reel 24, 9–10, MDB.

152. Interview with Dr. Rugatiri Mekacha, Dar es Salaam, 24 May 1996.

153. Interview with Jackson (Benedicto) Mang'oha Maginga, Mbiso, 18 March 1995.

154. Many elders were recommended to me as good informants on the basis of their nyangi titles.

155. Interviews with Mwita Magige, Mosongo, 9 September 1995; Morigo (Mchombocho) Nyarobi, 28 October 1995; Bokima Giringayi, Mbiso, 26 October 1995.

156. Interview with Mahiti Kwiro, Mchang'oro, 19 January 1996. Mzee Mahiti is a nyangi Omongibo for Nata. This story was told to me numerous times, always in the context of Ikizu and Nata unity and common origins. For other versions, interviews with Jackson (Benedicto) Mang'oha Maginga, Mbiso, 18 March 1995; Tetere Tumbo, Mbiso, 5 April 1995; Mang'oha Morigo, Bugerera, 24 June 1995; Riyang'ang'ara Nyang'urara, Sarawe, 20 July 1995.

157. Among the Ikizu the highest-ranking nyangi had more formal leadership status, while among the Ikoma little differentiation existed past the black-tail eldership title that most could attain. People feared the top eldership titles of the Ngoreme and looked to them for advice when community problems arose. Interview with Elfaresti Wambura Nyetonge, interview by Kemegesi, 29 September 1995.

158. Interview with Gabuso Shoka, Mbiso, 30 May 1995.

159. Interview with Kinanda Sigara and Nyawagamba Magoto, Mchang'oro, 19 January 1996.

160. R. Malcolm, System of Government, 2, 53, MDB.

161. Chief Secretary, notes on North Mara and agricultural development, about Luo and Kuria peoples, 24 July 1932; Agricultural Development, Mwanza Province, TNA.

162. Ibid.

163. District Book Agriculture, District Agricultural Officer's Report—compiled by D. Thula, inspected by H. C. Barlet, 6 May 1941, reel 24, 3–8, MDB.

164. Director, Department of Agriculture, Agriculture: Tanganyika Territory, TNA.

165. Investigation in Labour in Gold Mining Areas, Draft Final Report, Dodoma, February 1936, to Honorable Chief Secretary, DSM, Recruitment, Supply and Administration of Labour for the Mines, Secretariat Files, 23047, TNA.

166. Austen, *Northwest Tanzania*, 97–98; Agricultural Development, Lake Province, vol. 2, Secretariat Files, 22338, TNA; White, *Rediscovered Country*, 245; D. Sturdy, Senior Agricultural Officer, Lake Province, to Director of Agricultural Production, DSM, 28 September 1944, Agriculture, General, 215/909/II, TNA. Blue book figures do not show an absolute decrease in sesame acreage.

167. Anderson, "Depression, Dust Bowl," 323–26; Sir Roger Swynnerton, interview by Deborah Bryceson, on Tanganyika Colonial Agricultural Service, 1934–51, 22 April 1985, MSS. Afr. s. 1993, RHL, typescript.

168. G. F. Sayers, Acting Chief Secretary, Government Circular of 1938 from Secretariat, October 1938, Native Agriculture, Increased Production Campaign, vol. 1, Secretariat Files, 26298, TNA.

169. Department of Agriculture to Chief Secretary, 14 October 1938, Native Agriculture, Increased Production Campaign, vol. 1, Secretariat Files, 26298, TNA; Roger Swynnerton, former Director of Agriculture, Kenya, interview by Geoffrey Masefield, Lecturer in Tropical Agriculture at Oxford, 5 November, 1970, MSS. Afr. s. 1426, RHL.

170. United Africa Company (Tanganyika), Ltd., to Colonial Secretary, DSM, 4 October 1939; reply to above from Chief Secretary to Government, October 1939,

draft; A. J. Wakefield, Director of Agriculture and Food Controller, to all Provincial Commissioners, 27 October 1939; J. C. Muir, Director of Agriculture, to Deputy Chairman, Development Commission, 10 July 1948, vol. 2, Native Agriculture, Increased Production Campaign, Secretariat Files, 26298, TNA.

171. Senior Agricultural Officer, D. Sturdy, Lake Province, to Director of Agricultural Production, DSM, 28 September 1944, Agriculture, General, 215/909/II, TNA.

172. *Blue Books*, 1937–46. Original figures: 8,400 acres in 1937; 2,000 acres in 1938; 20,000 acres in 1941; 1,700 acres in 1943; 39,500 acres in 1945; 7,200 acres in 1946.

173. H. C. Barlet, District Book Agriculture—Agriculture Officer's general report, 7 February 1941, reel 24, MDB; Interview with Nyamaganda Magoto, Mbiso, 3 August 2003.

174. Interview with Nyamaganda Magoto, Mbiso, 3 August 2003.

175. *Blue Books*, 1930–31.

176. *Blue Books*, 1943–44.

177. H. C. Barlet, Trade—Marketing systems, 19 May 1941, reel 24, 1–2 C, MDB.

178. Interviews with Mang'oha Morigo and Paulo Machota Mongoreme, Kyandege, 8 August 2003; Nyamaganda Magoto, Mbiso, 3 August 2003.

179. Agricultural Development, Lake Province, vol. 2, TNA; Swynnerton, interview by Masefield, RHL; interview with David Maganya Masama, Kemegesi, 12 August 2003. For similar conclusions about the connection of cotton and poverty, see Allen F. Isaacman and Richard Roberts, *Cotton, Colonialism, and Social History in Sub-Saharan Africa* (Portsmouth, NH: Heinemann, 1995); Allen F. Isaacman, *Cotton Is the Mother of Poverty: Peasants, Work, and Rural Struggle in Colonial Mozambique, 1938–1961* (Portsmouth, NH: Heinemann, 1996).

180. Ghee Industry, vols. 1–3: Secretariat Files, 18638, TNA; Hides and Skins Industry, vols. 3–6: Secretariat Files, 12672, TNA.

181. On the history of gold mining in Tanzania, see Kaijage, *Labor Conditions*; Lemelle, "Capital, State."

182. Provincial Administration, Annual Report 1927, Mwanza Province, 1927–28, 246 PC 1/30, TNA.

183. Provincial Commissioner, Mwanza, Lake Province, to Chief Secretary, 20 April 1936, Labour Commission, Information and Statistics, Secretariat Files, 23202, TNA.

184. F. Longland, Report on Native Labour in Gold Mining Areas, Interim Report no. 1, Recruitment, Supply and Administration of Labour for the Mines, Secretariat Files, 23047, TNA.

185. Interviews with Joseph Sillery Magoto, Mbiso, 2 August 2003; Nyamaganda Magoto, Mbiso, 3 August 2003; Mang'oha Morigo, Kyandege, 8 August 2003; David Maganya Masama, Kemegesi, 12 August 2003.

186. Interviews with Joseph Sillery Magoto, Mbiso, 2 August 2003; David Maganya Masama, Kemegesi, 12 August 2003; Wambura Tonte, Kemegesi, 12 August 2003.

187. Interview with Paulo Machota Mongoreme, Kyandege, 8 August 2003; Fleisher, *Kuria Cattle Raiders*, 24.

188. Interviews with David Maganya Masama, Kemegesi, 12 August 2003; Wambura Tonte, Kemegesi, 12 August 2003.

189. Schnee, *Deutsches Kolonial-Lexikon*, 1920, 89–90; Mining Claims Book, 1914, 60, G 45/58 L R, TNA; Senior Inspector of Mines to District Officer, Musoma, 15 August 1932, 62; Senior Inspector of Mines, B. E. Frayling, to District Officer, Musoma, 21 July 1932, Labour Matters, Lake Province, vol. 1, Secretariat Files, 12019, TNA; Mwanza District Office, 1925–26, 246/PC 1/17, TNA; Office of Inspector of Mines, Mwanza, to Controller of Mines, DSM, 31 December 1925, Mwanza District Office, 1925–26, Head: Provincial Administration, 246, TNA.

190. H. C. Barlet, Labour Notes, 4 June 1941, reel 24, 1–6, MDB; Report on Musoma Gold Field, 27 February–23 March 1939, Labour Officer, Report by Labour Officer on Gold Mines in Musoma Area, Secretariat Files, 25944, TNA.

191. *Blue Books*, 1938–65; Report on Musoma Gold Field, 27 February–23 March 1939, Labour Officer, Report by Labour Officer on Gold Mines, TNA; notification of a trade dispute between Tanganyika Mine Workers Union and Buhemba Mines, Ltd., Musoma, Re: Suspension of Masaka Wambura, 14 September 1962, 460 724/196, TNA; Report on Native Labour in Gold Mining Areas, Interim Report no. 1, Musoma District, by F. Longland, Provincial Commissioner, 20 November 1935, Recruitment, Supply and Administration of Labour for the Mines, Secretariat Files, 23047, TNA.

192. Native Labour Advisor, memorandum to R. B. Smart, Esq., New Consolidated Gold Fields, Ltd., Geita, 16 January 1952, Recruitment, Supply and Administration of Labour for the Mines, Secretariat Files, 23047, TNA; Report on Native Labour in Gold Mining Areas, Interim Report no. 1, Musoma District, by F. Longland, Provincial Commissioner, 20 November 1935, and Investigation in Labour in Gold Mining Areas, Draft Final Report, Dodoma, February 1936, to Honorable Chief Secretary DSM, Recruitment, Supply and Administration of Labour for the Mines, Secretariat Files, 23047, TNA.

193. Office of Inspector of Mines, Mwanza, to Controller of Mines, DSM, 31 December 1925, Mwanza District Office Files, TNA; Report on Native Labour in Gold Mining Areas, Interim report no. 1, Musoma District, by F. Longland, Provincial Commissioner, 20 November 1935, Recruitment, Supply and Administration of Labour for the Mines, Secretariat Files, 23047, TNA.

194. Labour Department, Annual Report, 1926, Secretariat Files, AB 452 (3046/11), TNA.

195. Permit for labor recruiting, 11 January 1944, General Correspondence, Labour Matters of Lake Province, Department of Health, 87/24 III, TNA.

196. Native Labour Advisor, memorandum to R. B. Smart Esq., New Consolidated Gold Fields, Ltd., Geita, 16 January 1952, Recruitment, Supply and Administration of Labour, TNA.

197. *Blue Books*, 1940–43; Cereal Shortage: 1946–7, Secretariat Files, 34990, TNA; Anderson, " Depression, Dust-Bowl," 331.

198. Provincial Commissioner to District Officer, Musoma, 26 October 1933, 293, Shortage of Foodstuffs: Famine Relief, vol. 2, 13252, TNA.

199. J. Fairclough, District Commissioner, Chiefs Barazas, South Mara, 11 February 1931, reel 24, 1–3, MDB.

200. *Blue Books*, 1928, 1932. Original figures: 103,000 acres; 70,000 acres.

201. Provincial Commissioner to Chief Secretary, 30 April 1931; Provincial Commissioner, Lake, to Chief Secretary, 6 July 1932, Shortage of Foodstuffs: Famine Relief, vol. 1, Secretariat Files, 13252, TNA.

202. Acting District Officer, Musoma, to Provincial Commissioner, Mwanza, Monthly Report for May 1927, 8 June 1927, 1926–29 Provincial Administration, Monthly Reports Musoma District, 215/PC/1/7, TNA.

203. *Blue Books*, 1931.

204. *Blue Books*, 1933–34. Cattle numbers from 465,084 to 346,702; goats from 376,805 to 162,218.

205. *Blue Books*, 1938–43.

206. Schnee makes no mention of cassava and says that the Washashi of Chamuriho and Baridi grew eleusine millet, based on earlier reports of Baumann and Kollman. Schnee, *Deutsches Kolonial-Lexikon*, 680–81. See also Musoma District, notes on local tribes, EAF. Baumann reports Ikizu growing cassava. Baumann, *Durch Massailand*, 39–41.

207. Interview with David Maganya Masama, Kemegesi, 12 August 2003.

208. *Blue Books*, 1927. Original figures: 2,500 acres of cassava; 90,000 acres of millet.

209. Conservator of Forests to Honorable Acting Chief Secretary, DSM, 19 January 1934; Chief Secretary to Provincial Commissioner, Eastern Province, 31 January 1934, Locust Invasion 1934, Secretariat Files, 21965, TNA.

210. Provincial Commissioner, Lake Province, to Honorable Chief Secretary, DSM, 9 September 1932; A. Sillery, Acting District Officer, to Provincial Commissioner, Lake, 22 February 1934, Shortage of Foodstuffs: Famine Relief, vol. 2, 13252, TNA.

211. Provincial Commissioner, Lake Province, to Honorable Chief Secretary, DSM, 9 September 1932, Shortage of Foodstuffs: Famine Relief, vol. 2, 13252, TNA.

212. Provincial Commissioner Lake to the Honorable Chief Secretary, DSM, 17 August 1933, Shortage of Foodstuffs, Famine Relief, vol. 2, 13252, TNA.

213. District Officer, Musoma, to Provincial Commissioner, Lake, 21 October 1933; A. Sillery, Acting District Officer, to Provincial Commissioner, Lake, 22 February 1934, 291, Shortage of Foodstuffs: Famine Relief, vol. 2, 13252, TNA.

214. C. McMahon, District Officer, Musoma, to Provincial Commissioner, Lake, 27 July 1933 (266); Provincial Commissioner to District Officer, Musoma, 26 October 1933 (293) Shortage of Foodstuffs: Famine Relief, vol. 2, 13252, TNA.

215. C. McMahon, District Officer, Musoma, to Provincial Commissioner, Lake, 27 July 1933 (266); A. Sillery, Acting District Officer, to Provincial Commissioner, Lake, 22 February 1934, Shortage of Foodstuffs: Famine Relief, vol. 2, 13252, TNA.

216. Provincial Commissioner, Lake, to Honorable Chief Secretary, DSM, 26 October 1933, 289, 312; District Officer, Musoma, to Provincial Commissioner, Lake, 24 November 1933; A. Sillery, Acting District Officer, to Provincial Commissioner, Lake, 22 February 1934, Shortage of Foodstuffs: Famine Relief, vol. 2, 13252, TNA.

217. Interview with Joseph Sillery Magoto, Mbiso, 2 August 2003.

218. Interviews with Nyamaganda Magoto, Mbiso, 3 August 2003; Mang'oha Morigo and Paulo Machota Mongoreme, Kyandege, 8 August 2003.

219. Interview with Joseph Sillery Magoto, Mbiso, 2 August 2003.

220. Interviews with Joseph Sillery Magoto, 2 August 2003; Nyamaganda Magoto, Mbiso, 3 August 2003; Mang'oha Morigo and Paulo Machota Mongoreme, Kyandege, 8 August 2003.

221. C. McMahon, District Officer, Musoma, to Provincial Commissioner, Lake, 27 July 1933, 266; Provincial Commissioner, Lake, to the Honorable Chief Secretary, DSM, August 1933, 17, Shortage of Foodstuffs: Famine Relief, vol. 2, 13252, TNA.

222. Provincial Commissioner, Lake Province, Food Supply Reports for 21 October 1935, Shortage of Foodstuffs: Famine Relief, vol. 3, Secretariat Files, 13252, TNA.

223. Monthly report for July 1927, 16 August 1927, 1926–29 Provincial Administration, Monthly Reports, Musoma District, 215/PC/1/7, TNA.

224. E. C. Baker, District Officer, to Provincial Commissioner, Monthly report for the month of February 1928, 10 March 1928, 1926–29 Provincial Administration, Monthly Reports, Musoma District, 215/PC/1/7, TNA.

225. Report on Musoma District, District Office, Musoma, to Provincial Commissioner, Mwanza, 11 January 1929, Provincial Administration, Annual Report, 1928, 246/PC/1/53, TNA.

226. Amended and Consolidated Rules for the Control of the Movement and Disposal of Native Livestock (Enacted under Section 15 of the Native Authority Ordinance) Head: Native Administration, Subhead: Government Anthropologist's Report, Kuria, 544/C/28, TNA.

227. Agricultural Statistics, Secretariat Files, 35169, TNA; interviews with Joseph Sillery Magoto, Mbiso, 2 August 2003; Nyamaganda Magoto, Mbiso, 3 August 2003.

228. Month of February 1929, 12 March 1929, Provincial Administration, Monthly Reports, 1926–29, Musoma District, 215/PC/1/7, TNA.

229. E. C. Baker, District Officer, to Provincial Commissioner, Monthly report for the month of February 1928, 10 March 1928, Provincial Administration, Monthly Reports, 1926–29, Musoma District, 215/PC/1/7, TNA.

230. Acting Provincial Commissioner to Honorable Chief Secretary, DSM, 31 May 1929, 291, Game Regulations, 1923–29, 215/PC/14/I, TNA; also Provincial Commissioner responds to Chief Secretary, 31 May 1929, Killing of Game by Natives, vol. 1, TNA.

231. Acting Game Warden to Honorable Chief Secretary, 18 April 1934, Killing of Game by Natives, vol. 1, TNA; Provincial Commissioner, Mwanza, to Honorable Chief Secretary, 13 June 1934, Killing of Game by Natives, vol. 1, TNA.

232. Provincial Commissioner, Lake, to Honorable Chief Secretary, 4 May 1939, Killing of Game by Natives, vol. 2, TNA.

233. F. A. Montague, Chief Secretary, to Provincial Commissioner, Lake, 19 April 1941, Killing of Game by Natives, vol. 2, TNA; H. C. Barlet, System of Government—Native Administration in practice, 17 May 1941, reel 24, 23–27, MDB.

234. H. C. Barlet, District Book Natural History (Game), 16 December 1940, 1–2; Barlet, Illicit Hunting, 14 August 1941, reel 24, 3–4, MDS, microfilm.

235. Annual Report 1933, Musoma District, 150, 215/924/2, TNA; Annual Report 1934, Annual Reports, Native Affairs Section, Lake Province, 215, 1135, TNA.

236. A. M. D. Turnbull, Senior Commissioner, Mwanza District, 26 January 1926, Report on Mwanza District for the Year 1925, Mwanza District Office, 1925–26, TNA.

237. Interview with Wambura Tonte, Kemegesi, 12 August 2003.

238. A. M. D. Turnbull, Senior Commissioner, Mwanza District, 26 January 1926, Report on Mwanza District for the Year 1925, Mwanza District Office Files, 1925–26, TNA.

CHAPTER 6: THE CREATION OF SERENGETI NATIONAL PARK

1. For literature questioning the monolithic view of the colonial state, see Grove, *Green Imperialism*, 7; Cooper, "Conflict and Connection"; Frederick Cooper and Ann Laura Stoler, *Tensions of Empire Colonial Cultures in a Bourgeois World* (Berkeley: University of California Press, 1997).

2. Adams and McShane, *Myth of Wild Africa*, 113–21; Elizabeth Garland, "Park Managers, Biologists, and Camera Crews: Toward a Social History of Serengeti Conservationists," Paper presented at the African Studies Association Conference, Boston, 30 October–2 November 2003.

3. Grove, *Green Imperialism*, 3–4; Neumann, *Imposing Wilderness*, 19–21; Crandell, *Nature Pictorialized*; Dunlap, *Nature and the English Diaspora*.

4. Ankar, *Imperial Ecology*, 1–6; Lewis, *Inventing Global Ecology*, 13–24; Dunlap, *Nature and the English Diaspora*, 219–42.

5. Neumann, *Imposing Wilderness*, 18; Grove, *Green Imperialism*, 13–14, 478; David Anderson and Richard Grove, "The Scramble for Eden: Past, Present, and Future in African Conservation," in *Conservation in Africa: People, Policies and Practice*, ed. David Anderson and Richard Grove (Cambridge: Cambridge University Press, 1987), 4–5.

6. Neumann, *Imposing Wilderness*, 8–10, 22; Brockington, *Fortress Conservation*.

7. Interview with Mang'oha Morigo and Paulo Machota Mongoreme, Kyandege, 8 August 2003.

8. Grove, *Green Imperialism*, 480–81.

9. G. H. Swynnerton, Game Warden, to all Provincial Commissioners, 28 June 1952, Game, Including National Parks, 41 G 1/1, TNA.

10. D. J. Jardine, Chief Secretary to the Government, An Extract from Major R. W. G. Hingston, MC, Report on his Visit to Tanganyika Territory for the Purpose of

Investigating the Most Suitable Methods of Ensuring the Preservation of its Indigenous Fauna, sent out to all Provincial Commissioners, 6 June 1931, 8, National Game Parks, vol. 1, 215/350, TNA.

11. Grove, *Green Imperialism*, 485–86. Anderson, "Depression, Dust Bowl," 326–28. Anderson and Grove argue that the crisis was more perceptual than actual. Anderson and Grove, "Scramble for Eden," 7–8.

12. Neumann, *Imposing Wilderness*, 11.

13. For a similar analysis of methodological blind spots, see Melissa Leach and Robin Mearns, "Environmental Change and Policy: Challenging Received Wisdom in Africa," in *The Lie of the Land: Challenging Received Wisdom on the African Environment*, ed. Leach and Mearns (Portsmouth, NH: Heinemann, 1996), 15.

14. Acting District Officer, Musoma, to Provincial Commissioner, Mwanza, 12 July 1931; Response from District Officer, Mwanza, to Provincial Commissioner, Mwanza, 3 July 1931, National Game Parks, vol. 1, 215/350, TNA; note from F. J. Pedler, 7 July 1939; Protection of Fauna and Flora of the Empire, CO 323/1688/42, PRO.

15. Sinclair, "Serengeti Past and Present," 8–10, 14, 22–23; A. R. E. Sinclair and Peter Arcese, "Serengeti in the Context of Worldwide Conservation Efforts," in *Serengeti II: Dynamics, Management, and Conservation of an Ecosystem*, ed. Sinclair and Arcese (Chicago: University of Chicago Press, 1995), 34–43.

16. For Neumann's argument on the fallacy of making population increase the major cause of increased pressure on the park, see *Imposing Wilderness*, 9.

17. Stewart Edward White, *Lions in the Path: A Book of Adventure on the High Veldt* (Garden City, NY: Doubleday, Page and Co., 1926), 11–13.

18. Moore, *Serengeti*, 12–17. For the account of a filming trip in Serengeti among the lions, see Hoefler, *Africa Speaks*, 69–109; see also White, *Lions in the Path*, 124–26, 144.

19. Society for the Preservation of the Fauna of the Empire to Under Secretary of State for the Colonies, 31 July 1929, Game: Indiscriminate Slaughter, Newspaper Protest, 1929, Secretariat Files, 13582, TNA; M. Turner, *My Serengeti Years*, 35–37; Government Memorandum on the Legal History of SNP, TNA.

20. Director, Department of Agriculture, Agriculture: Tanganyika Territory, 41–43, TNA.

21. J. B. Davey (Principal Medical Officer, Tanganyika Territory) to Honorable Chief Secretary, 25 January 1922; C. F. M. Swynnerton, Game Warden, Memorandum on Game Preservation and Tsetse Control in the Tanganyika Territory (with Reference to Dr. Davey's Memorandum), 46, Tsetse Fly, vol. 1, 2702, TNA.

22. Confidential Circular no. 1 of 1935, The Killing of Game by Natives, from the Secretariat, 24 January 1935, vol. 1, TNA.

23. S. A. Walden, Provincial Commissioner, Lake Province, to District Commissioner, Musoma, 30 October 1954, in response to Grant report, National Game Parks, 215/350/III, TNA.

24. D. J. Jardine, Chief Secretary to the Government, Extract from the Hingston, Report on Visit to Tanganyika Territory, National Game Parks, TNA.

25. Protection of Fauna and Flora of the Empire, CO 323/1688/42, PRO.

26. Ref. 122–Comments on the necessity of the Serengeti National Park conforming to the ideals of the Convention and the need for a committee of qualified experts, minutes, 29 August 1939; Protection of Fauna and Flora of the Empire, CO 323/1689/23, PRO.

27. Adams and McShane, *Myth of Wild Africa*, 47.

28. Protection of Fauna and Flora of the Empire, CO 323/1689/23, PRO.

29. District Officer, Oldeani, Mbulu District, Mr. Gordon Russell's report on the National Park 1950, 1 January 1950, to the Honorable Member for Agriculture and Natural Resources, DSM, 6; E. G. Rowe, Provincial Commissioner, Lake Province, to Provincial Agricultural Officer, Provincial Veterinary Officer, Provincial Forestry Officer, 20 March 1950, National Game Parks, 215/350/II, TNA.

30. D. J. Jardine, Chief Secretary to the Government, Extract from Hingston, Report on his Visit to Tanganyika Territory National Game Parks, TNA.

31. Ibid.

32. Acting District Officer, Musoma, to Provincial Commissioner, Mwanza, 12 July 1931, National Game Parks, 215/350/I, TNA.

33. T. J. R. Dashwood, Member for Agricultural and Natural Resources, to Game Warden, Moshi, 25 November 1948, 232, National Game Parks, 215/350/I, TNA.

34. Lake Province, Report of the Serengeti Committee of Inquiry, 1957, FT3/599. Conservation Problems Overseas, Tanganyika, PRO; Case for Serengeti National Park Boundary Revision, 208, National Game Parks, vol. 1, 215/350, TNA.

35. Draft Recommendations of the National Park Boundaries Committee, Arusha, 8 October 1948, 120, National Game Parks, 215/350/I, TNA.

36. Memorandum on Masai History and Mode of Life, Prepared by H. St. J. Grant, District Officer in the Masai District, August 1950–December 1954, 10, National Game Parks, 215/350/IV, TNA.

37. District Commissioner, Musoma, to Provincial Commissioner, Lake Province, 18 December 1954, Confidential, National Game Parks, 215/350/III, TNA.

38. Governor, Tanganyika Territory, to A. B. Cohen, Colonial Office, London, 28 April 1950, Development in Tanganyika, Masai District, CO 691/208/6, PRO.

39. Provincial Commissioner, Northern, to Board of Trustees, letter of protest from Maasai for Being Removed from Crater, State Their Case, 2 August 1951, vol. 4, Secretariat Files, 24979, TNA. Some of the first visitors to the new park were representatives of the American Museum of Natural History who wanted to make a film of Maasai "habits and customs and life in general." Request for a filming safari to Ngorongoro Crater, Serengeti Plains, and Loliondo, vol. 5, Secretariat Files, 24979, TNA.

40. Game Warden, Chairman of Board of Management, Serengeti National Park, Memorandum for Consideration by the Trustees of the SNP regarding the position of and policy to be adopted towards the Masai Resident in the Park, 12 February 1952, Swynnerton, vol. 5, Secretariat Files, 24979, TNA.

41. Memorandum for Proposed Partition Plan for Serengeti National Park, 28 September 1955, National Game Parks, 215/350/III, TNA.

42. Oltimbau ole Masiaya, Memorandum on the Serengeti National Park, Signed on behalf of the Masai of the National Park, 1957, MSS. Afr. s. 1237, RHL.

43. Minutes of Meeting held at Office of Member for Agriculture and Natural Resources (MANR's Office), 14 June 1950, National Game Parks, 215/350/II, TNA.

44. Grant, Report on Human Habitation of the SNP, TNA.

45. Notes on Proposed Northern/Lakes Province Boundary Adjustment, National Game Parks, 215/350/III, TNA. (These notes are presumably by the director of national parks, Lt. Col. P. G. Molloy, from other responses to his comments, but no author and no date, found in papers responding to the Grant report. The quote is from a meeting of the park's administration to discuss Grant's proposal.)

46. Game Warden, Chairman of Board of Management, Serengeti National Park, Memorandum for Consideration by the Trustees of the SNP, TNA; District Commissioner, Musoma, to Provincial Commissioner, Lake, Confidential, 25 September 1954, National Game Parks, 215/350/III, TNA.

47. M. Turner, *My Serengeti Years*, 41–42. The Musoma district commissioner counted 140 adult male Maasai and their families in the Lake Province portion of the park in 1954. District Commissioner, Musoma, to Provincial Commissioner, Lake, Confidential, 25 September 1954, National Game Parks, 215/350/III, TNA.

48. District Commissioner, Musoma, to Provincial Commissioner, Lake Province, 18 December 1954, Confidential; S.A. Walden, Provincial Commissioner, Lake, to Honorable Member for Local Government, Secretariat, DSM, 11 August 1954, National Game Parks, 215/350/III, TNA.

49. Memorandum for Proposed Partition Plan for SNP, TNA.

50. Government Paper no. 5, Legislative Council of Tanganyika, Proposals for Reconstituting the Serengeti National Park, 1958, Government Printer, DSM, Game, Including National Parks, 41 G 1/1, TNA, 8; M. Turner, *My Serengeti Years*, 42–44; Memorandum for Proposed Partition Plan for SNP, TNA.

51. Government Paper no. 5, Legislative Council of Tanganyika, Proposals for Reconstituting the Serengeti National Park, 1958, Government Printer, DSM, Game, Including National Parks, 41 G 1/1, TNA; M. Turner, *My Serengeti Years*, 42–44; Serengeti National Park, Legislative Council of Tanganyika, Sessional Paper no.1, 1956, Government Printer, DSM, 1956.

52. Memorandum for Proposed Partition Plan for SNP, TNA.

53. W. David, "Must Such a Small Problem Cost Us the Serengeti?" *Sunday News*, 29 January 1957, 3 (629), National Game Parks, 215/350/III, TNA.

54. *East African Standard*, Nairobi, 20 April 1956, National Game Parks, 215/350/IV, TNA.

55. Kay Turner, *Serengeti Home* (New York: Dial Press/James Wade, 1977), 12.

56. G. H. Swynnerton, Memorandum, Game of the Serengeti National Park, Tanganyika, National Game Parks, 215/350/IV, TNA; also found in MSS Afr. s. 1238, RHL. See also M. Turner, *My Serengeti Years*, 44–47; Memorandum on Serengeti National Park, RHL; Proposals for Reconstituting the Serengeti National Park, Government Paper no. 5, National Game Parks, 215/350/IV, TNA.

57. M. Turner, *My Serengeti Years*, 47.

58. Report on an Ecological Survey, PRO; F. H. Page-Jones, Acting Chief Secretary to Government, DSM, Memorandum by Tanganyika Government with Reference to Report by Professor W. H. Pearsall, National Game Parks, 5 June 1957, 215/350/IV, TNA.

59. M. Turner, *My Serengeti Years*, 171–72.

60. K. Turner, *Serengeti Home*, 50–51.

61. R. de S. Stapledon, Chief Secretary, to all Provincial Commissioners, Fauna Conservation, 3 March 1956, Game, Including National Parks, 41 G 1/1, TNA.

62. Proposals for Reconstituting the Serengeti National Park, Government Paper no. 5, National Game Parks: 215/350/IV, TNA; M. Turner, *My Serengeti Years*, 48, 87.

63. Report of the Serengeti Committee of Enquiry, PRO.

64. East African Office, London, to Colonial House, London, 20 August 1958, appearance on BBC television, 21 August 1958, to discuss Serengeti; Future of the Serengeti National Park, Tanganyika, CO 847/65–64, PRO.

65. Confidential for Secretary of State's Signature, Sir R. Turnbull, Government House, 15 November 1958, from Governor, Tanganyika, to Secretary of State for Colonies, London; Future of Serengeti National Park, PRO.

66. Grzimek, *Serengeti Shall Not Die*, 169–73, 230.

67. F. R. Cowell of Thames and Hudson, Ltd., London, to J. K. Thompson, Colonial Office, 8 April 1959, Future of Serengeti National Park, PRO.

68. Grzimek, *Serengeti Shall Not Die*, 20–48, 153, 224, 245, 310; M. Turner, *My Serengeti Years*, 154.

69. Conservation Problems Overseas, PRO.

70. K. Turner, *Serengeti Home*, 94–99.

71. The assumed dominance of this view of the landscape is explored in Anderson and Grove, "Scramble for Eden," 3.

72. On the politics of rain in independent Tanzania, see Feierman, *Peasant Intellectuals*.

73. K. Turner, *Serengeti Home*, 142, 172.

74. Ibid., 202.

75. Julius K. Nyerere, *Freedom and Unity: Uhuru na umoja: A Selection from Writings and Speeches*, 1952–65 (London: Oxford University Press, 1967), xi–xii.

76. The conference was formally titled "The Conference on the Conservation of Nature and Natural Resources in Modern African States," held at Arusha, Tanzania, September 1961. Adams and McShane, *Myth of Wild Africa*, 113–14; Peter Rogers, "Global Governance/Governmentality, Wildlife Conservation, and Protected Area Management: A Comparative Study of Eastern and Southern Africa" (with permission of the author), paper presented at the African Studies Association, 45th Annual Meeting, Washington, DC, 2002, 17–19; Raymond Bonner, *At the Hand of Man: Peril and Hope for Africa's Wildlife* (New York: Knopf, 1993), 65.

77. Neumann, *Imposing Wilderness*, 142–44.

78. Fleisher, *Kuria Cattle Raiders*, 9, 71.

79. Ibid., 1–3.

80. Ibid., 115–28, 131–67.

81. Makacha, "Extending Robanda WMA," author's collection.

82. Neumann, *Imposing Wilderness*, 7, 166.

83. Thompson, *English Working Class*, 99.

84. Yusuf Q. Lawi, "Tanzania's Ujamaa Villagization and Local Ecological Consciousness: The Case of Eastern Iraqwland, 1974–1976." Paper presented at the African Studies Association conference, "Listening (Again) for the African Past: A Five-College Project on Producing Historical Knowledge," Smith College, 18–26 October 2003.

85. M. Turner, *My Serengeti Years*, 13, 20, 23, 169–71.

86. Ibid., 180, 183, 184, 188–89, 192, 194. For the militarization of the parks, see Neumann, *Imposing Wilderness*, 6–7.

87. K. Turner, *Serengeti Home*, 41, 42–45, 62.

88. M. Turner, *My Serengeti Years*, 167–69.

89. Grzimek, *Serengeti Shall Not Die*, 217–19.

90. Lt. Col. P. G. Molloy, Director of National Parks, Native Poaching on the Western Serengeti Boundaries, 1957, Fauna and Flora International Archives, London.

91. Audrey Lousada, *Poachers in the Serengeti* (New York: Walker, 1965), 139.

92. M. Turner, *My Serengeti Years*, 167–69, 187.

93. K. Turner, *Serengeti Home*, 49.

94. Hoefler, *Africa Speaks*, 64, 115, 145.

95. A. Moore, *Serengeti*, 176, 178–80. See also Molloy, Native Poaching on the Western Serengeti Boundaries.

96. M. Turner, *My Serengeti Years*, 185, 192.

97. K. Turner, *Serengeti Home*, 50, 53–54.

98. M. Turner, *My Serengeti Years*, 193, 195, 196–97.

99. Ibid., 174–75, 177, 184, 185, 188, 192–93, 195.

100. Ibid., 67.

101. K. Turner, *Serengeti Home*, 61.

102. Political District Commissioner, Tarime, to Game Ranger, Musoma, 26 May 1956, 230, General Correspondence, Game, Including National Parks, 436/G.1, TNA.

103. Grzimek, *Serengeti Shall Not Die*, 160.

104. On the dynamics of postcolonial development, see Goran Hyden, *Beyond Ujamaa in Tanzania: Underdevelopment and an Uncaptured Peasantry* (Berkeley: University of California Press, 1980); David A. McDonald, and Eunice Njeri Sahle, *The Legacies of Julius Nyerere: Influences on Development Discourse and Practice in Africa* (Trenton: Africa World Press, 2002); Dean E. McHenry, *Limited Choices: The Political Struggle for Socialism in Tanzania* (Boulder: L. Rienner, 1994).

105. R. R. Matango, "Operation Mara: The Paradox of Democracy," *Majimaji* (TANU Youth League, DSM) 20 (1975): 17–29.

106. Lawi, "Tanzania's Ujamaa Villagization." On ujamaa, see Dean E. McHenry, *Ujamaa Villages in Tanzania: A Bibliography* (Uppsala: Scandinavian Institute of African Studies, 1981). For the Mara region, see Lenin Bega Kasoga," An Evaluation of the Ujamaa Village Policy: A Case Study of Musoma Vijijini District in Tanzania, 1974–1987," 2 vols. (PhD diss., Michigan State University, 1990); Mustafa, "Development of Ujamaa."

107. Neumann, *Imposing Wilderness*, 145.

108. Interviews with Nyamaganda Magoto, Bugerera, 3 March 1995; Mbiso, 2 August 2003.

109. Interview with Keneti Mahembora, Sang'anga, 17 February 1996.

110. Interview with Joseph Sillery Magoto, Mbiso, 2 August 2003.

111. Makacha, "Extending Robanda WMA," author's collection.

112. Interview with Joseph Sillery Magoto, Mbiso, 2 August 2003.

113. K. Turner, *Serengeti Home*, 53.

114. Interviews with Kisenda Mwita, Hezekia Marwa Sarya, Philemon Mbota, Mugumu/Matare, 15 March 1996; Peter Mgosi Siwa, Morotonga, 23 August 2003.

115. District Commissioner, Musoma, to Provincial Commissioner, Lake, 19 March 1955, National Game Parks, 215/350/III, TNA.

116. Kuroz Feroz, "Human Population Densities and Their Changes around Major Conservation Areas of Tanzania," Bureau of Resource Assessment and Land Use Planning, University of Dar es Salaam, n.d., 65, EAF.

117. Issa G. Shivji, *Not Yet Democracy: Reforming Land Tenure in Tanzania* (Dar es Salaam: IIED/HAKIARDHI/Faculty of Law, University of Dar es Salaam, 1998), 2–7. For the literature on land tenure in Africa and Tanzania, see R. W. James, *Land Tenure and Policy Tanzania* (Toronto: University of Toronto Press, 1971); Elizabeth Colson, "The Impact of the Colonial Period on the Definition of Land Rights," in *Profiles of Change, African Society and Colonial Rule*, ed. Victor Turner (Cambridge: Cambridge University Press, 1971); Thomas J. Bassett and Donald E. Crummey, eds., *Land in African Agrarian Systems* (Madison: University of Wisconsin Press, 1991); Parker Shipton and Mitzi Goheen, "Understanding African Land-Holding: Power, Wealth, and Meaning," *Africa* 62, no. 3 (1992): 307–25.

118. Mtaala wa Sera na Sheria za Ardhi, Msitu, na Wanyamapori, SRCP, Ardi, Sehemu ya Kwanza, n.d., author's collection.

119. Shivji, *Not Yet Democracy*, 109.

120. Interview with Simion Hunga Nason, agricultural extension officer and his wife Bhoke Mtoka, Kemegesi, 14 August 2003.

121. See E. P. Thompson, *Whigs and Hunters: The Origin of the Black Act* (New York: Pantheon Books, 1975).

122. G. B. Mitchell, Secretary, Board of Trustees, Tanganyika National Parks, to Provincial Commissioner, Lake Province, 7 July 1952; Provincial Commissioner, Lake, to Board of Trustees, TNP, 4 September 1952; Government Paper no. 5, Legislative Council of Tanganyika, Proposals for Reconstituting the SNP, 1958, Government Printer, DSM, Game, Including National Parks, 41 G 1/1, TNA; Game

Ranger, Banagi Hill, to Park Warden, Ngorongoro, 21 May 1953, National Game Parks, 215/350/II, TNA.

123. Veterinarian of Sukumaland Development, to Provincial Commissioner, Lake, 26 February 1949, 275, National Game Parks, 215/350/I, TNA; District Commissioner, Maswa, to Provincial Commissioner, Lake, 24 August 1951. 359, National Game Parks, 215/350/II, TNA; Scout G. R. Phiri, Banagi Hill, to Mtemi Makongoro Ikizu, 17 February 1947, 129; M. M. Matutu (Makongoro) Mtemi wa Ikizu to District Commissioner, Musoma, 11 April 1948; District Commissioner, Musoma, to Game Warden, Moshi, 12 October 1948, 205; District Commissioner, Musoma to Provincial Commissioner, Lake, 30 December 1948, 247; Field Officer—Safari, 9 January 1948, Handa Jega to Simba Kopjes, 12 March 1948; Game Warden to Provincial Commissioner, Lake, National Game Parks, 215/350/I, TNA; Game Warden, Lyamungu, Moshi, to Provincial Commissioner, Lake, 17 November 1950, 343; G. R. Phiri, Game Scout, Banagi Hill, to Political Officer, Maswa, 13 September 1950 (341), National Game Parks, 215/350/II, TNA.

124. District Commissioner, Musoma, to Game Warden, Arusha, 18 July 1957, re: your letter 20 June 1957 to Honorable Member for Agriculture and Natural Resources; Provincial Commissioner, Lake, to Game Warden, 1 August 1957, Proposed Extension to SNP, National Game Parks, 215/350/III, TNA.

125. Report of Game Ranger, Banagi Hill, to Game Warden, Tengeru, Arusha, 9 March 1953, 418, National Game Parks, 215/350/II, TNA; Col. E. C. T. Wilson, District Commissioner, North Mara, to J. Blower, Game Ranger, Banagi Hill, 6 July 1953; John Blower, Game Ranger, Banagi Hill, to Wilson, District Commissioner, North Mara, 11 August 1953, 37; Game Warden to District Commissioner, Tarime, 11 June 1954, 178, Game, Including National Parks, General Correspondence, 436/G.1, TNA.

126. M. Turner, *My Serengeti Years*, 50–53; Mara Region, Annual Report, 1964, 11, TNA.

127. Extract from monthly letter from Member for Agriculture and Natural resources file 2460/A/24 from Secretariat, 24 April 1953, National Game Parks, 215/350/II, TNA. For an evaluation of buffer zones, see Neumann, "Primitive Ideas," 559–82.

128. G. H. Swynnerton, Game Warden, to all Provincial Commissioners, 28 June 1952, Game, Including National Parks, 41 G 1/1, TNA.

129. Game Ranger, Banagi to Provincial Commissioner, Mwanza, 3 November 1952, Game, Including National Parks, 41 G 1/1, TNA.

130. Report of the Serengeti Committee of Enquiry, 1957, Government Printer, DSM, 1957, 24, FT 3/599; Conservation Problems Overseas, PRO.

131. Mtoni, "Involve Them," 39; interview with Joseph Tareta Masina, SEPDA director, Issenye, 7 August 2003.

132. Mara Region, Annual Report, 1964, 11, TNA.

133. Interview with Joseph Sillery Magoto, Mbiso, 2 August 2003.

134. M. Turner, *My Serengeti Years*, 53–54.

135. Feroz, "Human Population Densities," 65, EAF.

136. Occasional Paper no. 1, 1956, Legislative Council of Tanganyika, "The Serengeti National Park," Government Printer, DSM booklet, Game, Including National Parks, 41 G 1/1, TNA.

137. Interview with Wambura Tonte, Kemegesi, 12 August 2003.

138. Interviews with George Wambura Gehamba, Samweli Muya Mongita, Morotonga, 21 August 2003.

139. Interviews with Mang'ombe Makuru, Morotonga, 28 August 2003; Tetere Tumbo, Mbiso, 11 August 2003.

140. Interview with Peter Mgosi Siwa, Morotonga, 23 August 2003.

141. Interviews with George Wambura Gehamba, Samweli Muya Mongita, Morotonga, 21 August 2003.

142. Hill, "Zimbabwe's Wildlife Utilization," 103–4.

143. Stefano Ponte, *Farmers and Markets in Tanzania: How Policy Reforms Affect Rural Livelihoods in Africa* (Oxford: James Currey, 2002); Aili Mari Tripp, *Changing the Rules: The Politics of Liberalization and the Urban Informal Economy in Tanzania* (Berkeley: University of California Press, 1997); Julius Edo Nyang'oro and Timothy M. Shaw, *Beyond Structural Adjustment in Africa: The Political Economy of Sustainable and Democratic Development* (New York; Praeger, 1992).

144. Serengeti National Park Management Plan, 1991–95, 46, 60, author's collection; Sinclair, "Serengeti Past and Present," 8–9.

145. Neumann, *Imposing Wilderness*, 144.

146. Sinclair, "Serengeti Past and Present," 8–10, 14, 22–23; Sinclair and Arcese, "Serengeti in Context," 34–43.

147. Sinclair, "Serengeti Past and Present," 8–10, 14, 22–3; Sinclair and Arcese, "Serengeti in Context," 34–43; interview with Herbert Kiyata, SRCP, Morotonga, 18 August 2003.

148. Makacha, "Extending Robanda WMA," author's collection.

149. SRCP, Lengo la Serkikali kuwashirikisha wananchi katika Uhifadhi was Maliasili: Warsha ya uongozi na usimamizi bora wa shughuli za uhifadhi wa maliasili vijijini, 18–20 May 2000, author's collection. On Community Conservation philosophy, see Kiss, *Living with Wildlife*; Edmund Barrow, *Community Conservation Approaches and Experiences from East Africa*, Rural Extension Bulletin no. 10, Theme Issue on Community Conservation, University of Reading Agricultural Extension and Rural Development Department, 1996; P. Bergin, *Reforming a Conservation Bureaucracy: TANAPA and Community Conservation*, Community Conservation in Africa Working Paper, Institute for Development Policy and Management, University of Manchester, 1998.

150. Sinclair and Arcese, "Serengeti in Context," 40–43; B. N. N. Mbano, R. C. Malpas, M. K. S. Maige, P. A. K. Symonds, and D. M. Thompson, "The Serengeti Regional Conservation Strategy," in *Serengeti II: Dynamics, Management, and Conservation of an Ecosystem*, ed. A. R. E. Sinclair and Peter Arcese (Chicago: Chicago University Press, 1995), 605–16; Wim Olthof, "Wildlife Resources and

Local Development: Experiences from Zimbabwe's CAMPFIRE Programme," in *Local Resource Management in Africa*, ed. J. P. M. van den Breemer, C. A. Drijver, and L. B. Venema (New York: Wiley, 1995): 111–28.

151. Lucy Emerton and Iddi Mfunda, *Making Wildlife Economically Viable for Communities Living around the Western Serengeti, Tanzania* (London: International Institute for Environment and Development, 1999).

152. Edmund Barrow, Helen Gichohi, and Mark Infield, "The Evolution of Community Conservation Policy and Practice in East Africa," in *African Wildlife and Livelihoods: The Promise and Performance of Community Conservation*, ed. David Hulme and Marshall Murphree (Oxford: James Currey, 2001): 59–73; Songorwa, "Community-Based Wildlife Management," 2061–79.

153. Kiss, *Living with Wildlife*, 183; Wells, Brandon, and Hannah, *People and Parks*, 47.

154. Clark C. Gibson, "The Consequences of Institutional Design: The Impact of 'Community-Based' Wildlife Management Programs at the Local Level," in *Politicians and Poachers: The Political Economy of Wildlife Policy in Africa*, ed. Clark C. Gibson (Cambridge: Cambridge University Press, 1999), 147–50.

155. Patrick Bergin, "Accommodating New Narratives in a Conservation Bureaucracy: TANAPA and Community Conservation," in *African Wildlife and Livelihoods: The Promise and Performance of Community Conservation*, ed. David Hulme and Marshall Murphree (Oxford: James Currey, 2001): 88–105; A. Kauzeni and H. Kiwasila, "Serengeti Regional Conservation Strategy: a Socio-Economic Study," Institute of Resource Assessment, University of Dar es Salaam, 1994; I. Mfunda, *The Economics of Wildlife for Communities Living around the Serengeti National Park, Tanzania* (Fort Ikoma: Serengeti Regional Conservation Strategy, 1998).

156. SRCP, Mwelekeo wa uvunaji wa wanyamapori kwa jamii katika kanda ya Serengeti: Warsha—Usimamizi bora wa shughuli za uhifadhi was maliasili vijijini, Bunda, 18–20 May 2000, author's collection; Sinclair and Arcese, "Worldwide Conservation Efforts," 39; interviews with John Muya, SRCP, Fort Ikoma, 22 August 2003; Herbert Kiyata, SRCP, Morotonga, 18 August 2003.

157. SCRP, Mradi wa Hifadhi ya Kanda ya Serengeti SRCP, Mwongozo wa muundo na utaratibu wa utekelezaji wa shughuli za uhifadhi kwanjia ya ushirikishaji wananchi katika ngazi ya kijiji, Idara ya Wanyamapori, Wizara ya Maliasili na Utalii, March 1999, author's collection; Wildlife Conservation Act, 1974, DSM, Government Printer, 27 December 2002, author's collection.

158. Interview with Peter Mkome Mahembora, Mbiso, 4 August 2003.

159. SCRP, Mradi wa Hifadhi ya Kanda ya Serengeti; SRCP, Uzoefu uliopatikana kutokanana na matumizi ya mfuko wa uhifadhi kwa jamii, vijiji teule, chini ya mradi wa Hifadhi Kanda ya Serengeti (SRCP) (Warsha), Bunda, 18–20 May 2000, author's collection; interview with Peter Mkome Mahembora, Mbiso, 4 August 2003.

160. SRCP, Mwelekeo wa uvunaji.

161. SRCP, Uzoefu Uliopatikana kutokanana.

162. For a review of the forest laws, see SRCP, Andrew B. Maregesi, Afisa Misitu, Msaidizi Mwandamizi, Sera Mpya ya Misitu: Usimamizi shirikishi wa misitu kama njia pekee ya kuhifadhi misitu: Mada iliyotolewa katika mafunzo ya Kamati za Hifadhi za Vijiji vilivyoko china ya Mradi wa Hifadhi kanda ya Serengeti (SRCP) Katika Wilaya ya Tarime, Serengeti, na Bunda, June 2002, author's collection.

163. SRCP, Lengo la Serkikali, author's collection; SRCP, Sera ya Wanyamapori: Mada iliyoandaliwa na Ndugu Mokiri Warento, Afisa Wanyamapori (W) Serengeti, katika Semina ya Uhifadhi iliyofanyika Ukumbi wa Lutheran, Mjini Mugumu, 25–27 July 2002, author's collection.

164. Maeneo ya Jumuiya ya Hifadhi ya Wanyama Pori (WMAs): Mhutasari wa Sera na Kanuni, 2003, author's collection; Vernon Booth, George Nangale, Hamudi Majamba, "Procedures for Communities to Enter into Joint Ventures in Wildlife Management Areas," report prepared for Tanzanian government and financed by EPIQ-USAID/Tanzania and GTZ-Tanzania, Ministry of Natural Resources and Tourism, United Republic of Tanzania, 2000 (Dar es Salaam: United Republic of Tanzania, Ministry of Natural Resources and Tourism, 2000); Markus Borner, "Conservation with the People—For the People: Establishment of Community Wildlife Management Areas (WMAs) around the Serengeti National Park," Serengeti National Park Service, http://www.serengeti.org/download/Conservation.pdf.

165. Maeneo ya Jumuiya ya Hifadhi; Wildlife Conservation Act, 1974, author's collection; Workshop Results of the Stakeholders Meeting for Establishment of Ikona Wildlife Management Area, workshop held at Mugumu, Serengeti District, a Programme to Establish Ikona Wildlife Management Area, May 1999, in possession of participant, author's collection.

166. Makacha, "Extending Robanda WMA," author's collection.

167. Minutes of the Interim Committee Meeting for IKONA Wildlife Management Area, Mugumu, 4 October 1999, in possession of member, author's collection.

168. Makacha, "Extending Robanda WMA."

169. Workshop proceedings, WMA: Policy and Regulations Facilitated by the Wildlife Department and Frankfurt Zoological Society, notes taken and recorded by Yannick Ndoinyo, 2003, author's collection.

170. Workshop results of Stakeholders Meeting for Ikona.

171. Makacha, "Extending Robanda WMA."

172. Neumann, Imposing Wilderness, 8; Hill, "Zimbabwe's Wildlife Utilization," 103–21.

173. Vincent Shauri, "The New Wildlife Policy in Tanzania: Old Wine in a New Bottle?" LEAT (Lawyers' Environmental Action Team), http://www.leat.or.tz/publications/wildlife.policy/index.php.

174. Interviews with Jackson Mteba Mabura, Morotonga, 23 August 2003; Raymond Nyamasagi, Morotonga, 27 August 2003; Mang'ombe Makuru, Morotonga, 28 August 2003.

175. Interview with Jackson Mteba Mabura, Morotonga, 23 August 2003.

176. Interview with David Maganya Masama, Kemegesi, 12 August 2003.

177. Interview with Nyamuko Soka, Morotonga, 28 August 2003.

178. Interview with Jackson Mteba Mabura, Morotonga, 23 August 2003.

179. Interview with Raymond Nyamasagi, Morotonga, 27 August 2003.

180. Shauri, "New Wildlife Policy in Tanzania."

181. Interviews with John Muya, SRCP, Fort Ikoma, 22 August 2003; Herbert Kiyata, Morotonga, 18 August 2003.

182. Edward Paul Machumu, Leadership and Problems of Policy Implementation: A Case Study of Two Villages in Musoma District" (MA thesis, University of Dar es Salaam, 1987); D. W. O. Oming'o, "Self-Help Schemes in South Mara District," Political Science Paper 7 (PhD diss., University of East Africa, University College DSM, 1970).

183. Makacha, "Extending Robanda WMA," author's collection; interview with Peter Mkome Mahembora, Mbiso, 4 August 2003.

184. Workshop Results of Stakeholders Meeting for Ikona, author's collection. Although VIP refused to bring their title to the 1999 meeting as ordered, their lease included 1,958 hectares in plots 96 and 97, of which 1,917 hectares were for the construction of a lodge or tented camp and 4.16 hectares were for a residential area.

185. Interview with Peter Mkome Mahembora, Mbiso, 4 August 2003; Robert M. Poole, "Heartbreak on the Serengeti," *National Geographic*, February 2006, 16.

186. Interviews with Faustine Wanzagi Nyerere, Public Relations Officer, VIP, Mbiso, 6 August 2003. Barbara Schachenmann, Community Officer, VIP, Sasakwa, 6 August 2003.

187. Poole, "Heartbreak," 16.

188. Interviews with Joseph Sillery Magoto, Mbiso, 2 August 2003; Mang'oha Morigo and Paulo Machota Mongoreme, Kyandege, 8 August 2003.

189. Various informants who remain anonymous. Quote from interview with George Wambura Gehamba and Samuweli Muya Mongita, Morotonga, 21 August 2003.

190. Interview with Jackson Mteba Mabura, Morotonga, 23 August 2003.

191. Interview with Mang'oha Morigo and Paulo Machota Mongoreme, Kyandege, 8 August 8 2003.

192. Interviews with David Maganya Masama, Wambura Tonte, Kemegesi, 12 August 2003.

193. Feroz, "Human Population Densities," 65. A 1999 FZS survey found that the recruitment rate for Nata was 6 percent, Makundusi 4 percent, and the rest 3 percent. However, between 1988 and 1997, 446 people moved out of Nata, 85 out of Nyichoka, and 613 into Makundusi, while the natural rate of population increase in the districts making up the Serengeti-Maswa conservation area was the lowest, at 2.7 percent.

194. Interview with Mang'oha Morigo and Paulo Machota Mongoreme, Kyandege, 8 August 2003.

195. Naomi Zakayo Kaitira, Taarifa ya Kikao, Rogoro Cultural and Museum Group, Mugumu, 23 June 2003, author's collection.

196. Also argued by Hofmeyr in *We Spend Our Years*.

197. Fentress and Wickham, *Social Memory*, 11–14.

198. Anderson and Grove, *Conservation in Africa*, 3–4, 9–10. For an anthropological argument see, Croll and Parkin, *Bush Base*, 3–9.

199. Cronon, "Trouble with Wilderness," 24.

200. Ibid., 24–25.

201. Cronon, "Place for Stories," 1375.

202. For a similar argument about the Amazon, see Slater, *Entangled Edens*, 187–88, 190, 204.

203. For an argument about the need for historical research and against one-size-fits-all solutions see, Giles-Vernick, *Cutting the Vines*, 6–7, 201–2. One of the criticisms of the model CAMPFIRE community conservation program in Zimbabwe was that it did not take into consideration various local histories or local interests and attitudes about wildlife. Jocelyn Alexander and Jo Ann McGregor, "Wildlife and Politics: CAMPFIRE in Zimbabwe," *Development and Change* 31 (November 2000): 625.

Bibliography

Author's Collection

Haimati, P., and P. Houle. "Mila na matendo ya Wangoreme." Iramba Mission, 1969. Unpublished mimeo.

Kaitira, Naomi Zakayo. Taarifa ya Kikao. Rogoro Cultural and Museum Group, Mugumu, 23 June 2003.

Kikao cha mila, desturi na asili ya kabila la Waishenyi kilichokutana tarehe 6/6/1990. Nyiberekera, Ishenyi.

Maeneo ya Jumuiya ya Hifadhi ya Wanyama Pori (WMAs): Mhutasari wa Sera na Kanuni. 2003.

Magoto, Nyamaganda. Cultural vocabulary list, Mbiso and Bugerera, 1995–96.

Makacha, Stephen. IKONA. "Extending Robanda Wildlife Management Area Now Called IKONA to Cover Four Villages, Mugumu District, Mara Region, Tanzania (Robanda, Nyichoka, Makundusi/Nyakitono, and Nattambiso)." Report to Pia Zimmerman c/o Frankfurt Zoological Society, May 1999, SNP, TZ.

Maregesi, Andrew B., Afisa Misitu, and Msaidizi Mwandamizi. Lengo la Serkikali kuwashirikisha wananchi katika Uhifadhi was Maliasili: Warsha ya uongozi na usimamizi bora wa shughuli za uhifadhi wa maliasili vijijini, 18–20 May 2000.

———. Mtaala wa sera na sheria za ardhi, msitu, na wanyamapori. SRCP. Ardi, n.d.; Sehemu ya Kwanza, n.d.

———. Mwelekeo wa uvunaji wa wanyamapori kwa jamii katika kanda ya Serengeti: Warsha—Usimamizi bora wa shughuli za uhifadhi was maliasili vijijini, Bunda, 18–20 May 2000.

———. Sera Mpya ya Misitu: Usimamizi shirikishi wa misitu kama njia pekee ya kuhifadhi misitu. Mada iliyotolewa katika mafunzo ya Kamati za Hifadhi za Vijiji vilivyoko china ya Mradi wa Hifadhi kanda ya Serengeti (SRCP) katika Wilaya ya Tarime, Serengeti, na Bunda, June 2002.

———. Sera ya Wanyamapori: Mada iliyoandaliwa na Ndugu Mokiri Warento, Afisa Wanyamapori (W) Serengeti, katika Semina ya Uhifadhi iliyofanyika Ukumbi wa Lutheran Mjini, Mugumu, 25–27 July 2002.

———. Uzoefu uliopatikana kutokanana na matumizi ya mfuko wa uhifadhi kwa jamii, vijiji teule, chini ya mradi wa Hifadhi Kanda ya Serengeti (SRCP) (Warsha). Bunda, 18–20 May 2000.

Maryknoll Fathers, Iramba Parish. "Ngoreme-English Dictionary." n.d.

Minutes of the Interim Committee Meeting for IKONA Wildlife Management Area. Held on 4 October 1999, Mugumu. In possession of member.

Mradi wa Hifadhi ya Kanda ya Serengeti SRCP. Mwongozo wa muundo na utaratibu wa utekelezaji wa shughuli za uhifadhi kwanjia ya ushirikishaji wananchi katika ngazi ya kijiji, Idara ya Wanyamapori, Wizara ya Maliasili na Utalii, March 1999.

Ngokolo, Mtemi Seni. "Historia ya Utawala wa Nchi ya Kanadi ilivyo andikwa na marahemu Mtemi Seni Ngokolo mnamo tarehe 10 June 1928." Typescript provided by his son, Mtemi Mgema Seni, 20 May 1971, to Buluda Itandala.

Serengeti National Park Management Plan. 1991–95.

Siso, Zedekia Oloo. Oral traditions of North Mara collected by Zedekia Oloo Siso, Buturi, Tanzania, 1965–90. Unpublished accounts.

Wildlife Conservation Act, 1974. Dar es Salaam: Government Printer, 27 December 2002.

Workshop Proceedings, WMA: Policy and Regulations Facilitated by the Wildlife Department and Frankfurt Zoological Society. Notes taken and recorded by Yannick Ndoinyo, 2003.

Workshop Results of the Stakeholders Meeting for Establishment of Ikona Wildlife Management Area. Workshop held at Mugumu, Serengeti District, A Programme to Establish Ikona Wildlife Management Area, May 1999. In possession of participant.

E. C. Baker Papers

Baker, Edward Conway. *Tanganyika Papers*. Oxford: Oxford University Press, 1935. Microfilm.

East Africana Library, University of Dar es Salaam (EAF)

Cory, Hans. "Report on the Pre-European Tribal Organization in Musoma (South Mara District) and . . . Proposals for Adaptation of the Clan System to Modern Circumstances." 1945. CORY 173.

———. "Sukuma Secret Societies." 1938. CORY 191.

Feroz, Kuroz. "Human Population Densities and their Changes Around Major Conservation Areas of Tanzania." Bureau of Resource Assessment and Land Use Planning, University of Dar es Salaam.

Fosbrooke, H. A. Senior Sociologist, Tanganyika. "Masai History in Relation to Tsetse Encroachment." Arusha, 1954. CORY 254.

———. "Sections of the Masai in Loliondo Area." 1953. CORY 259.

Musoma District Book. Notes from Musoma district book on local tribes and chieftains, in German. [1912]. CORY 348.

President's Office, Planning Commission, Bureau of Statistics. Mara Regional Statistical Abstract, 1993. Dar es Salaam, June 1995.

Fauna and Flora International Archives, London

Molloy, Lt. Col. P. G. Director of National Parks. Native Poaching on the Western Serengeti Boundaries. Typescript, 1957.

Mara Region Office of Culture, Musoma

Anacleti, Odhiambo. Kijiji cha Butiama: Mila na desturi zinazohusiana na mahari na ndoa: Tarafa ya Makongoro, Wilaya ya Musoma, Mara, 26 March 1979, Utafiti na Mipango/Utamaduni, Dar es Salaam, UTV/M 13/22.

Musoma District Books, Microfilm

Baker, Edward Conway, comp. Tribal History and Legends.
Baumann, Oskar. Musoma Neighbourhood in 1892. A translation of Dr. Baumann's account of his safari through South Mara and Ukara Island with an introduction and notes by G. W. Y. Hucks.
Koritschoner, Hans. Laws, Manners and Customs. Society of the Bayeye, 14–3/7, reel 24.
Malcolm, R. S. W. System of Government. Extracts from a report, 1937.

Public Records Office, London (PRO)

Annual Report of the Game Department 1959. CO 822/2662.
Conservation Problems Overseas: Serengeti National Park, Tanganyika. FT 3/587.
Development in Tanganyika, Masai District. CO 691/208/6.
Future of the Serengeti National Park, Tanganyika. CO 847/65–64.
Protection of Fauna and Flora of the Empire. CO 323/1688/42, CO 323/1689/23.
Report of the Serengeti Committee of Enquiry, 1957. Conservation Problems Overseas, Tanganyika. FT 3/599.
Report on an Ecological Survey of the Serengeti National Park Tanganyika, November and December 1956. Conservation Problems Overseas, Tanganyika. FT 3/599.
Tanganyika Territory Blue Book, CO 726/2–30.

Rhodes House Library, Oxford (RHL)

Malcolm, D. W. Report on Land Utilization in Sukuma, 1938. MSS. Afr. s. 1445/5.
Marchant, P. J. C. Personal Letters, 1950–52. MSS. Afr. s. 1701 (1).
Memorandum on the Serengeti National Park, signed on behalf of the Masai of the National Park by Oltimbau ole Masiaya, 1957. MSS. Afr. s. 1237.

Swynnerton, Roger. Interview by Deborah Bryceson, on Tanganyika Colonial Agricultural Service 1934–51 (typescript). 22 April 1985. MSS. Afr. s. 1993.

———. Former Director of Agriculture, Kenya. Interview by Geoffrey Masefield, Lecturer in Tropical Agriculture. Oxford, 5 November 1970. MSS. Afr. s. 1426.

Winnington-Ingram, C. Administrative Officer. North Mara District (Tanzania) Survey of the African Farming and Land Utilization Problem in North Mara, Annual Report, 9 March 1950. MSS. Afr. s. 1749.

Sukuma Museum, Bujora, Mwanza, Tanzania

Bourget, L. Trip Diary, 1904. "Report of a Trip in 1904 from Bukumbi to Mwanza, Kome? Ukerewe, Kibara, Ikoma—Mara Region, together with some stories." N.p., n.d. M-SRC54, Sukuma Archives.

Tanzania National Archives (TNA)

Administration of Tanganyika Territory, 1926. Secretariat Files, 1733/4–12.

African Population by Chiefdom, 1948. Secretariat Files, 40641.

Agricultural Development, Lake Province. Vol. 2. Secretariat Files, 22338.

Agricultural Development, Mwanza Province. Secretariat Files, 19080.

Agricultural Statistics. Secretariat Files, 35169.

Agriculture: General, 1934–48. 215/909/II.

Agriculture: Tanganyika Territory. Report AB 1025 (7013).

Annual Reports, Native Affairs Section, Lake Province. 215, 1135.

Compilation of Economic Maps, 1934–36. Secretariat Files, 23275.

Game. 215/13/I.

Game. Secretariat Files, AB 914 (3733) I.

Game, Including National Parks. 41 G 1/1.

———. Vermin Destruction, 41 G 1/2.

Game: Indiscriminate Slaughter. Newspaper Protest. Secretariat Files, 13582.

Game Licenses. 215/1153/I.

Game Preservation Department. Annual Report, 1936, Tanganyika Territory.

Game Regulations. Vol. 1, 1923–29. 215/PC/14/I.

Ghee Industry. Vols. 1–3. Secretariat Files, 18638.

Government Memorandum on the Legal History of the Area Now Forming the Serengeti National Park. Government, 10 June 1957. Distributed by the Secretariat in Reference to Serengeti Committee of Enquiry, National Game Parks. 215/350/IV.

Government Notice no. 368 of 1942.

Grant, H. St. J. Report on Human Habitation of the Serengeti National Park. Secretariat Files, AB 915 (3733) II and National Game Parks, 215/350/III.

Hides and Skins Industry. Vols. 3–6. Secretariat Files, 12672.

Hingston, Maj. R. W. G., MC. Report on his Visit to Tanganyika Territory for the Purpose of Investigating the Most Suitable Methods of Ensuring the Preservation of Its Indigenous Fauna, Sent out to All Provincial Commissioners, 6 June 1931. National Game Parks. Vol. 1. 215/350.

Karte von Deutsch-Ostafrika. A.4, Ikoma. German Maps. GM 30/3.

Killing of Game by Natives. Vols. 1–2. Secretariat Files, 13371.

Labour Commission. Information and Statistics. Secretariat Files, 23202.

Labour Department. Annual Report, 1926. Secretariat Files, AB 452 (3046/11).

Labour Matters, Lake Province. Vol. 1. Secretariat Files, 12019.

———. Department of Health. 87/24/III.

Land and Mines. Chiefdoms' Boundary Dispute Files, North Mara District. 83, 3/127.

Legislative Council of Tanganyika. Proposals for Reconstituting the Serengeti National Park, 1958. Government Paper no. 5. Game, Including National Parks. 41 G 1/1.

———. The Serengeti National Park. Occasional Paper no. 1. Game, Including National Parks, 41 G 1/1.

Mara Region. Annual Report, 1964.

Memorandum for Proposed Partition Plan for Serengeti National Park, 28 September 1955. National Game Parks, 215/350/III.

Memorandum on Masai History and Mode of Life. Prepared by H. St. J. Grant, a District Officer in the Masai District from August 1950 to December 1954. National Game Parks. 215/350/IV.

Mining Claims Book, 1914. G 45/58 L R.

National Game Parks. 215/350/I–IV.

National Park, Serengeti. Vols. 4–5. Secretariat Files, 24979.

Native Administration Successions in Sultanates, 1926–31. 246, PC/3/3.

Native Administration. Subhead: Government Anthropologist's Report. Kuria, 544/C/28.

Native Affairs Census, 1926–29. 246/PC/3/21.

Native Affairs. Collective Punishments and Prosecution of Chiefs, 1926–31. 246/PC/3/4.

Native Agriculture. Increased Production Campaign. Vols. 1–2. Secretariat Files, 26298.

Native Courts. Mwanza, 1924. 2860/32.

Native Settlement in Game Reserves. 7227.

Notification of a Trade Dispute. 460 724/196.

Outbreak of Sleeping Sickness in Mwanza Province. Secretariat Files, 20909.

Provincial Administration, 1925–26. Head: Provincial Administration. Sub-Head Annual Report. 246/PC/ 1/17.

Provincial Administration, 1927–28. 246 PC/1/30.

———. Annual Report, 1928. 246/PC/1/53.

———. Monthly Reports, 1926–29. Musoma District. 215/PC/1/7.

Raids by Masai, 1936. Vol. 1. Secretariat Files, 23384.

Recruitment, Supply and Administration of Labour for the Mines. Secretariat Files, 23047.

Report by Labour Officer on Gold Mines in Musoma Area. Secretariat Files, 25944.
Report on Economic Conditions in the Island of Ukara with Special Reference to
Soil Erosion, 1934. 22425.
Russell, Gordan. Report on the National Park 1950 . . . National Game Parks,
215/350/II.
Shortage of Foodstuffs, Famine Relief. Vols. 1–3. 13252.
Sleeping Sickness, Musoma District. 215/463.
Tsetse Fly. Vols. 1–2. 2702.
Veterinary Department. Annual Report, 1921. 3046/22.
White Paper: Serengeti National Park, Legislative Council of Tanganyika, Ses-
sional Paper no. 1. 1956.

White Fathers Regionals' House, Nyegezi, Mwanza

Société des Missionnaires d'Afrique (Pères-Blancs). *Rapports annuels.* 1910–11,
1911–12, 1915–16.
———. *Chronique trimestrielle de la Société des Missionnaires d'Afrique.* 1902, 1905.
———. "Visitations Book." Nyegina, Mwanza I, 1931–32.

INTERVIEWS

In the western Serengeti and throughout Tanzania, a first name is an individual's personal name, and a second name is his or her father's personal name,
not the family name. If the individual has a Christian (western) name, it will be
listed first, followed by the personal name and father's name. The title "mzee"
has been used when I do not know the father's name.

Adamu Matutu and Kihumbo elders, Kihumbo and Gaka, 31 August 1995 (Ikizu).
Ali Maro Wambura, Masinki, 30 September 1995 (Ngoreme).
Apolinari Maro Makore, Megasa, 29 September 1995 (Ngoreme).
Atanasi Kebure Wambura, Maburi, 7 October 1995 (Ngoreme).
Baginyi Mutani, Sanzate, 8 September 1995 (Ikizu).
Barichera Machage Barichera, Nyichoka to Riyara, 7 March 1996 (Nata).
Bhoke Mtoka, Kemegesi, 14 August 2003.
Bhoke Rotegenga, Motokeri, 13 March 1995, 4 August 2003 (Nata).
Bhoke Wambura, Maburi, 7 October 1995 (Ngoreme).
Bhosa Rugatiri, Mbiso, 17 June 1995 (Nata).
Bita Makuru, Bugerera, 11 February 1995 (Nata).
Bokima Giringayi, Mbiso, 26 October 1995 (Ikoma).
Chamuriho, Mchang'oro, 19 January 1996.
Charles Nyamaganda, Burunga, 31 March 1995 (Nata).
Charwe Matiti, Nyeboko, 22 September 1995 (Ngoreme).
Chengero, Mzee, Robanda, 29 August 2003.
Daniel Kitaro Wambura, Motokeri, 2 April 1996 (Nata).

Daudi Katama Maseme, Bwai, 11 November 1995 (Ruri).
David Maganya Masama, Kemegesi, 12 August 2003 (Ngoreme).
Efaristi Bosongo Gikaro, Masinki, 30 September 1995 (Ngoreme).
Elfaresti Wambura Nyetonge, Kemegesi, 20 September 1995 (Ngoreme).
Elia Masiyana Mchanake, Borenga, 21 September 1995 (Ngoreme).
Emmanuel Ndenu, Sale, 6 December 1995 (Sonjo/Temi).
Faini Magoto, Mbiso, 6 March and 19 August 1995, 3 August 2003 (Nata).
Faustine Wanzagi Nyerere, VIP, Mbiso, 6 August 2003 (Zanaki).
Francis Sabayi Maro, Masinki, 6 October 1995 (Ngoreme).
Frederick Mochogu Munyera, Judge, Maji Moto, 28 September 1995 (Ngoreme).
Gabuso Shoka, Mbiso, 30 May 1995 (Nata).
Gejera Ginanani, Kyandege, 26 July 1995 (Tatoga Rotigenga).
George Wambura Gehamba, Morotonga, 21 August 2003 (Ikoma).
Gesura Mwatagu, Issenye, 5 August 1995 (Tatoga Rotigenga).
Getara Mwita, Mesaga, 29 September 1995 (Ngoreme).
Ghamarhizisiji (Uyayehi) Nuaasi, Issenye, 5 August 1995 (Tatoga Rotigenga).
Gilumughera Gwiyeya, Issenye, 28 July 1995 (Tatoga Rotigenga).
Ginanani Chokora, Kyandege, 26 July 1995 (Tatoga Rotigenga).
Girihoida Masaona, Issenye, 28 July 1995 (Tatoga Rotigenga).
Girimanda Mwarhisha Gishageta, Issenye, 27 July 1995 and 28 March 1996 (Tatoga
    Rotigenga).
Giruchani Masanja, Mariwanda, 6 July 1995 (Tatoga Rotigenga).
Gisuge Chabwasi, Mariwanda, 22 June 1995; Sanzate, 22 June 1995 (Ikizu).
Gorobani Gesura, Issenye, 28 July 1995 (Tatoga Rotigenga).
Guti Manyeni Nyabwango, Sanzate, 15 June 1995 (Ikizu).
Herbert Kiyata, Morotonga, 18 August 2003.
Hezekia Marwa Sarya, Mugumu/Matare, 15 March 1996 (Kuria Nyabasi).
Ibrahim Mutatiro, Maji Moto, 29 September 1995 (Ngoreme).
Ikizu elders at historical places, Kilinero, 16 August 1995.
Ikota Mwisagija, Kihumbo, 5 July 1995 (Ikizu).
Isaya Charo Wambura, Buchanchari, 22 September 1995 (Ngoreme).
Jacob Mugaka, Bunda, 10 March 1995 (Sizaki).
Jackson (Benedicto) Mang'oha Maginga, Mbiso, 18 March and 13 May 1995 (Nata).
Jackson Mteba Mabura, Morotonga, 23 August 2003 (Ikoma).
Jackson Witari, Mariwanda, 8 July 1995 (Ikizu).
Jasper Malewo, Swala Safaris Robanda, Ikoma Bush Camp, 22 August 2003.
John Muya, Fort Ikoma, 22 August 2003.
Joseph Mashohi, Nyeberekera, 16 February 1996 (Ishenyi).
Joseph Sillery Magoto, Mbiso, 2 August 2003 (Nata).
Joseph Tareta Masina, Issenye, 7 August 2003.
Juana Masanja, Manawa, 24 February 1996 (Tatoga Isimajega).
Katani Magori Nyabunga, Sanzate, 10 June 1995 (Ikizu).
Keneti Mahembora, Sang'anga, 17 February 1996; Gitaraga and Nyichoka, 9 Feb-
    ruary 1996 (Nata).

Kenyatta Mosoka, Robanda, 12 July 1995 (Ikoma).

Kihenda Manyorio, Kurasanga, 3 August 1995 (Ikizu).

Kinanda Sigara (Itara), Bugerera, 21 May and 23 May 1995; Mchang'oro, 19 January 1996 (Ikizu).

Kirigiti Ng'orogita, Mbiso, 8 June 1995, 17 June 1995 (Nata).

Kisenda Mwita, Mugumu/Matare, 15 March 1996 (Kuria Nyabasi).

Kitang'ita Robi, Busawe, 22 September 1995 (Ngoreme).

Kiyarata Mzumari, Kihumbo, 5 July 1995; Mariwanda, 8 July 1995 (Ikizu).

Mabenga Nyahega, Mbiso, 1 September 1995; Bugerera, 5 September 1995 (Ikoma).

Machaba Nyahega, Mbiso, 1 September 1995 (Ikoma).

Machota Nyantitu, Morotonga/Mugumu, 28 May 1995 (Ikoma).

Machota Sabuni, Issenye, 14 March 1996 (Ikoma).

Maguye Maginga, Nyeketono, 21 June 1995 (Nata).

Mahewa Timanyi, Robanda, 27 May 1995 (Ikoma).

Mahiti Gamba, Bugerera, 1 November 1995, 4 February and 3 March 1996 (Nata).

Mahiti Kwiro, Mchang'oro, 19 January 1996 (Nata).

Makanda Magige, Bumangi, 10 November 1995 (Zanaki).

Makang'a Magigi, Bisarye, 9 November 1995 (Zanaki).

Makongoro Nyemwitweka, Rubana, 4 April 1996 (Nata).

Makuru Magambo, Geteku, 8 March 1996 (Nata).

Makuru Maro, Kemegesi, 13 August 2003 (Ngoreme).

Makuru Moturi, Maji Moto, 29 September 1995 (Ngoreme).

Makuru Nyang'aka, Motokeri, 8 June 1995; Bwanda, Nyichoka, and Gitaraga, 9 February and 16 February 1996; Nyichoka to Riyara, 7 March 1996 (Nata).

Mang'oha Machunchuriani, Mbiso, 24 March 1995 (Nata).

Mang'oha Morigo, Bugerera, 24 June 1995; Kyandege, 8 August 2003 (Nata).

Mang'ombe Makuru, Morotonga, 28 August 2003 (Ikoma).

Mang'ombe Morimi, Issenye/Iharara, 26 August 1995 (Ishenyi).

Maria Maseke, Busawe, 22 September 1995 (Ngoreme).

Mariam Mturi, Nyamuswa, Makongoro Secondary School, 30 June 1995 (Ikizu).

Mariko Romara Kisigiro, Burunga, 31 March 1995 (Nata).

Marimo Nyamakena, Sanzate, 10 June 1995 (Ikizu).

Marindaya Sanaya, Samonge, 5 December 1995 (Sonjo/Temi).

Maro Mchari Maricha, Maji Moto, 28 September 1995 (Ngoreme).

Maro Mugendi, Busawe, 22 September 1995 (Ngoreme).

Maronyi Bwana, Kemegesi, 12 August 2003 (Ngoreme).

Marunde Godi, Manawa, 24 February 1996 (Tatoga Isimajega).

Mashauri Ng'ana, Issenye, 2 November 1995 (Ishenyi).

Masosota Igonga, Ring'wani, 6 October 1995 (Ngoreme).

Maswe Makore, Mesaga, 28 September 1995 (Ngoreme).

Matias Mahiti Kebumbeko, Torogoro, 2 April, 1996 (Nata).

Mayani Magoto, Bugerera, 18 February and 4 April 1995, 3 March 1996 (Nata).

Mayenye Nyabunga, Sanzate, 8 September 1995 (Ikizu).

Mayera Magondora, Manawa, 24 February 1996 (Tatoga Isimajega).

Mechara Masauta, Robanda, 22 August 2003 (Ikoma).

Megasa Mokiri, Motokeri, 4 March and 13 March 1995 (Nata).

Merekwa Masunga, Mariwanda, 6 July 1995 (Tatoga Rotigenga).

Mgoye Magutachuba Rotegenga Megasa, Motokeri, 13 March 1995; Mbiso, 1 May 1995 (Nata).

Mikael Magessa Sarota, Issenye, 25 August 1995 (Ishenyi).

Mnyengere (Bhoke) Magoto, Mbiso, 13 May 1995 (Nata).

Mogusuhi Nyasarigoko, Kemegesi, 13 August 2003 (Ngoreme).

Mohere Mogoye, Bugerera, 25 March 1995 (Nata).

Moremi Mwikicho, Robanda, 12 July 1995 (Ikoma).

Morigo (Mchombocho) Nyarobi, Issenye, 28 September 1995 (Ishenyi).

Mossi Chagana, Nyeberekera, 16 February 1996 (Ishenyi).

Mswaga, Mzee, Bugerera, 29 March 1995.

Musa Matabarwa, Mariwanda, 8 July 1995 (Ikizu).

Mwamedi Hassan, Bugerera, 3 May 1995.

Mwenge Elizabeth Magoto, Mbiso, 6 May 1995 (Nata).

Mwikwabe Maro, Busawe, 22 September 1995 (Ngoreme).

Mwinoki Munyewa, Bugerera, 21 May 1995 (Ikizu).

Mwita Magige, Mosongo, 9 September 1995 (Ngoreme).

Mwita Maro, Maji Moto, 29 September 1995 (Ngoreme).

Nata elders meeting, Mbiso, 14 August 1995 (Nata).

Nata singers and dancers, Bugerera, 19 August 1995 (Nata).

Nchota Chachamogohe, Kemegesi, 13 August 2003 (Ngoreme).

Njaga Nyasama, Kemegesi, 2 September and 14 September 1995, 29 March 1996 (Ngoreme).

Nyabori Marwa, Kemegesi, 14 September 1995 (Ngoreme).

Nsaho Maro, Kenyana, 14 September 1995 (Ngoreme).

Nyabusogesi Nying'asa, Nyamuswa, Makongoro Secondary School, 30 June 1995 (Ikizu).

Nyakaho Magambo, Kemegesi, 31 August 2003 (Ngoreme).

Nyakerenge Nyamusaki, Morotonga, 25 August 2003 (Ikoma).

Nyamaganda Magoto, Bugerera, 3 March and 4 October 1995; Mbiso, 4 July 2002, 3 August 2003; cultural vocabulary, numerous days (Nata).

Nyambeho Marangini, Issenye, 7 September 1995 (Ishenyi).

Nyambureti Morumbe, Robanda, 27 May 1995 (Ikoma).

Nyamuko Soka, Morotonga, 28 August 2003 (Ikoma).

Nyanchiwa Mesika, Morotonga, 27 August 2003 (Ikoma).

Nyawagamba Magoto, Site, Bugerera, 27 February, 3 September, 2 October, 4 October, 6 October, and 1 November 1995; Kikongoti, 2 April 1996; M'chongoro, 19 January 1996; Rubana, 4 April 1996 (Nata).

Paulina Wambura, Bugerera, 16 April 1995 (Kuria).

Paulo Machota Mongoreme, Kyandege, 8 August 2003 (Nata).

Paulo Maitari Nyigana and Ibrahim Mutatiro, Maji Moto, 29 September 1995 (Ngoreme).

Peter Mgosi Siwa, Morotonga, 23 August 2003 (Ikoma).

Peter Mkome Mahembora, Mbiso, 4 August 2003 (Ikoma).

Peter Nabususa, Samonge, 5 December 1995 (Sonjo/Temi).

Philemon Mbota, Mugumu/Matare, 15 March 1996 (Kuria).

Philipo Haimati, Iramba, 15 September 1995 (Ngoreme).

Raheli Wanchota Nyanchiwa, Mugumu/Morotonga, 16 March 1996 (Ikoma).

Ramadhani Masaigana, Motokeri, 8 June 1995, 4 August 2003 (Nata).

Raphael Machogote, Issenye, 14 March 1996 (Ishenyi).

Raymond Nyamasagi, Morotonga, 27 August 2003 (Nata).

Reterenge Nyigena, Maji Moto, 23 September 1995 (Ngoreme).

Riyang'ang'ara Nyang'urara, Sarawe, 20 July 1995 (Ikizu).

Robert Rashidi Rotiginga, Motokeri, 4 August 2003 (Nata).

Robi Chacha, Kemegesi, 13 August 2003 (Ngoreme).

Robi Nykisokoro, Borenga, 21 September 1995 (Ngoreme).

Rugatiri Mekacha, Dr., Dar es Salaam, 24 May 1996 (Nata).

Rugayonga Nyamohega, Mugeta, 27 October 1995 (Ishenyi).

Sagochi Nyekipegete, Robanda, 12 July 1995 (Ikoma).

Samueli Buguna Katama, Bwai, 11 November 1995 (Ruri).

Samweli Ginduri, Samonge, 6 December 1995 (Sonjo/Temi).

Samweli M. Kiramanzera, Kurasanda, 3 August 1995 (Ikizu).

Samweli Muya Mongita, Morotonga, 21 August 2003 (Ikoma).

Sarah Wanchota Noku, Morotonga, 25 August 2003 (Ikoma).

Sarya Nyamuhandi, Bumangi, 10 November 1995 (Zanaki).

Schachenmann, Barbara, VIP, Sasakwa, 6 August 2003.

Senteu Maghanye, Issenye, 14 March 1996 (Ikoma).

Silas King'are Magori, Kemegesi, 21 September 1995 (Ngoreme).

Simion Hunga Nason, Kemegesi, 14 August 2003.

Sira Masiyora, Nyerero, 17 November 1995 (Kuria/Nyabasi).

Sochoro Kabati, Nyichoka, 2 June 1995; Bwanda, 16 February 1996; Riyara, 7 March 1996 (Nata).

Songoro Sasora, Nyamuswa, Makongoro Secondary School, 30 June 1995 (Ikizu).

Stephano Makondo Karamanga, Robanda, 22 August 2003 (Ikoma).

Stephen Gojat Gishageta, Issenye, 27 July 1995, 28 March 1996 (Tatoga Rotigenga).

Sumwa Nyamutwe, Mugeta, 9 March 1996; Mbiso, 4 April 1996 (Nata).

Surati Wambura, Morotonga, 13 July 1995 (Ikoma).

Taranka, Mzee, Bugerera, 10 May 1995.

Tatoga elders and singers, Kyandege, 18 August 1995 (Tatoga Rotigenga).

Tetere Tumbo, Mbiso, 5 April and 1 May 1995, 11 August 2003 (Nata).

Thomas Kubini, Bunda, 10 March 1995, Sizaki.

Tirani Wankunyi, Issenye, 7 April 1995 (Nata).

Wambura Edward Kora, Morotonga, 14 and 19 August 2003; Robanda, 29 August 2003 (Ikoma).

Wambura Nyikisokoro, Sang'anga Buchanchari, 23 September 1995 (Ngoreme).

Wambura Tonte, Kemegesi, 12 August 2003 (Ngoreme).

Warioba Mabusi Nyangabara, Sarawe, 20 July 1995 (Ikizu).
Webiro Ginyewe, Sarawe, 20 July 1995 (Ikizu).
Webiro Zeze, Mariwanda, 8 July 1995 (Ikizu).
Weigoro Mincha, Kemegesi, 29 March 1996 (Ngoreme).
Wilson Shanyangi Machota, Morotonga, 12 July 1995, 16 March 1996, 18 August 2003; Robanda, 29 August 2003 (Ikoma).
Yohana Kitena Nyitanga, Makundusi, 1 May 1995 (Nata).
Zabron Kisubundo Nyamamera, Bisarye, 9 November 1995 (Zanaki).
Zakaria Gereta, Kemegesi, 12 August 2003 (Ngoreme).
Zamberi Manyeni, Sanzate, 15 June 1995 (Ikizu).
Zamberi Masambwe, Mariwanda, 22 June 1995 (Ikizu).

PUBLISHED SOURCES

Abrahams, R. G. "Aspects of Labwor Age and Generation Grouping and Related Systems." In *Age, Generation, and Time: Some Features of East African Age Organizations*, edited by P. T. Baxter and Uri Almagor, 37–67. New York: St. Martin's, 1978.

Abuso, Paul Asaka. *A Traditional History of the Abakuria, c. A.D. 1400–1914.* Nairobi: Kenya Literature Bureau, 1980.

Adams, Jonathan S., and Thomas O. McShane. *The Myth of Wild Africa: Conservation without Illusion.* New York: Norton, 1992.

Ahmed, Christine Choi. "Before Eve Was Eve: 2200 Years of Gendered History in East-Central Africa." PhD dissertation, UCLA, 1996.

Akyeampong, Emmanuel Kwaku. *Between the Sea and the Lagoon: An Eco-social History of the Anlo of Southeastern Ghana, c. 1850 to Recent Times.* Athens: Ohio University Press, 2001.

Alexander, Jocelyn, and JoAnn McGregor. "Wildlife and Politics: CAMPFIRE in Zimbabwe." *Development and Change* 31 (November 2000): 605–27.

Allman, Jean, and Victoria Tashjian. *"I Will Not Eat Stone": A Women's History of Colonial Asante.* Portsmouth, NH: Heinemann, 2000.

Alpers, Edward A. *Ivory and Slaves in East Central Africa: Changing Patterns of International Trade to the Later Nineteenth Century.* London: Heinemann, 1975.

———. "The Ivory Trade in Africa: An Historical Overview." In *Elephant: The Animal and Its Ivory in African Culture*, edited by Doran H. Ross, 349–63. Los Angeles: Fowler Museum of Cultural History, UCLA, 1992.

Ambler, Charles H. *Kenyan Communities in the Age of Imperialism: The Central Region of the Late Nineteenth Century.* New Haven: Yale University Press, 1988.

Ambrose, Stanley H. "Archaeology and Linguistic Reconstructions of History in East Africa." In *The Archaeological and Historical Reconstruction of African History*, edited by Christopher Ehret and Merrick Posnansky, 104–57. Berkeley: University of California Press, 1982.

———. "Hunter-Gatherer Adaptations to Non-marginal Environments: An Ecological and Archaeological Assessment." *Sprache und Geschichte in Afrika* 7, no. 2 (1986): 11–42.

———. "The Introduction of Pastoral Adaptations to the Highlands of East Africa." In *From Hunters to Farmers: The Causes and Consequences of Food Production in Africa*, edited by J. Desmond Clark and Steven A. Brandt, 212–39. Berkeley: University of California Press, 1984.

Anacleti, A. Odhiambo. "Pastoralism and Development: Economic Changes in Pastoral Industry in Serengeti, 1750–1961." MA thesis, University of Dar es Salaam, 1975.

———. "Serengeti: Its People and Their Environment." *Tanzania Notes and Records*, nos. 81/82 (1977): 23–34.

Anderson, Benedict. *Imagined Communities: Reflections on the Origin and Spread of Nationalism*. London: Verso, 1983.

Anderson, David. "Depression, Dust Bowl, Demography, and Drought: The Colonial State and Soil Conservation in East Africa during the 1930s." *African Affairs* 83 (July 1984): 321–43.

Anderson, David, and Richard Grove. "The Scramble for Eden: Past, Present and Future African Conservation." Introduction to *Conservation in Africa: People, Policies and Practice*, edited by Anderson and Grove, 1–12. Cambridge: Cambridge University Press, 1987.

———, eds. *Conservation in Africa: People, Policies and Practice*. Cambridge: Cambridge University Press, 1987.

Anker, Peder. *Imperial Ecology: Environmental Order in the British Empire, 1895–1945*. Cambridge, MA: Harvard University Press, 2001.

Ardener, Shirley. *Women and Space: Ground Rules and Social Maps*. New York: St. Martin's, 1981.

Atkins, Peter, Ian Simmons, and Brian Roberts. *People, Land and Time: An Historical Introduction to the Relations between Landscape, Culture and Environment*. London: Arnold, 1998.

Atwell, C. A. M., and F. P. D. Cotterill. "Postmodernism and African Conservation Science." *Biodiversity and Conservation* 9, no. 5 (2000): 559–77.

Austen, Ralph. "Memoirs of a German D.C. in Mwanza, 1907–16: Extracts from the Memoirs of Theodor Gunzert." *Tanzania Notes and Records* 66 (1966): 171–79.

———. *Northwest Tanzania under German and British Rule: Colonial Policy and Tribal Politics, 1889–1939*. New Haven: Yale University Press, 1968.

Bachelard, Gaston. *The Poetics of Space*. New York: Orion Press, 1964.

Baker, Alan R. H. "On Ideology and Landscape." Introduction to *Ideology and Landscape in Historical Perspective: Essays on the Meanings of some places in the past*, edited by Alan R. H. Baker and Gideon Biger, 1–14. Cambridge: Cambridge University Press, 1992.

Baker, E. C. "Age-Grades in Musoma District, Tanganyika Territory." *Man* 27 (December 1927): 221–24.

———. *Tanganyika Papers*. Oxford: Oxford University Press, 1935.

Barber, Karin. *I Could Speak until Tomorrow: Oriki, Women and the Past in a Yoruba Town*. Edinburgh: Edinburgh University Press, 1991.

Barrow, Edmund. *Community Conservation Approaches and Experiences from East Africa*. Rural Extension Bulletin no. 10: Theme Issue on Community Conservation, University of Reading Agricultural Extension and Rural Development Department (1996).

Barrow, Edmund, Helen Gichohi, and Mark Infield. "The Evolution of Community Conservation Policy and Practice in East Africa." In *African Wildlife and Livelihoods: The Promise and Performance of Community Conservation*, edited by David Hulme and Marshall Murphree, 59–73. Oxford: James Currey, 2001.

Barth, Fredrik, ed. *Ethnic Groups and Boundaries: The Social Organization of Culture Difference*. Bergen: Universitetsforlaget; London: George Allen and Unwin, 1969.

Bassett, Thomas J., and Donald E. Crummey, eds. *Land in African Agrarian Systems*. Madison: University of Wisconsin Press, 1993.

Baumann, Oskar. *Durch Massailand zur Nilquelle: Reisen und Forschungen der Massai-Expedition des deutschen Antisklaverei-Komite in den Jahren 1891–1893*. Berlin: Geographische Verlagshandlung Dietrich Reimer, 1894.

———. *Ngorongoro's First Visitor, Being an Annotated and Illustrated Translation from Dr. O. Baumann's* Durch Mas[s]ailand zur Nilquelle. Kampala: East African Literature Bureau, 1963.

Baxter, P. T. W., and Uri Almagor. Introduction to *Age, Generation, and Time: Some Features of East African Age Organizations*, edited by P. T. W. Baxter and Uri Almagor. New York: St. Martin's, 1978.

Beachy, R. W. *The Slave Trade of Eastern Africa*. New York: Barnes and Noble, 1976.

Beattie, John. "Group Aspects of the Nyoro Spirit Mediumship Cult." *Rhodes-Livingstone Institute Journal* (1961): 11–35.

Beinart, William. "African History and Environmental History." *African Affairs* 99 (2000): 269–302.

Beinart, William, and Peter A. Coates. *Environment and History: The Taming of Nature in the USA and South Africa*. Historical Connections. New York: Routledge, 1995.

Beinart, William, and JoAnn McGregor. Introduction to *Social History and African Environments*, edited by William Beinart and JoAnn McGregor. Athens: Ohio University Press, 2003.

Bennett, Norman Robert. *Mirambo of Tanzania, 1840–1884*. New York: Oxford University Press, 1971.

Bentley, Jerry H., and Herbert Ziegler. *Traditions and Encounters: A Global Perspective on the Past*. 2nd ed. 2 vols. Boston: McGraw-Hill, 2003.

Bergin, Patrick. "Accommodating New Narratives in a Conservation Bureaucracy: TANAPA and Community Conservation." In *African Wildlife and Livelihoods: The Promise and Performance of Community Conservation*, edited by David Hulme and Marshall Murphree, 88–105. Oxford: James Currey, 2001.

———. *Reforming a Conservation Bureaucracy: TANAPA and Community Conservation*. Community Conservation in Africa Working Paper, Manchester: Institute for Development Policy and Management, University of Manchester, 1998.

Berntsen, John L. "The Maasai and Their Neighbors: Variables of Interaction." *African Economic History* 2 (Fall 1976): 1–11.

———. "Pastoralism, Raiding, and Prophets: Maasailand in the Nineteenth Century." PhD dissertation, University of Wisconsin, Madison, 1979.

Birley, Martin H. "Resource Management in Sukumaland, Tanzania." *Africa* 52, no. 2 (1982): 1–29.

Bischofberger, Otto. *The Generation Classes of the Zanaki (Tanzania).* Fribourg, Switzerland: University Press, 1972.

Bloch, Marc L. B. *The Historian's Craft.* Translated from the French by Peter Putnam. Manchester: Manchester University Press, 1954.

Bloch, Maurice. *From Blessing to Violence: History and Ideology in the Circumcision Ritual of the Merina of Madagascar.* Cambridge: Cambridge University Press, 1986.

Boddy, Janice. *Wombs and Alien Spirits: Women, Men, and the Zar Cult in Northern Sudan.* Madison: University of Wisconsin Press, 1989.

Bonner, Raymond. *At the Hand of Man: Peril and Hope for Africa's Wildlife.* New York: Knopf, 1993.

Booth, Vernon, George Nangale, and Hamudi Majamba. "Procedures for Communities to Enter into Joint Ventures in Wildlife Management Areas." Dar es Salaam: United Republic of Tanzania, Ministry of Natural Resources and Tourism, 2000.

Borner, Markus. "Conservation with the People—For the People: Establishment of Community Wildlife Management Areas (WMAs) around the Serengeti National Park." Serengeti National Park Service. http://www.serengeti.org/download/Conservation.pdf.

Bouniol, J. *The White Fathers and Their Missions.* London: Sands and Co., 1929.

Bourdieu, Pierre. *Outline of a Theory of Practice.* Translated from the French by Richard Nice. 1972. Cambridge: Cambridge University Press, 1977.

Bower, John. "The Pastoral Neolithic of East Africa." *Journal of World Prehistory* 5, no. 1 (1991): 49–82.

Brandstrom, Per. "Seeds and Soil: The Quest for Life and the Domestication of Fertility in Sukuma-Nyamwezi Thought and Reality." In *The Creative Communion: African Folk Models of Fertility and the Regeneration of Life,* edited by Anita Jacobson-Widding and Walter van Beek, 167–86. Uppsala: Almqvist and Wiksell International, 1990.

Brantley, Cynthia. "Through Ngoni Eyes: Margaret Read's Matrilineal Interpretations from Nyasaland." *Critique of Anthropology* 17, no. 2 (June 1997): 147–69.

Braudel, Fernand. *The Mediterranean and the Mediterranean World in the Age of Philip II.* Translated from the French by Siân Reynolds. Edited by Richard Ollard. London: Harper Collins, 1992.

Brockington, Dan. *Fortress Conservation: The Preservation of the Mkomazi Game Reserve, Tanzania.* Bloomington: Indiana University Press, 2002.

Brown, Michael F. "On Resisting Resistance." *American Anthropologist* 98, no. 4 (1996): 729–35.

Buchanan, Carole A. "Perceptions of Ethnic Interaction in the East African Interior: The Kitara Complex." *International Journal of African Historical Studies* 11, no. 3 (1978): 410–28.

Burton, Richard F. *The Lake Regions of Central Africa: A Picture of Exploration.* 2 vols. London: Longman, Green, Longman, and Roberts, 1860.

Campbell, Ken, and Markus Borner. "Population Trends and Distribution of Serengeti Herbivores: Implications for Management." In *Serengeti II: Dynamics, Management, and Conservation of an Ecosystem,* edited by A. R. E. Sinclair and Peter Arcese, 117–45. Chicago: University of Chicago Press, 1995.

Carruthers, Jane. "Dissecting the Myth: Paul Kruger and the Kruger National Park." *Journal of Southern African Studies* 20, no. 2 (1994): 263–83.

———. *The Kruger National Park: A Social and Political History.* Pietermaritzburg: University of Natal Press, 1995.

———. "Past and Future Landscape Ideology: The Kalahari Gemsbok National Park." In *Social History and African Environments,* edited by William Beinart and JoAnn McGregor, 255–66. Oxford: James Currey, 2003.

Carruthers, Mary J. *The Book of Memory: A Study of Memory in Medieval Culture.* Cambridge: Cambridge University Press, 1990.

Chacker, E. A. "Early Arab and European Contacts with Ukerewe." *Tanzania Notes and Records* 68 (1968): 75–86.

Chatwin, Bruce. *The Songlines.* New York: Viking, 1987.

Chouin, G. "Sacred Groves in History: Pathways to the Social Shaping of Forest Landscapes in Coastal Ghana." *IDS Bulletin* (Institute for Development Studies, Brighton, England) 33, no. 1 (2002): 39–46.

Clyde, David F. *History of the Medical Services of Tanganyika.* Dar es Salaam: Government Press, 1962.

Cohen, David W. "Doing Social History from Pim's Doorway." In *The African Past Speaks: Essays on Oral Tradition and History,* edited by Joseph C. Miller, 191–235. Folkestone, Eng.: Dawson, 1980.

———. "The Face of Contact: A Model of a Cultural and Linguistic Frontier in Early Eastern Uganda." In *Nilotic Studies. Part 2.* Proceedings of the international symposium on languages and history of the Nilotic peoples, Cologne, 4–6 January 1982. Edited by Bechhaus-Gerst Rainer and Marianne Vossen, 339–56. Berlin: Dietrich Reimer Verlag, 1983.

———. "Food Production and Food Exchange in the Precolonial Lakes Plateau Region." In *Imperialism, Colonialism, and Hunger: East and Central Africa,* edited by Robert I. Rotberg, 1–18. Lexington, MA: D. C. Heath, 1983.

———. "Reconstructing a Conflict in Binafu: Seeking Evidence outside the Narrative Tradition." In *The African Past Speaks: Essays on Oral Tradition and History,* edited by Joseph C. Miller, 201–20. Folkestone, Eng.: Dawson, 1989.

———. *Towards a Reconstructed Past: Historical Texts from Busoga, Uganda.* Oxford: Oxford University Press, 1986.

———. *Womunafu's Bunafu: A Study of Authority in a Nineteenth-Century African Community.* Princeton: Princeton University Press, 1977.

Cohen, David William, and E. S. Atieno Odhiambo. *Siaya: The Historical Anthropology of an African Landscape.* Athens: Ohio University Press, 1989.

Cohen, Ronald. "Ethnicity: Problem and Focus in Anthropology." *Annual Review of Anthropology* 7 (1978): 379–403.

Collett, D. P. "Models of the Spread of the Early Iron Age." In *The Archaeological and Linguistic Reconstruction of African History,* edited by Christopher Ehret and Merrick Posnansky, 182–95. Berkeley: University of California Press, 1982.

Colson, Elizabeth. "The Impact of the Colonial Period on the Definition of Land Rights." In *Profiles of Change: African Society and Colonial Rule,* edited by Victor Turner, 193–215. Cambridge: Cambridge University Press, 1971.

———. "Places of Power and Shrines of the Land." *Paideuma* 43 (1997): 47–59.

Connerton, Paul. *How Societies Remember.* Cambridge: Cambridge University Press, 1989.

Cooper, Frederick. "Conflict and Connection: Rethinking Colonial African History." *American Historical Review* 99, no. 5 (1994): 1516–45.

———. *Plantation Slavery on the East Coast of Africa in the Nineteenth Century.* New Haven: Yale University Press, 1974.

Cooper, Frederick, and Ann L. Stoler. "Tensions of Empire: Colonial Control and Visions of Rule." Introduction. *American Ethnologist* 16, no. 4 (1989): 609–21.

———. *Tensions of Empire: Colonial Cultures in a Bourgeois World.* Berkeley: University of California, 1997.

Cory, Hans. *The Indigenous Political System of the Sukuma and Proposals for Political Reform.* Dar es Salaam: East African Institute of Social Research/Eagle Press, 1954.

———. "Land Tenure in Bukuria." *Tanganyika Notes and Records* 23 (1947): 70–79.

Cosgrove, Denis E. *Social Formation and Symbolic Landscape.* Totowa, NJ: Barnes and Noble, 1985.

Crandell, Gina. *Nature Pictorialized: "The View" in Landscape History.* Baltimore: Johns Hopkins University Press, 1993.

Croll, Elisabeth, and David Parkin, eds. *Bush Base: Forest Farm: Culture, Environment, and Development.* New York: Routledge, 1992.

Cronon, William. "A Place for Stories: Nature, History, and Narrative." *Journal of American History* 78, no. 4 (March 1992): 1347–76.

———. "The Trouble with Wilderness, or, Getting Back to the Wrong Nature." In *Uncommon Ground: Toward Reinventing Nature,* edited by William Cronon, 1–55. New York: Norton, 1995.

Crosby, Alfred W. "The Past and Present of Environmental History." *American Historical Review* 100, no. 4 (1995): 1177–89.

Curtin, Philip, Steven Feierman, Leonard Thompson, and Jan Vansina. *African History.* New York: Longman, 1978.

Davis, Mike. *Late Victorian Holocausts: El Niño Famines and the Making of the Third World.* London: Verso, 2001.

Deleuze, Gilles. "Rhizome." Introduction to *A Thousand Plateaus: Capitalism and Schizophrenia*, edited by Felix Guattari, 1–19. Minneapolis: University of Minnesota Press, 1987.

Dobson, Andy. "The Ecology and Epidemiology of Rinderpest Virus in Serengeti and Ngorongoro Conservation Area." In *Serengeti II: Dynamics, Management, and Conservation of an Ecosystem*, edited by A. R. E. Sinclair and Peter Arcese, 485–505. Chicago: University of Chicago Press, 1995.

Dobson, E. B. "Comparative Land Tenure of Ten Tanganyika Tribes." *Tanganyika Notes and Records* 38 (1955): 31–39.

Dublin, Holly T. "Dynamics of the Serengeti-Mara Woodlands: An Historical Perspective." *Forest and Conservation History* 35, no. 4 (October 1991): 169–78.

———. "Vegetation Dynamics in the Serengeti-Mara Ecosystem: The Role of Elephants, Fire, and Other Factors." In *Serengeti II: Dynamics, Management, and Conservation of an Ecosystem*, edited by A. R. E. Sinclair and Peter Arcese, 71–90. Chicago: University of Chicago Press, 1995.

Dunlap, Thomas R. *Nature and the English Diaspora: Environment and History in the United States, Canada, Australia, and New Zealand*. Cambridge: Cambridge University Press, 1999.

Durkheim, Emile. *Elementary Forms of Religious Life*. London: Allen and Unwin, 1954.

Durkheim, Emile, and Marcel Mauss. *Primitive Classification*. Translated from the French and edited by Rodney Needham. 1903. Chicago: University of Chicago Press, 1963.

East African Literature Bureau. *Zamani mpaka siku hizi*. Dar es Salaam: East African Literature Bureau, 1930.

Ehret, Christopher. *An African Classical Age: Eastern and Southern Africa in World History, 1000 B.C. to A.D. 400*. Oxford: James Currey, 1998.

———. *Southern Nilotic History: Linguistic Approaches to the Study of the Past*. Evanston: Northwestern University Press, 1971.

Emerton, Lucy, and Iddi Mfunda. *Making Wildlife Economically Viable for Communities Living around the Western Serengeti, Tanzania*. London: International Institute for Environment and Development, 1999.

Evans-Pritchard, E. E. *The Nuer: A Description of the Modes of Livelihood and Political Institutions of a Nilotic People*. Oxford: Clarendon Press, 1940.

Feierman, Steven. "The Myth of Mbegha." In Feierman, *Shambaa Kingdom*, 40–64.

———. *Peasant Intellectuals: Anthropology and History in Tanzania*. Madison: University of Wisconsin Press, 1990.

———. *The Shambaa Kingdom: A History*. Madison: University of Wisconsin Press, 1974.

Fentress, James, and Chris Wickham. *Social Memory*. Oxford: Blackwell, 1992.

Finch, F. G. "Hambageu: Some Additional Notes on the God of the Wasonjo." *Tanganyika Notes and Records* 47/48 (1957): 203–8.

Fleisher, Michael L. *Kuria Cattle Raiders: Violence and Vigilantism on the Tanzania/Kenya Frontier.* Ann Arbor: University of Michigan Press, 2000.

Ford, John. *The Role of the Trypanosomiases in African Ecology: A Study of the Tsetse Fly Problem.* Oxford: Clarendon Press, 1971.

Ford, V. C. R. *The Trade of Lake Victoria.* Kampala: East African Institute of Social Research, 1955.

Fortes, Meyer. *Kinship and the Social Order.* London: Routledge and Kegan Paul, 1969.

Fosbrooke, H. A. "Hambageu, the God of the Wasonjo." *Tanganyika Notes and Records* 35 (1953): 38–42.

———. "A Rangi Circumcision Ceremony: Blessing a New Grove." *Tanganyika Notes and Records* 50 (1958): 30–38.

Galaty, John G. "Maasai Expansion in the New East African Pastoralism." In *Being Maasai: Ethnicity and Identity in East Africa,* edited by Thomas Spear and Richard Waller, 61–86. Athens: Ohio University Press, 1993.

Garland, Elizabeth. "Park Managers, Biologists, and Camera Crews: Toward a Social History of Serengeti Conservationists." Paper presented at the African Studies Association Conference, Boston, 30 October–2 November 2003.

Geary, Patrick J. *Phantoms of Remembrance: Memory and Oblivion at the End of the First Millennium.* Princeton: Princeton University Press, 1994.

George, Lemi G. *Report Paper on Ft. Ikoma Monument Research Project.* Dar es Salaam: TANAPA, 2001.

Geographical Section of the Naval Intelligence Division, Naval Staff, Admiralty. *A Handbook of German East Africa.* 1920. Reprint, New York: Negro University Press, 1969.

Gerresheim, Klaus. *The Serengeti Landscape Classification.* Serengeti Research Institute Publication no. 165, Serengeti Monitoring Programme, African Wildlife Leadership Foundation, 1974.

Giblin, James L. *The Politics of Environmental Control in Northeastern Tanzania, 1840–1940.* Philadelphia: University of Pennsylvania Press, 1992.

———. "Trypanosomiasis Control in African History: An Evaded Issue?" *Journal of African History* 31, no. 1 (1990): 59–80.

Gibson, Clark C. "The Consequences of Institutional Design: The Impact of 'Community-Based' Wildlife Management Programs at the Local Level." In *Politicians and Poachers: The Political Economy of Wildlife Policy in Africa,* edited by Clark C. Gibson, 119–53. Cambridge: Cambridge University Press, 1999.

Gifford-Gonzalez, Diane. "Animal Disease Challenges to the Emergence of Pastoralism in Sub-Saharan Africa." *African Archaeological Review* 17, no. 3 (2000): 95–139.

———. "Early Pastoralists in East Africa: Ecological and Social Dimensions." *Journal of Anthropological Archaeology* 17 (1998): 166–200.

Giles-Vernick, Tamara. *Cutting the Vines of the Past: Environmental Histories of the Central African Rain Forest.* Charlottesville: University Press of Virginia, 2002.

Gissibl, Bernhard. "The Nature of Colonialism: Being and Elephant in German East Africa." Paper presented at the Animals in History Conference, Cologne, May 2005.

Glassie, Henry. *Passing the Time in Ballymenone: Culture and History of an Ulster Community.* Philadelphia: University of Pennsylvania Press, 1982.

Glassman, Jonathon. *Feasts and Riots: Revelry, Rebellion, and Popular Consciousness on the Swahili Coast, 1856–1888.* Portsmouth, NH: Heinemann, 1995.

Gluckman, Max. *Custom and Conflict in Africa.* Oxford: Basil Blackwell, 1963.

———. *Rituals of Rebellion in South-East Africa.* Manchester: Manchester University Press, 1954.

Gould, Peter, and Rodney White. *Mental Maps.* Baltimore: Penguin, 1974.

Gray, Christopher. "The Disappearing District? Territorial Transformation in Southern Gabon 1850–1950." In *The Spatial Factor in African History: The Relationship of the Social, Material, and Perceptual,* edited by Allen M. Howard and Richard M. Shain, 221–44. Leiden: Brill, 2005.

———. *Modernization and Cultural Identity: The Creation of National Space in Rural France and Colonial Space in Rural Gabon.* Occasional Paper no. 21. Bloomington: Indiana Center on Global Change and World Peace, February 1994.

Gray, Robert F. "Sonjo Lineage Structure and Property." In *The Family Estate in Africa: Studies in the Role of Property in Family Structure and Lineage Continuity,* edited by Robert F. Gray and Philip H. Gulliver, 231–61. London: Routledge and Kegan Paul, 1964.

———. *The Sonjo of Tanganyika: An Anthropological Study of an Irrigation-Based Society.* London: Oxford University Press, 1963.

Greene, Sandra E. *Sacred Sites and the Colonial Encounter: A History of Meaning and Memory in Ghana.* Bloomington: Indiana University Press, 2002.

Gregory, Derek. *Geographical Imaginations.* Cambridge: Blackwell, 1994.

Gregory, Derek, and John Urry. *Social Relations and Spatial Structures.* New York: St. Martin's, 1985.

Griffiths, Tom, and Libby Robin, eds. *Ecology and Empire: Environmental History of Settler Societies.* Seattle: University of Washington Press, 1997.

Grove, Richard H. *Green Imperialism: Colonial Expansion, Tropical Island Edens, and the Origins of Environmentalism, 1600–1860.* Cambridge: Cambridge University Press, 1995.

Grzimek, Bernhard. *Serengeti darf nicht sterben* [Serengeti Shall Not Die]. 1959. Berlin: Universal Family Entertainment, 2004.

———. *Serengeti Shall Not Die.* Translated from the German by E. L. Rewald and D. Rewald. 1st American ed. New York: E. P. Dutton, 1961.

Grzimek, Bernhard, and Michael Grzimek. *Kein Platz für wilde Tiere* [No Room for Wild Animals]. 1956. Berlin: Universal Family Entertainment, 2004.

Guthrie, Malcolm. *Comparative Bantu: An Introduction to the Comparative Linguistics and Prehistory of the Bantu Languages.* 4 vols. Westmead: Farnborough, Gregg, 1967–71.

Guyer, Jane I. "Household and Community in African Studies." *African Studies Review* 24, nos. 2–3 (1981): 87–137.

———. "Wealth in People, Wealth in Things—Introduction." *Journal of African History* 36, no. 1 (1995): 83–90.

Guyer, Jane I., and Samuel M. Eno Belinga. "Wealth in People as Wealth in Knowledge: Accumulation and Composition in Equatorial Africa." *Journal of African History* 36, no. 1 (1995): 91–121.

Hakansson, Thomas. "Family Structure, Bridewealth, and Environment in Eastern Africa: A Comparative Study of the House-Property Systems." *Ethnology* 28, no. 2 (1989): 117–35.

Halbwachs, Maurice. *On Collective Memory.* Edited and translated by Lewis A. Coser. The Heritage of Sociology. Chicago: University of Chicago Press, 1992.

Hanson, Holly. *Landed Obligation: The Practice of Power in Buganda.* Portsmouth, NH: Heinemann, 2003.

Harms, Robert W. *Games against Nature: An Eco-cultural History of the Nunu of Equatorial Africa.* Cambridge: Cambridge University Press, 1987.

Hartwig, Gerald W. *The Art of Survival in East Africa: The Kerebe and Long-Distance Trade, 1800–1895.* New York: Africana Publishing, 1976.

———. "Oral Traditions Concerning the Early Iron Age in Northwestern Tanzania." *African Historical Studies* 4, no. 1 (1971): 93–114.

Harvey, David. *The Condition of Postmodernity: An Enquiry into the Origins of Cultural Change.* Oxford: Basil Blackwell, 1989.

Hay, Margaret Jean. "Local Trade and Ethnicity in Western Kenya." *Economic History Review* 2, no. 1 (1975): 7–12.

Henige, David P. "Oral Tradition and Chronology." *Journal of African History* 12, no. 3 (1971): 371–89.

Herlocker, Dennis. *Woody Vegetation of the Serengeti National Park.* College Station: Caesar Kleberg Research Program in Wildlife Ecology/Texas A&M University, 1973.

Hill, Kevin A. "Zimbabwe's Wildlife Utilization Programs: Grassroots Democracy or an Extension of State Power?" *African Studies Review* 39, no. 1 (April 1996): 103–21.

Hirsch, Eric, and Michael O'Hanlon, eds. *The Anthropology of Landscape: Perspectives on Place and Space.* Oxford: Clarendon Press, 1995.

Hoefler, Paul L. *Africa Speaks: A Story of Adventure.* Chicago: John C. Winston Co., 1931.

Hofmeyr, Isabel. *We Spend Our Years as a Tale That Is Told: Oral Historical Narrative in a South African Chiefdom.* Portsmouth, NH: Heinemann, 1993.

Holmes, C. F. "Zanzibar Influence at the Southern End of Lake Victoria: The Lake Route." *African Historical Studies* 4, no. 3 (1971): 479–503.

Hoskins, W. G. *The Making of the English Landscape.* Harmondsworth: Penguin, 1970.

Howard, Allen M. "Nodes, Networks, Landscapes, and Regions: Reading the Social History of Tropical Africa, 1700s–1920." In *The Spatial Factor in African His-*

tory: The Relationship of the Social, Material, and Perceptual, edited by Allen M. Howard and Richard M. Shain, 21–140. Leiden: Brill, 2005.

———. "The Relevance of Spatial Analysis for African Economic History: The Sierra Leone–Guinea System." *Journal of African History* 17, no. 3 (1976): 365–88.

Howard, Allen M., and Richard M. Shain. "African History and Social Space in Africa." Introduction to *The Spatial Factor in African History: The Relationship of the Social, Material, and Perceptual,* edited by Allen M. Howard and Richard M. Shain, 1–20. Leiden: Brill, 2005.

Huber, Hugo. *Marriage and Family in Rural Bukwaya (Tanzania).* Fribourg, Switzerland: University Press, 1973.

Hulme, David, and Marshall Murphree, eds. *African Wildlife and Livelihoods: The Promise and Performance of Community Conservation.* Oxford: James Currey, 2001.

Hyden, Goran. *Beyond Ujamaa in Tanzania: Underdevelopment and an Uncaptured Peasantry.* Berkeley: University of California Press, 1980.

Iliffe, John. *Africans: The History of a Continent.* Cambridge: Cambridge University Press, 1995.

———. *A Modern History of Tanganyika.* Cambridge: Cambridge University Press, 1979.

Illius, A. W., and T. G. O'Connor. "On the Relevance of Nonequilibrium Concepts to Arid and Semiarid Grazing Systems." *Ecological Applications* 9, no. 9 (1999): 798–813.

Irwin, Paul. *Liptako Speaks: History from Oral Traditions in Africa.* Princeton: Princeton University Press, 1981.

Isaacman, Allen F. *Cotton Is the Mother of Poverty: Peasants, Work, and Rural Struggle in Colonial Mozambique, 1938–1961.* Portsmouth, NH: Heinemann, 1996.

———. "Peasants and Rural Social Protest in Africa." *African Studies Review* 33, no. 2 (1990): 1–120.

Isaacman, Allen F., and Richard Roberts. *Cotton, Colonialism, and Social History in Sub-Saharan Africa.* Portsmouth, NH: Heinemann, 1995.

Itandala, A. Buluda. "A History of the Babinza of Usukuma, Tanzania, to 1890." PhD dissertation, Dalhousie University, 1983.

Jacobs, Alan H. "A Chronology of the Pastoral Maasai." In *Hadith: Proceedings of the Annual Conference of the Historical Association of Kenya.* Vol. 1. Nairobi: East African Publishing House, 1968, 10–31.

———. "The Irrigation Agricultural Maasai of Pagasi: A Case of Maasai-Sonjo Acculturation." Paper presented at the Dar es Salaam Social Science Conference, 1968.

———. "The Traditional Political Organization of the Pastoral Maasai." PhD dissertation, Oxford University, 1965.

Jacobson-Widding, Anita. "The Encounter in the Water Mirror." In *Body and Space: Symbolic Models of Unity and Division in African Cosmology and*

Experience, edited by Anita Jacobson-Widding, 177–216. Uppsala: Almqvist and Wiksell International, 1991.

——, ed. *Body and Space: Symbolic Models of Unity and Division in African Cosmology and Experience*. Uppsala: Almqvist and Wiksell International, 1991.

——, ed. *The Creative Communion: African Folk Models of Fertility and the Regeneration of Life*. Uppsala: Almqvist and Wiksell International, 1990.

Jager, Tj. *Soils of the Serengeti Woodlands, Tanzania*. Serengeti Research Institute Publications, no. 301. Wageningen, Netherlands: Centre for Agricultural Publishing and Documentation, 1982.

James, R. W. *Land Tenure and Policy in Tanzania*. Toronto: University of Toronto Press, 1971.

Janzen, John. *Ngoma: Discourses of Healing in Central and Southern Africa*. Berkeley: University of California Press, 1992.

Johnson, George. *In the Palaces of Memory: How We Build the Worlds inside Our Heads*. New York: Knopf, 1991.

Jones, Laird Revis. "The District Town and the Articulation of Colonial Rule: The Case of Mwanza, Tanzania, 1890–1945." 2 vols. PhD dissertation, Michigan State University, 1992.

July, Robert W. *A History of the African People*. 4th ed. Prospect Heights, IL: Waveland Press, 1992.

Jungblut, Carl. *Vierzig Jahre Afrika, 1900–1940*. Berlin-Friedenau: Spiegel Verlag Paul Lippa, 1941.

Kaggwa, Apolo. *The Kings of Buganda* [Basekabaka be Buganda]. Translated by M. S. M. Kiwanuka. Nairobi: East African Publishing House, 1971.

Kaijage, Fred J. *Labor Conditions in the Tanzanian Mining Industry, 1930–1960*. Working Papers, no. 83. Boston: African Studies Center, Boston University, 1983.

Kasoga, Lenin Bega. "An Evaluation of the Ujamaa Village Policy: A Case Study of Musoma Vijijini District in Tanzania, 1974–1987." 2 vols. PhD dissertation, Michigan State University, 1990.

Kauzeni, A., and H. Kiwasila. *Serengeti Regional Conservation Strategy: A Socio-Economic Study*. University of Dar es Salaam: Institute of Resource Assessment, 1994.

Kenny, Michael G. "Mirror in the Forest: The Dorobo Hunter-gatherers as an Image of the Other." *Africa* 51, no. 1 (1981): 477–94,

——. "The Stranger from the Lake: A Theme in the History of the Lake Victoria Shorelands." *Azania* 17 (1982): 1–26.

Kershaw, Greet. *Mau Mau from Below*. Athens: Ohio University Press, 1997.

Kimambo, Isaria N. *Penetration and Protest in Tanzania: The Impact of the World Economy on the Pare, 1860–1960*. Dar es Salaam: Tanzania Publishing House, 1991.

——. *A Political History of the Pare of Tanzania, c. 1500–1900*. Nairobi: East African Publishing House, 1969.

Kirwen, Michael. *The Missionary and the Diviner: Contending Theologies of Christian and African Religions.* Maryknoll, NY: Orbis Books, 1987.

Kisigiro, Augustine N. M. *Kamusi ndogo ya Kinata-Kiswahili.* Edited by Jan Shetler. Goshen, IN: Goshen College Printing Services, 2001.

Kiss, Agnes. *Living with Wildlife: Wildlife Resource Management with Local Participation in Africa.* Washington, DC: World Bank, 1990.

Kitereza, Aniceti. *Mr. Myombekere and His Wife, Bugonoka, Their Son Ntulanalwo and Daughter Bulihwali: The Story of an Ancient African Community.* Translated from the Kikerewe by Gabriel Ruhumbika. Dar es Salaam: Mkuki na Nyota Publishers, 2002.

Kjekshus, Helge. *Ecology Control and Economic Development in East African History: The Case of Tanganyika, 1850–1950.* Portsmouth, NH: Heinemann Educational Books, 1977.

Kjerland, Kirsten Alsaker. "Cattle Breed; Shillings Don't: The Belated Incorporation of the abaKuria into Modern Kenya." PhD thesis, University of Bergen, 1995.

Klein, Kerwin Lee. "On the Emergence of Memory in Historical Discourse." *Representations* 69 (Winter 2000): 127–50.

Klieman, Kairn A. *"The Pygmies Were Our Compass": Bantu and Batwa in the History of West Central Africa, Early Times to c. 1900 C.E.* Portsmouth, NH: Heinemann, 2003.

Klima, George J. *The Barabaig: East African Cattle-Herders.* New York: Holt, Rinehart and Winston, 1970.

Klumpp, Donna, and Corinne Kratz. "Aesthetics, Expertise, and Ethnicity: Okiek and Maasai Perspectives on Personal Ornament." In *Being Maasai: Ethnicity and Identity in East Africa,* edited by Richard Waller and Thomas Spear, 195–221. Athens: Ohio University Press, 1993.

Kollmann, Paul. *The Victoria Nyanza: The Land, the Races and Their Customs, with Specimens of Some of the Dialects.* London: Swan Sonnenschein and Co., 1899.

Koponen, Juhani. *Development for Exploitation: German Colonial Policies in Mainland Tanzania, 1884–1914.* Studia Historica, no. 49. Helsinki: Finnish Historical Society, 1994.

Kopytoff, Igor. "The Internal African Frontier: The Making of African Political Culture." In *The African Frontier: The Reproduction of Traditional African Societies,* edited by Igor Kopytoff, 3–84. Bloomington: Indiana University Press, 1987.

Kramer, Fritz W. *The Red Fez: Art and Spirit Possession in Africa.* Translated from the German by Malcolm Green. 1987. London: Verso, 1993.

Kuper, Adam. "The 'House' and Zulu Political Structure in the Nineteenth Century." *Journal of African History* 34, no. 3 (1993): 469–87.

——. "Lineage Theory: A Critical Retrospect." *Annual Review of Anthropology* 11 (1982): 71–95.

Lamphear, John. "Aspects of 'Becoming Turkana': Interactions and Assimilation between Maa- and Ateker-Speakers." In *Being Maasai: Ethnicity and Identity in East Africa*, edited by Thomas Spear and Richard Waller, 87–104. Athens: Ohio University Press, 1993.

Lamprey, Richard, and Richard Waller. "The Loita-Mara Region in Historical Times: Patterns of Subsistence, Settlement and Ecological Change." In *Early Pastoralists of South-western Kenya*, edited by Peter Robertshaw, 16–35. Nairobi: British Institute in Eastern Africa, 1990.

Lan, David. *Guns and Rain: Guerillas and Spirit Mediums in Zimbabwe*. Oxford: James Currey, 1985.

Lawi, Yusuf Q. "Tanzania's Ujamaa Villagization and Local Ecological Consciousness: The Case of Eastern Iraqwland, 1974–1976." Paper presented at the African Studies Association, Listening (Again) for the African Past: A Five-College Project on Producing Historical Knowledge, Smith College, 18–26 October 2003.

———. "Where Physical and Ideological Landscapes Meet: Landscape Use and Ecological Knowledge in Iraqw, Northern Tanzania, 1920s–1950s." *International Journal of African Historical Studies* 32, nos. 2–3 (1999): 281–310.

Lawrence, Denise L., and Setha M. Low. "The Built Environment and Spatial Form." *Annual Review of Anthropology* 19 (1990): 453–505.

Leach, Melissa, and Robin Mearns. "Environmental Change and Policy: Challenging Received Wisdom in Africa." In *The Lie of the Land: Challenging Received Wisdom on the African Environment*, edited by Melissa Leach and Robin Mearns, 1–33. Portsmouth, NH: Heinemann, 1996.

———, eds. *The Lie of the Land: Challenging Received Wisdom on the African Environment*. Portsmouth, NH: Heinemann, 1996.

Leakey, L. S. B. "The Newest Link in Human Evolution: The Discovery by L. S. B. Leakey of *Zinjanthropus boisei*." *Current Anthropology* 1, no. 1 (1960): 76–77.

———. "Preliminary Report on Examination of Engaruka Ruins." *Tanganyika Notes and Records* 1 (1936): 57–60.

Leakey, L. S. B., and Vanne Morris Goodall. *Unveiling Man's Origins: Ten Decades of Thought about Human Evolution*. Cambridge, MA: Schenkman Publishing, 1969.

Leakey, L. S. B., P. V. Tobias, and J. R. Napier. "A New Species of Genus *Homo* from Olduvai Gorge." *Current Anthropology* 6, no. 4 (October 1965): 424–27.

Lemelle, Sidney J. "Capital, State, and Labor: A History of the Gold Mining Industry in Colonial Tanganyika, 1890–1942." PhD dissertation, UCLA, 1986.

LeVine, Robert A. "The Gusii Family." In *The Family Estate in Africa: Studies in the Role of Property in Family Structure and Lineage Continuity*, edited by Robert F. Gray and Philip H. Gulliver, 63–82. London: Routledge and Kegan Paul, 1964.

LeVine, Robert A., and Sarah E. LeVine. "House Design and the Self in an African Culture." In *Body and Space: Symbolic Models of Unity and Division in*

African Cosmology and Experience, edited by Anita Jacobson-Widding, 155–76. Uppsala: Almqvist and Wiksell International, 1991.

Lewis, Michael L. *Inventing Global Ecology: Tracking the Biodiversity Ideal in India, 1947–1997*. Athens: Ohio University Press, 2004.

Listowel, Judith. *The Making of Tanganyika*. London: Chatto and Windus, 1965.

Lomnitz-Adler, Claudio. "Concepts for the Study of Regional Culture." *American Ethnologist* 18, no. 2 (May 1991): 195–214.

Long, Charles H. *Alpha: The Myths of Creation*. Chico, CA: Scholars Press, 1983.

Lord, A. B. *The Singer of Tales*. Cambridge, MA: Harvard University Press, 1964.

Lousada, Audrey. *Poachers in the Serengeti*. New York: Walker, 1965.

Luck, Anne. *Charles Stokes in Africa*. Nairobi: East African Publishing House, 1972.

Luig, Ute, and Achim von Oppen. "Landscape in Africa: Process and Vision, an Introductory Essay." *Paideuma* 43 (1997): 7–45.

Machumu, Edward Paul. "Leadership and Problems of Policy Implementation: A Case Study of Two Villages in Musoma District." MA thesis, University of Dar es Salaam, 1987.

MacKenzie, John M. *The Empire of Nature: Hunting, Conservation, and British Imperialism*. Manchester: Manchester University Press, 1988.

Maddox, Gregory H. "Leave Wagogo, You Have No Food: Famine and Survival in Ugogo, Tanzania, 1916–1961." PhD dissertation, Northwestern University, 1988.

Maddox, Gregory, James L. Giblin, and Isaria N. Kimambo. *Custodians of the Land: Ecology and Culture in the History of Tanzania*. London: James Currey, 1996.

Maguire, R. A. J. "Il-Torobo." *Tanganyika Notes and Records* 25 (1948): 1–26.

Maine, Sir Henry Sumner. *Ancient Law: Its Connection with the Early History of Society and Its Relation to Modern Ideas*. London: James Murray, 1861.

Mandala, Elias. *Work and Control in a Peasant Economy: A History of the Lower Tchiri Valley in Malawi, 1859–1960*. Madison: University of Wisconsin Press, 1990.

Marean, Curtis W. "Hunter to Herder: Large Mammal Remains from the Hunter-Gatherer Occupation at Enkapune Ya Muto Rock-Shelter, Central Rift Kenya." *African Archaeological Review* 10 (1992): 65–127.

Marks, Stuart A. *The Imperial Lion: Human Dimensions of Wildlife Management in Central Africa*. Boulder: Westview Press, 1984.

Marwa, Sebastiani Muraza. *Mashujaa wa Tanzania: Mtemi Makongoro wa Ikizu: Historia ya Mtemi Makongoro na kabila lake la Waikizu, mwaka 1894 hadi 1958*. Peramiho, Tanzania: Benedictine Publications Ndanda, 1988.

Matango, R. R. "Operation Mara: The Paradox of Democracy." *Majimaji* (TANU Youth League, Dar es Salaam) 20 (1975): 17–29.

Mauss, Marcel. *The Gift: Forms and Functions of Exchange in Archaic Societies*. Translated by Ian Cunnison. New York: Norton, 1967.

Mayer, Iona. "From Kinship to Common Descent: Four Generation Genealogies among the Gusii." *Africa* 35, no. 4 (October 1965): 366–84.

Mbano, B. N. N., R. C. Malpas, M. K. S. Maige, P. A. K. Symonds, and D. M. Thompson. "The Serengeti Regional Conservation Strategy." In *Serengeti II: Dynamics, Management, and Conservation of an Ecosystem*, edited by A. R. E. Sinclair and Peter Arcese, 605–16. Chicago: University of Chicago Press, 1995.

McCann, James. *From Poverty to Famine in Northeast Ethiopia: A Rural History, 1900–1935*. Philadelphia: University of Pennsylvania Press, 1987.

——. *Green Land, Brown Land, Black Land: An Environmental History of Africa, 1800–1990*. Portsmouth, NH: Heinemann, 1999.

McDonald, David A., and Eunice Njeri Sahle. *The Legacies of Julius Nyerere: Influences on Development Discourse and Practice in Africa*. Trenton: Africa World Press, 2002.

McGregor, JoAnn. "Living with the River: Landscape and Memory in the Zambezi Valley, Northwest Zimbabwe." In *Social History and African Environments*, edited by William Beinart and JoAnn McGregor, 87–105. Athens: Ohio University Press, 2003.

McHenry, Dean E. *Limited Choices: The Political Struggle for Socialism in Tanzania*. Boulder: L. Rienner, 1994.

——. *Ujamaa Villages in Tanzania: A Bibliography*. Uppsala: Scandinavian Institute of African Studies, 1981.

McIntosh, Susan Keech. "Pathways to Complexity: An African Perspective." In *Beyond Chiefdoms: Pathways to Complexity in Africa*, edited by Susan Keech McIntosh, 1–30. Cambridge: Cambridge University Press, 1999.

Mellars, Paul. "Fire Ecology, Animal Populations and Man: a Study of Some Ecological Relationships in Prehistory." *Proceedings of the Prehistoric Society* 42 (1976): 15–45.

Merchant, Carolyn. *The Death of Nature: Women, Ecology, and the Scientific Revolution*. San Francisco: Harper and Row, 1980.

Mfunda, I. *The Economics of Wildlife for Communities Living around the Serengeti National Park, Tanzania*. Fort Ikoma: Serengeti Regional Conservation Strategy, 1998.

Miller, Joseph C. "Listening for the African Past." Introduction to *The African Past Speaks: Essays on Oral Tradition and History*, edited by Joseph C. Miller, 1–12. Folkestone, Eng.: Dawson, 1980.

——, ed. *The African Past Speaks: Essays on Oral Tradition and History*. Folkestone, Eng.: Dawson, 1980.

Mitchell, W. J. T. "Imperial Landscape." In *Landscape and Power*, edited by W. J. T. Mitchell, 5–34. Chicago: University of Chicago Press, 1994.

——. Introduction to *Landscape and Power*, edited by W. J. T. Mitchell, 1–4. Chicago: University of Chicago Press, 1994.

Mkirya, Benjamin. *Historia, mila, na desturi ya Wazanaki*. Peramiho, Tanzania: Benedictine Publications, Ndanda Mission Press, 1991.

Moore, Audrey. *Serengeti*. London: Country Life, 1938.

Moore, Henrietta L. *Space, Text, and Gender: An Anthropological Study of the Marakwet of Kenya.* Cambridge: Cambridge University Press, 1986.

Moore, Henrietta L., and Megan Vaughan. *Cutting Down Trees: Gender, Nutrition, and Agricultural Change in the Northern Province of Zambia, 1890–1990.* Portsmouth, NH: Heinemann: 1994.

Morgan, Lewis Henry. *Ancient Society.* New York: World Publishing, 1877.

Mtoni, Paul Enock. "Involve Them or Lose Both: Local Communities Surrounding Serengeti National Park in Bunda and Serengeti Districts in Relation to Wildlife Conservation." MA Thesis, Agricultural University of Norway, 1999.

Muniko, S. M., B. Muita oMagige, and Malcolm J. Ruel, eds. *Kuria-English Dictionary.* London: International African Institute, 1996.

Murray, Martyn G. "Specific Nutrient Requirements and Migration of Wildebeest." In *Serengeti II: Dynamics, Management, and Conservation of an Ecosystem,* edited by A. R. E. Sinclair and Peter Arcese, 231–56. Chicago: University of Chicago Press, 1995.

Mustafa, Kemal. "The Concept of Authority and the Study of African Colonial History." *Kenya Historical Review Journal* 3, no. 1 (1975): 55–83.

———. "The Development of Ujamaa in Musoma: A Case Study of Butiama Ujamaa Village." MA thesis, University of Dar es Salaam, 1975.

Napper, D. M. *Grasses of Tanganyika.* Dar es Salaam: Ministry of Agriculture, Forests and Wildlife, Tanzania, 1965.

Neumann, Roderick P. *Imposing Wilderness: Struggles over Livelihood and Nature Preservation in Africa.* Berkeley: University of California Press, 1998.

———. "Primitive Ideas: Protected Area Buffer Zones and the Politics of Land in Africa." *Development and Change* 28 (July 1997): 559–82.

———. "Ways of Seeing Africa: Colonial Recasting of African Society and Landscape in Serengeti National Park." *Ecumene* 2, no. 2 (1995): 149–69.

Newbury, Catharine. *The Cohesion of Oppression: Clientship and Ethnicity in Rwanda, 1860–1960.* New York: Columbia University Press, 1988.

Newbury, David. *Kings and Clans: Ijiwi Island and the Lake Kivu Rift, 1780–1840.* Madison: University of Wisconsin Press, 1991.

Nicholson, Sharon Elaine. "A Climatic Chronology for Africa: Synthesis of Geological, Historical, and Meteorological Information and Data." PhD dissertation, University of Wisconsin, Madison, 1976.

Norton-Griffiths, M. "The Influence of Grazing, Browsing, and Fire on the Vegetation Dynamics of the Serengeti." In *Serengeti: Dynamics of an Ecosystem,* edited by A. R. E. Sinclair and M. Norton-Griffiths, 310–52. Chicago: University of Chicago Press, 1979.

Norton-Griffiths, M., D. Herlocker, and Linda Pennycuick. "The Patterns of Rainfall in the Serengeti Ecosystem, Tanzania." *East African Wildlife Journal* 13 (1975): 347–74.

Nurse, Derek, and Franz Rottland. "The History of Sonjo and Engaruka: A Linguist's View." *Azania* 28 (1993): 1–5.

———. "Sonjo: Description, Classification, History." *Sprache und Geschichte in Afrika* 12–13 (1991–92): 171–289.

Nyang'oro, Julius Edo, and Timothy M. Shaw. *Beyond Structural Adjustment in Africa: The Political Economy of Sustainable and Democratic Development.* New York: Praeger, 1992.

Nyerere, Julius K. *Freedom and Unity: Uhuru na umoja: A Selection from Writings and Speeches, 1952–65.* London: Oxford University Press, 1967.

Oboler, Regina Smith. "The House-Property Complex and African Social Organization." *Africa* 64, no. 3 (1994): 342–58.

Ochieng', William R. *A Pre-colonial History of the Gusii of Western Kenya.* Kampala: East African Literature Bureau, 1974.

Ofcansky, Thomas P. "The 1889–97 Rinderpest Epidemic and the Rise of British and German Colonialism in Eastern and Southern Africa." *Journal of African Studies* 8, no. 1 (1981): 31–38.

Ogot, B. A. *Migration and Settlement, 1500–1900.* Vol. 1 of *A History of the Southern Luo.* Nairobi: East African Publishing House, 1967.

Okpewho, Isidore. *African Oral Literature: Backgrounds, Character, and Continuity.* Bloomington: Indiana University Press, 1992.

Olthof, Wim. "Wildlife Resources and Local Development: Experiences from Zimbabwe's CAMPFIRE Programme." In *Local Resource Management in Africa,* edited by J. P. M. van den Breemer, C. A. Drijver, and L. B. Venema, 111–28. New York: Wiley, 1995.

Oming'o, D. W. O. "Self-Help Schemes in South Mara District." Political Science Paper 7. PhD dissertation, University of East Africa, University College, Dar es Salaam, 1970.

Ortner, Sherry B. "Resistance and the Problem of Ethnographic Refusal." *Comparative Studies in Society and History* 37, no. 1 (1995): 173–93.

Packard, Randall M. *Chiefship and Cosmology: An Historical Study of Political Competition.* Bloomington: Indiana University Press, 1981.

———. "The Study of Historical Process in African Traditions of Genesis: The Bashu Myth of Muhiyi." In *The African Past Speaks: Essays on Oral Tradition and History,* edited by Joseph C. Miller, 157–77. Folkestone, Eng.: Dawson, 1980.

Parkin, David. *Sacred Void: Spatial Images of Work and Ritual among the Giriama of Kenya.* Cambridge: Cambridge University Press, 1991.

Peel, J. D. Y. *Ijeshas and Nigerians: The Incorporation of a Yoruba Kingdom, 1890s–1970s.* Cambridge: Cambridge University Press, 1983.

Peires, J. B. *The House of Phalo: A History of the Xhosa People in the Days of Their Independence.* Berkeley: University of California Press, 1982.

Poewe, Karla O. *Matrilineal Ideology: Male-Female Dynamics in Luapula, Zambia.* London: Academic Press, 1981.

Ponte, Stefano. *Farmers and Markets in Tanzania: How Policy Reforms Affect Rural Livelihoods in Africa.* Oxford: James Currey, 2002.

Poole, Robert M. "Heartbreak on the Serengeti." *National Geographic*, February 2006, 2–29.

Povinelli, Elizabeth A. *Labor's Lot: The Power, History, and Culture of Aboriginal Action*. Chicago: University of Chicago Press, 1993.

Prazak, Miroslava. "Cultural Expressions of Socioeconomic Differentiation among the Kuria of Kenya." PhD dissertation, Yale University, 1992.

Pred, Allan. *Making Histories and Constructing Human Geographies: The Local Transformation of Practice, Power Relations, and Consciousness*. Boulder: Westview Press, 1990.

Presley, Cora Ann. "The Mau Mau Rebellion, Kikuyu Women and Social Change." *Canadian Journal of African Studies* 22, no. 3 (1988): 502–27.

Pudup, Mary Beth. "Arguments within Regional Geography." *Progress in Human Geography* 12, no. 3 (September 1989): 369–91.

Purseglove, J. M. *Tropical Crops: Monocotyledons*. London: Longman, 1972.

Radcliffe-Brown, A. R. *Structure and Function in Primitive Society: Essays and Addresses*. New York: Free Press, 1965.

Ranger, Terence. "Taking Hold of the Land: Holy Places and Pilgrimages in Twentieth-Century Zimbabwe." *Past and Present* 117 (November 1987): 158–94.

———. *Voices from the Rocks: Nature, Culture and History in the Matopos Hills of Zimbabwe*. Bloomington: Indiana University Press, 1999.

———. "Women and Environment in African Religion: The Case of Zimbabwe." In *Social History and African Environments*, edited by William Beinart and JoAnn McGregor, 72–86. Athens: Ohio University Press, 2003.

Raymond, W. D. "Tanganyika Arrow Poisons." *Tanganyika Notes and Records* 23 (1947): 49–65.

Rempel, Ruth. "Trade and Transformation: Participation in the Ivory Trade in Late 19th Century East and Central Africa." *Canadian Journal of Development Studies* 19, no. 3 (1998): 529–552.

Richards, Paul. "Ecological Change and the Politics of African Land Use." *African Studies Review* 26, no. 2 (1983).

Rigby, Peter. *Cattle and Kinship among the Gogo: A Semi-pastoral Society of Central Tanzania*. Ithaca: Cornell University Press, 1969.

Roberts, Andrew. "The Nyamwezi." In *Tanzania before 1900*, edited by Andrew Roberts, 117–50. Nairobi: East African Publishing House, 1968.

———. "Nyamwezi Trade." In *Pre-colonial African Trade: Essays on Trade in Central and Eastern Africa before 1900*, edited by Richard Gray and David Birmingham, 39–74. New York: Oxford University Press, 1976.

Robertson, Claire C. *Trouble Showed the Way: Women, Men, and Trade in the Nairobi Area, 1890–1990*. Bloomington: Indiana University Press, 1997.

Rockel, Stephen J. "A Nation of Porters: The Nyamwezi and their Labor Market in Nineteenth-Century Tanzania." *Journal of African History* 41, no. 2 (2000): 173–95.

Rogers, Peter. "Global Governance/Governmentality, Wildlife Conservation, and Protected Area Management: A Comparative Study of Eastern and Southern

Africa (with permission)." Paper presented at the African Studies Association, 45th Annual Meeting, Washington, DC, 2002.

Rosaldo, Renato. *Ilongot Headhunting, 1883–1974: A Study in Society and History.* Stanford: Stanford University Press, 1980.

Roscoe, John. *The Baganda; An Account of Their Native Customs and Beliefs.* 2nd ed. New York: Barnes and Noble, 1966.

Rose, Deborah Bird. *Nourishing Terrains: Australian Aboriginal Views of Landscape and Wilderness.* Canberra: Australian Heritage Commission, 1996.

Ruel, Malcolm J. "Kuria Generation Classes." *Africa* 32, no. 1 (1962): 14–36.

———. "Kuria Generation Sets." Paper presented at the Makerere Institute of Social Research Conference, Kampala, 1954–58.

———. "Piercing: Ritual Actions of the Kuria, East Africa." Paper presented at the Makerere Institute of Social Research Conference, Kampala, 1954–58.

———. "Religion and Society among the Kuria of East Africa." *Africa* 35, no. 3 (1965): 295–306.

Sack, Robert David. *Human Territoriality: Its Theory and History.* Cambridge: Cambridge University Press, 1986.

Sacks, Karen. *Sisters and Wives: The Past and Future of Sexual Equality.* Westport, CT: Greenwood Press, 1979.

Said, Edward. *Orientalism.* New York: Vintage Books, 1994.

Schama, Simon. *Landscape and Memory.* New York: Knopf, 1995.

Schlobach, Von Hauptmann, "Die Volksstämme der deutschen Ostkuste des Victoria-Nyansa." *Mitteilungen aus den deutschen Schutzgebieten* (Berlin: U. Usher) 14 (1901): 183–93.

Schmidt, Heike. "'Penetrating' Foreign Lands: Contestations over African Landscapes: A Case Study from Eastern Zimbabwe." *Environment and History* 1, no. 3 (1995): 351–76.

Schmidt, Peter. *Historical Archaeology: A Structural Approach in an African Culture.* Westport, CT: Greenwood Press, 1978.

Schnee, Heinrich. *Deutsches Kolonial-Lexikon.* 3 vols. Leipzig: Quelle und Meyer, 1920.

Schoenbrun, David Lee. "Early History in Eastern Africa's Great Lakes Region: Linguistic, Ecological, and Archaeological Approaches, ca. 500 B.C. to ca. A.D. 1000." PhD dissertation, UCLA, 1990.

———. "Gendered Histories between the Great Lakes: Varieties and Limits." *International Journal of African Historical Studies* 29, no. 3 (1996): 461–92.

———. *A Green Place, A Good Place: Agrarian Change, Gender, and Social Identity in the Great Lakes Region to the 15th Century.* Portsmouth, NH: Heinemann, 1998.

———. *The Historical Reconstruction of Great Lakes Bantu Cultural Vocabulary: Etymologies and Distributions.* Cologne: Rüdiger Köppe Verlag, 1996.

———. "We Are What We Eat: Ancient Agriculture between the Great Lakes." *Journal of African History* 34, no. 1 (1993): 1–31.

Schoffeleers, J. M. Introduction to *Guardians of the Land: Essays on Central African Territorial Cults*, edited by J. M. Schoffeleers, 1–45. Gwelo, Zimbabwe: Mambo Press, 1978.

——. "Oral History and the Retrieval of the Distant Past: On the Use of Legendary Chronicles as Sources of Historical Information." In *Theoretical Explorations in African Religion*, edited by Wim van Binsbergen and Matthew Schoffeleers, 164–88. London: KPI, 1985.

——. *River of Blood: The Genesis of a Martyr Cult in Southern Malawi, c. A.D. 1600*. Madison: University of Wisconsin Press, 1992.

Scott, Jonathan. *The Great Migration*. London: Elm Tree Books, 1988.

Scott, James C. *The Moral Economy of the Peasant: Rebellion and Subsistence in Southeast Asia:* New Haven: Yale University Press, 1977.

——. *Weapons of the Weak: Everyday Forms of Peasant Resistance*. New Haven: Yale University Press, 1985.

Searing, James F. *West African Slavery and Atlantic Commerce: The Senegal River Valley, 1700–1860*. Cambridge: Cambridge University Press, 1993.

Sen, Amartya Kumar. *Poverty and Famines: An Essay on Entitlement and Deprivation*. Oxford: Clarendon Press, 1981.

Senior, H. S. "The Sukuma Homestead." *Tanganyika Notes and Records* 9 (1940): 42–4.

Shauri, Vincent. "The New Wildlife Policy in Tanzania: Old Wine in a New Bottle?" LEAT (Lawyers' Environmental Action Team). http://www.leat.or.tz/publications/wildlife.policy/index.php.

Sheriff, Abdul. *Slaves, Spices, and Ivory in Zanzibar: Integration of an East African Commercial Empire into the World Economy, 1770–1873*. Athens: Ohio University Press, 1987.

Shetler, Jan Bender. "The Gendered Spaces of Historical Knowledge: Women's Knowledge and Extraordinary Women in the Serengeti District, Tanzania." *International Journal of African Historical Studies* 36, no. 2 (2003): 283–307.

——. "'A Gift for Generations to Come': A Kiroba Popular History from Tanzania and Identity as Social Capital in the 1980s." *International Journal of African Historical Studies* 28, no. 1 (1995): 69–112.

——. "Interpreting Rupture in Oral Memory: The Regional Context for Changes in Western Serengeti Age Organization (1850–1895)." *Journal of African History* 44, no. 3 (2003): 385–412.

——. "The Landscapes of Memory: A History of Social Identity in the Western Serengeti, Tanzania." PhD dissertation, University of Florida, 1998.

——. "The Politics of Publishing Oral Sources from the Mara Region, Tanzania." *History in Africa* 29 (2002): 413–26.

——. "'Region' as Historical Production: Narrative Maps from the Western Serengeti, Tanzania." In *The Spatial Factor in African History: The Relationship of the Social, Material, and Perceptual*, edited by Allen M. Howard and Richard M. Shain, 141–76. Leiden: Brill, 2005.

———. *Telling Our Own Stories: Local Histories from South Mara, Tanzania*. Leiden: Brill, 2003.

Shipton, Parker M. "Strips and Patches: A Demographic Dimension in Some African Land-Holding and Political Systems." *Man*, n.s., 19, no. 4 (1984): 613–34.

Shipton, Parker, and Mitzi Goheen. "Understanding African Land-Holding: Power, Wealth, and Meaning." Introduction. *Africa* 62, no. 3 (1992): 307–25.

Shivji, Issa G. *Not Yet Democracy: Reforming Land Tenure in Tanzania*. Dar es Salaam: IIED/HAKIARDHI/Faculty of Law, University of Dar es Salaam, 1998.

Simenauer, E. "The Miraculous Birth of Hambageu, Hero-God of the Sonjo." *Tanganyika Notes and Records* 38 (1955): 23–30.

Simonse, Simon, and Eisei Kurimoto. Introduction to *Conflict, Age and Power in North East Africa: Age Systems in Transition*, edited by Eisei Kurimoto and Simon Simonse, 1–28. Athens: Ohio University Press, 1998.

Sinclair, A. R. E. "Dynamics of the Serengeti Ecosystem." In Sinclair and Norton-Griffiths, *Serengeti: Dynamics of an Ecosystem*, 1–30.

———. "Equilibria in Plant-Herbivore Interactions." In Sinclair and Arcese, *Serengeti II*, 91–113.

———. "The Serengeti Environment." In Sinclair and Norton-Griffiths, *Serengeti: Dynamics of an Ecosystem*, 31–45.

———. "Serengeti Past and Present." In Sinclair and Arcese, *Serengeti II*, 3–30.

Sinclair, A. R. E., and Peter Arcese. "Serengeti in the Context of Worldwide Conservation Efforts." In Sinclair and Arcese, *Serengeti II*, 31–46.

———, eds. *Serengeti II: Dynamics, Management and Conservation of an Ecosystem*. Chicago: University of Chicago Press, 1995.

Sinclair, A. R. E., and M. Norton-Griffiths, eds. *Serengeti: Dynamics of an Ecosystem*. Chicago: University of Chicago Press, 1979.

Skully, R. T. K. "Fort Sites of East Bukusu, Kenya." *Azania* 4 (1969): 105–14.

———. "Nineteenth Century Settlement Sites and Related Oral Traditions from the Bungoma Area, Western Kenya." *Azania* 14 (1979): 81–96.

Slater, Candace. *Entangled Edens: Visions of the Amazon*. Berkeley: University of California Press, 2002.

Smith, Andrew B. "The Kalahari Bushman Debate: Implications for Archeology in Southern Africa." *South African Historical Journal* 35 (1996): 1–15.

Smith, Carol A. "Regional Economic Systems: Linking Geographical Models and Socioeconomic Problems." In *Economic Systems*, vol. 1 of *Regional Analysis*, 3–63. New York: Academic Press, 1976.

Smith, R. S. *Kingdoms of the Yoruba*. 3rd ed. Madison: University of Wisconsin Press, 1988.

Smith, William Edgett. *Nyerere of Tanzania*. London: Victor Gollancz, 1973.

Soja, Edward W. *Postmodern Geographies: The Reassertion of Space in Critical Social Theory*. London: Verso, 1989.

Songorwa, Alexander N. "Community-Based Wildlife Management (CWM) in Tanzania: Are the Communities Interested?" *World Development* 27, no. 12 (1999): 2061–79.

Southall, Aiden. *Alur Society: A Study in Processes and Types of Domination.* Cambridge: W. Heffer, 1956.

——. "The Segmentary State: From the Imaginary to the Material Means of Production." In *Early State Economics*, edited by Henri J. M. Claessen and Pieter van de Velde, 75–96. New Brunswick, NJ: Transaction Publishers, 1991.

Spear, Thomas. *Kenya's Past: An Introduction to Historical Method in Africa.* London: Longman, 1981.

——. "Oral Traditions: Whose History?" *History in Africa* 8 (1981): 165–81.

Spear, Thomas, and Richard Waller, eds. *Being Maasai: Ethnicity and Identity in East Africa.* Athens: Ohio University Press, 1993.

Spence, Jonathan D. *The Memory Palace of Matteo Ricci.* New York: Viking, 1984.

Spencer, Paul. "Age Systems and Modes of Predatory Expansion." In *Conflict, Age and Power in North East Africa: Age Systems in Transition*, edited by Eisei Kurimoto and Simon Simonse, 168–85. Athens: Ohio University Press, 1998.

——. "Becoming Maasai, Being in Time." In *Being Maasai: Ethnicity and Identity in East Africa*, edited by Thomas Spear and Richard Waller, 140–56. Athens: Ohio University Press, 1993.

Steinhart, Edward I. *Black Poachers, White Hunters: A Social History of Hunting in Colonial Kenya.* Athens: Ohio University Press, 2006.

——. *Conflict and Collaboration: The Kingdoms of Western Uganda, 1890–1907.* Princeton: Princeton University Press, 1977.

——. "Elephant Hunting in 19th Century Kenya: Kamba Society and Ecology in Transformation," *International Journal of African Historical Studies* 33, no. 2 (2001): 335–49.

——. "Hunters, Poachers and Gamekeepers: Towards a Social History of Hunting in Colonial Kenya." *Journal of African History* 30, no. 2 (1989): 247–64.

Sunseri, Thaddeus. "Reinterpreting a Colonial Rebellion: Forestry and Social Control in German East Africa, 1874–1915." *Environmental History* 8, no. 3 (2003): 430–51.

——. *Vilimani: Labor Migration and Rural Change in Early Colonial Tanzania.* Portsmouth, NH: Heinemann, 2002.

Sutton, John E. G. *The Archaeology of the Western Highlands of Kenya.* Nairobi: British Institute in Eastern Africa, 1973.

——. "Becoming Maasailand." In *Being Maasai: Ethnicity and Identity in East Africa*, edited by Thomas Spear and Richard Waller, 38–60. Athens: Ohio University Press, 1993.

——. "Engaruka and Its Waters." *Azania* 13 (1978): 37–70.

——. "Engaruka etc." *Tanzania zamani* 10 (1972): 7–10.

——. "The Irrigation and Manuring of the Engaruka Field System: Further Observations and Historical Discussion of a Later Iron Age Settlement in the Northern Tanzanian Rift." *Azania* 21 (1986): 27–51.

——. "Some Reflections on the Early History of Western Kenya." In *Hadith: Proceedings of the Annual Conference of the Historical Association of Kenya* 2 (Nairobi: East African Publishing House, 1970), edited by Bethwell A. Ogot, 2:17–29.

———. *A Thousand Years of East Africa*. Nairobi: British Institute in Eastern Africa, 1990.

———. "Towards a History of Cultivating the Fields." *Azania* 24 (1989): 99–112.

Swainson, Nicola. *The Development of Corporate Capitalism in Kenya, 1918–77*. Berkeley: University of California Press, 1980.

Tanner, R. E. S. "Cattle Theft in Musoma, 1958–59." *Tanzania Notes and Records* 65 (1966): 31–45.

Tantala, Renee Louise. "The Early History of Kitara in Western Uganda: Process Models of Religious and Political Change." PhD dissertation, University of Wisconsin, Madison, 1989.

———. "Verbal and Visual Imagery in Kitara (Western Uganda): Interpreting 'The Story of Isimbwa and Nyinamwiru.'" In *Paths toward the Past: African Historical Essays in Honor of Jan Vansina*, edited by Robert W. Harms, 223–43. Atlanta: African Studies Association Press, 1994.

Taylor, Christopher C. *Milk, Honey, and Money: Changing Concepts in Rwandan Healing*. Washington, DC: Smithsonian Institution Press, 1992.

Tedlock, Dennis, and Bruce Mannheim. *The Dialogic Emergence of Culture*. Urbana: University of Illinois Press, 1995.

Thompson, E. P. *The Making of the English Working Class*. New York: Vintage Books, 1966.

———. *Whigs and Hunters: The Origin of the Black Act*. New York: Pantheon Books, 1975.

Thomson, Joseph. *Through Masai Land: A Journey of Exploration among the Snow-clad Volcanic Mountains and Strange Tribes of Eastern Equatorial Africa*. London: Sampson Low, Marston, Searle, and Rivington, 1887. Reprint, London: Frank Cass, 1968.

Thornton, John K. *Africa and Africans in the Making of the Atlantic World, 1400–1800*. 2nd ed. Cambridge: Cambridge University Press, 1998.

Thornton, Robert J. *Space, Time, and Culture among the Iraqw of Tanzania*. New York: Academic Press, 1980.

Thurnwald, Richard C. *Black and White in East Africa: The Fabric of a New Civilization: A Study in Social Contact and Adaptation of Life in East Africa*. London: George Routledge and Sons, 1935.

Tilley, Helen. "African Environments and Environmental Sciences: The African Research Survey, Ecological Paradigms and British Colonial Development, 1920–1940." In *Social History and African Environments*, edited by William Beinart and JoAnn McGregor, 109–30. Oxford: James Currey, 2003.

Tobisson, Eva. *Family Dynamics among the Kuria: Agro-Pastoralists in Northern Tanzania*. Göteborg, Sweden: Acta Universitatis Gothoburgensis, 1986.

Tonkin, Elizabeth. *Narrating Our Pasts: The Social Construction of Oral History*. Cambridge: Cambridge University Press, 1992.

Tooke, D. W. Hammond. "In Search of the Lineage: The Cape Nguni Case." *Man* 19 (1984): 77–93.

Toppenberg, Valdemar E. *Africa Has My Heart*. Mountain View, CA: Pacific Press Publishing Association, 1958.

Tripp, Aili Mari. *Changing the Rules: The Politics of Liberalization and the Urban Informal Economy in Tanzania*. Berkeley: University of California Press, 1997.

Turner, Kay. *Serengeti Home*. New York: Dial Press/James Wade, 1977.

Turner, Myles. *My Serengeti Years: The Memoirs of an African Game Warden*. Edited by Brian Jackman. New York: Norton, 1987.

Turner, Victor. *The Forest of Symbols: Aspects of Ndembu Ritual*. Ithaca: Cornell University Press, 1967.

Uhlig, Carl. *Die ostafrikanische Bruchstufe und die angrenzenden Gebiete zwischen den Seen Magad und Lawa ja Mweri sowie dem Westfluss des Meru*. Part 1, *Die Karte*. Berlin: Ernst Siegfried Mittler und Sohn, 1909.

Vail, Leroy, ed. *The Creation of Tribalism in Southern Africa*. Berkeley: University of California Press, 1991.

Vansina, Jan. *The Children of Woot: A History of the Kuba Peoples*. Madison: University of Wisconsin Press, 1978.

———. *Oral Tradition as History*. Madison: University of Wisconsin Press, 1985.

———. *Paths in the Rainforests: Toward a History of Political Tradition in Equatorial Africa*. Madison: University of Wisconsin Press, 1990.

Van Young, Eric. "Are Regions Good to Think?" Introduction to *Mexico's Regions: Comparative History and Development*, 1–36. San Diego: Center for U.S.-Mexican Studies, UCSD, 1992.

Vaughan, Megan. *The Story of an African Famine: Gender and Famine in Twentieth-Century Malawi*. Cambridge: Cambridge University Press, 1987.

Wagner, Michele. "Environment, Community and History: 'Nature in the Mind' in Nineteenth- and Early Twentieth-Century Buha, Western Tanzania." In *Custodians of the Land: Ecology and Culture in the History of Tanzania*, edited by Gregory Maddox, James Giblin, and Isaria Kimambo, 175–99. Athens: Ohio University Press, 1996.

———. "Whose History Is History? A History of the Baragane People of Buragane, Southern Burundi, 1850–1932." 2 vols. PhD dissertation, University of Wisconsin, Madison, 1991.

Wakefield, Rev. T. "Native Routes through the Masai Country, from information obtained by the Rev. T. Wakefield." *Proceedings of the Royal Geographical Society*, n.s. 4 (1882): 742–47.

———. "Wakefield's Notes on the Geography of Eastern Africa, Routes of Native Caravans from the Coast . . ." *Journal of the Royal Geographical Society* 40 (1870): 303–39.

Waller, Richard. "Ecology, Migration, and Expansion in East Africa." *African Affairs* 84 (July 1985): 347–70.

———. "Economic and Social Relations in the Central Rift Valley: The Maa-Speakers and Their Neighbors in the Nineteenth Century." In *Kenya in the 19th Century (Hadith 8)*, edited by Bethwell A. Ogot, 83–151. Kisumu, Kenya: Anyange Press, 1985.

———. "The Lords of East Africa: The Maasai in the Mid-nineteenth Century (c. 1840–c. 1885)." PhD dissertation, Cambridge University, 1979.

———. "The Maasai and the British, 1895–1905: The Origins of Alliance." *Journal of African History* 17, no. 4 (1976): 529–53.

———. "Tsetse Fly in Western Narok, Kenya." *Journal of African History* 31, no. 1 (1990): 81–101.

Walter, Bob J. *Territorial Expansion of the Nandi of Kenya, 1500–1905*. Papers in International Studies, Africa Series, no. 9. Athens: Ohio University Center for International Studies, 1970.

Watts, Michael. *Silent Violence: Food, Famine, and Peasantry in Northern Nigeria*. Berkeley: University of California Press, 1983.

Weiss, Brad. *The Making and Unmaking of the Haya Lived World: Consumption, Commoditization, and Everyday Practice*. Durham: Duke University Press, 1996.

Weiss, Max. *Die Völkerstämme im norden Deutsch-Ostafrikas*. Berlin: Carl Marschner, 1910.

Wells, Michael, Katrina Brandon, and Lee Hannah. *People and Parks: Linking Protected Area Management with Local Communities*. Washington DC: World Bank, 1992.

Werbner, Richard P. *Regional Cults*. New York: Academic Press, 1977.

White, Luise, Stephan F. Miescher, and David William Cohen, eds. *African Words, African Voices: Critical Practices in Oral History*. Bloomington: Indiana University Press, 2001.

White, Stewart Edward. *Lions in the Path: A Book of Adventure on the High Veldt*. Garden City, NY: Doubleday, Page and Co., 1926.

———. *The Rediscovered Country*. Garden City, NY: Doubleday, Page and Co., 1915.

Wilks, Ivor. *Asante in the Nineteenth Century: The Structure of Evolution of a Political Order*. Cambridge: Cambridge University Press, 1975.

Wilmsen, Edwin N. *Land Filled with Flies: A Political Economy of the Kalahari*. Chicago: University of Chicago Press, 1989.

Wilmsen, Edwin N., and James R. Denbow. "Paradigmatic History of San-Speaking Peoples and Current Attempts at Revision." *Current Anthropology* 31, no. 5 (December 1990): 489–524.

Wilson, G. McL. "The Tatoga of Tanganyika, Part One." *Tanganyika Notes and Records* 33 (1952): 34–47.

———. "The Tatoga of Tanganyika, Part Two." *Tanganyika Notes and Records* 34 (1953): 35–56.

Wright, Marcia. *Strategies of Slaves and Women: Life-Stories from East and Central Africa*. New York: L. Barber Press, 1993.

Wrigley, Christopher. *Kingship and State: The Buganda Dynasty*. Cambridge: Cambridge University Press, 1996.

Yates, Frances A. *The Art of Memory*. Chicago: University of Chicago Press, 1966.

Index

asimoka, 29, 40–44, 50, 56, 60–61, 64. *See also* origin traditions and places of
askari, 184, 289n50
Australopithecus, 33
authority, 11, 66, 70, 126, 139, 141, 145, 151, 152, 155, 183, 207, 232, 235; chiefs, 173–77; colonial, 170, 173–74, 185, 214 (*see also* colony/colonial: rule); communal, 119–20; ecological, 31, 39, 50, 175, 185, 210; gendered, 56–59, 61; generational, 118–19, 122, 129–31, 158, 162, 281n76; government, 220–21; homestead, 61; Native Authority, 182–83; pastoralist, 52–53; political, 123, 157, 169, 172–74, 184; ritual, 53–54, 58, 85, 90, 96–97, 101, 103, 112–14, 120

baboon, 94
Babu, David Stevens, 212
Bachelard, Gaston, 20
Bagini, 176
Bajuta, 146
Banagi, 44, 92, 208, 219, 222; Hill, 125; Ranger Station, 182, 185, 198, 216
Bangwesi Mountain, 42–44, 74, 104, 124–25, 245n15. *See also* Mangwesi Mountain
Bantu languages, 10, 22, 31, 46–48, 50, 55, 67, 91, 96, 106, 117, 119, 124, 128–29
Bantu speakers, 40, 46–49, 52–53, 55, 90, 107, 117, 123–24, 135, 138, 146, 150–51; earliest in western Serengeti, ix, 30, 46, 48; East Nyanza, 7, 31, 46–49, 61, 65–66, 69, 76, 83, 117, 123; Great Lakes, 69, 77, 96, 117, 123–24; Sukuma farmers, 8
bao, 52
baragumu, 122
baraza, 195
battle: of age-sets, 163; of Ndabaka, 148–50, 160; in oral tradition, 11, 97, 114, 136; against poachers, 24; in WWI, 170
Battle of Ndabaka, 148–50, 160
Batwa, 42
Baumann, Oskar, 140–41, 146, 156, 186
beer: parties, 11, 75, 190–91, 235; preparation of, 76, 127, 196
bees, 39, 42, 85, 137
big man, 42, 96, 172, 189, 191
bilateral descent, 69–70
binding (core spatial image), 103, 119, 123, 127–29
Bischofberger, Otto, 118, 274n129
black-cotton soil, 50, 196, 219. *See also* mbuga

blacksmith, 50, 94, 128, 274n129
bless, 48, 112, 122, 127–28, 150
blessing, 91, 96, 109–10, 112, 114, 121–23, 128–29, 265
blood, 73–74, 77, 82–83, 85, 90, 107, 129, 147, 215
blood brotherhood, 49, 94–95, 122, 127, 275n140
Bongirate, 121, 160, 163–64, 203; age-set cycles, 138, 158, 164, 219, 280n69; age-set territory, 177, 184, 196
Borumarancha, 138, 148, 163–64
boundaries, 64; clan, 94; colonial, 56; crossing, 49, 117; ecological, 2, 7, 9, 28, 37, 124–26, 145; ethnic group, 68, 91, 94, 117, 124–26, 130, 154, 164–65, 191; park, 30, 43, 74, 103, 125, 181–82, 200–208, 210, 214, 218, 222–23, 225–26, 229; ritual, 110, 115, 119, 121, 123–30, 136; territorial, 101–4, 124–25, 130, 163–65, 175
bracelet, 141, 147, 186, 305n144
Braudel, Fernand, 22
bridewealth, 55, 78, 83–84, 130, 151, 186, 189, 194, 197, 213
British, 70, 85, 161, 180–85, 188–89, 191–92, 194–96, 202, 206, 210, 220, 223; indirect rule, 9, 169–70, 173–74, 176–78; officers, 38, 80, 85, 176–77; view of the land, 4, 17, 85, 180–81
Buchanan, Carole, 90
buffalo, 34, 39, 42, 109, 123, 205, 222, 227, 282n94
buffer zone, 124, 202, 218, 222–23
Buganda, 94, 141, 176
Bugerera, 10, 12
Buhemba, 87, 91–92, 194
Buhoro, 128
Bukiroba, 82
Bumare, 93, 123, 280n76
Bunda, 148; District, 7, 226
Bunyoro, 94, 274n128
Buri, 154
burning, 57, 234; bush, 33–37, 72, 178; houses, 143, 156, 174. *See also* fire
Burundi, 16
Burunga, 55
Busaai, 138, 158, 160, 163–64
Busegwe, 82, 298n28
bush, 6, 32, 40–41, 43, 58, 72, 79, 81, 110, 136, 150–51, 171–72, 178, 182, 184, 188, 203, 218, 230; encroachment of tsetse fly, 34, 36–38, 124–25, 138–39, 145–46, 178–79; land, 51, 53–54, 164–65

bushbuck, 58, 77, 93
Busoga, 94
Butiama, 93, 131, 212
Bwanda, 45, 54, 64
Bwinamoki, 136
Bwiregi, 91
Bwiro, 94, 274nn130–31
Bwitenge, 74

capitalist, 15, 17, 68, 169, 214, 217, 224
captive, 143, 163, 171
caravan trade, 138–45, 148, 152, 157, 162,
    186, 188, 287n22, 297n7
Carruthers, Jane, 15
cash: crops, 170, 186, 192–93, 195, 198, 233;
    economy, 192, 194, 198–99, 203, 213,
    224, 231
cassava, 15, 95, 195–97, 199, 219, 309n206
catfish, 38
Catholic, 131
cattle, 4, 14, 52–53, 74, 85, 136, 177, 183–84,
    225, 232; clan, 89, 91, 93; corral, 59,
    76–77, 80, 122; disease, 34, 36–37, 45,
    72, 139–40, 142–43, 197, 218–19; raiding,
    126–27, 138–43, 145–49, 154, 162, 179,
    186, 197, 207–9, 213, 219, 232; trustees,
    53, 71, 189; wealth, 64, 68, 70–71, 97,
    130, 173–74, 186, 188–99
cave, 19, 41, 45, 57–58
census, 68, 189, 211, 220
central Africa, 16, 205
central Kenya, 48
central plains, 208–10, 212
ceremony, 102, 115, 119–23, 126, 128, 191,
    277n22, 283n115, 283n120
chama, 121–22, 129–30, 164
Chamuriho Mountain, 42, 45, 93, 113–14,
    119, 245n15
chanderema, 74
charcoal, 203
charm, 128
Chengero, 43–44
chesiri, 75
chiefs, 5, 7, 15, 57, 85, 96, 110, 123, 187, 189,
    195–97, 199, 207, 222, 230; Butiama, 131;
    colonial, 172–77, 182–84; German, 161,
    171, 174; Ikizu, 57–58, 70, 151–53;
    Ikoma, 170, 184; Ishenyi, 136; Kerewe,
    142; Kwaya clan, 90, 152; Nata, 85, 156,
    171–72, 175, 185
cholera, 140
chronology, 9, 14, 21, 23, 65, 158–59, 287n22
Chuuma, 116–18, 120, 176

circumcision, 117, 128, 147, 190, 294n131,
    305n144; ceremony, 82, 112, 118; female,
    84, 123; sets, 159–60
clan, 4, 66–67, 69–70, 81–83, 116, 118, 150,
    271n88; adoption and, 148, 154; black-
    smith, Turi, 94; chief, Kwaya, 152; con-
    flict, 213, 219; Hemba, 64, 86–89; histo-
    ries, 7, 11, 19, 22, 57, 63–65, 67, 75–77,
    81, 85–86, 90, 97, 101, 103, 109, 123, 137,
    139, 147, 154–55, 175, 235; hunter, 48,
    54–55, 90, 109, 261n81; identity, 80, 83,
    909, 121; migration, 19, 63, 86, 154–55;
    networks, 91, 95–96, 98, 123, 138,
    151–52, 233; residential (hamate), 77,
    82–83; resource use, 73–74, 79, 82–83,
    90, 98, 102, 118–20, 126–28, 150, 164,
    173, 184, 217; spirits, 106, 108; territory
    (ekyaro), 82–83; totem or praise name
    or avoidance, 38, 91–94, 108, 122
client, 96, 146, 180, 206, 227, 290n62
coevolution, 33
coffee, 289n50
Cohen, David William, 4, 16, 262n87
collective memory, 2, 235
colonialism, 15, 169, 178, 217, 233
colonial period, 7–9, 17, 37–38, 50–51, 68,
    70–73, 82, 85, 96, 145, 147, 169–70, 177,
    180, 185, 188, 190–91, 195, 198–99, 211,
    214, 217, 230
colony/colonial, 2, 15–17, 50–51, 53, 68, 97,
    125–26, 141, 169–70, 175, 183–84, 186,
    194, 214, 217–18; boundaries, 56, 125;
    chief, 131, 172–73, 175–77, 199, 230; de-
    mands, 75, 170, 172, 185; discourse, 15,
    165, 170, 173, 199; economy, 191–92, 195,
    204; government, 5, 17, 34, 75, 106, 145,
    168, 170, 172–73, 178–79, 185, 188, 192–93,
    195, 201, 204, 211, 219, 229; office/officer,
    15, 34, 36–37, 71, 80, 85, 143, 157, 169,
    173, 177, 179–80, 182, 184–85, 191, 195,
    201, 206, 209; policy, 170, 176–80, 212,
    220; rule, ix, 9, 169–70, 175, 180, 184–85,
    199, 201; system, 166, 176
community, 140, 143, 183–84, 186, 191,
    215–16, 222, 235, 306n157; conservation,
    16, 24, 202, 204, 212, 223, 225–27, 229–33,
    323n203; ecology, 36, 40–41, 46, 49, 68;
    founder, 39, 43, 45, 57, 64, 103; identity,
    115, 117, 126–28, 130–31; inclusion,
    81–86, 147; leaders, 105, 111–12, 174–78;
    moral economy and networks, 73,
    75–76, 79, 85, 213, 227; residential, 83;
    ritual, 96–97, 103, 110, 112, 128, 130;

diversification, 22, 63, 65–68, 70, 74–76, 97, 151, 184, 194, 237
dog: hunting, 181, 187, 216; wild, 205, 225
Dorobo. *See* Ndorobo
drought, 34, 39, 63, 68, 110, 135, 162, 179, 208, 212, 216, 219, 221; and abagore, 84; farming and grazing during, 37, 67, 71–72; of late nineteenth century, 138–40, 142, 144, 170, 185; of 1930s–40s, 192–93, 195–98; and rainmaker, 97, 137; survival during, 74, 151
Dublin, Holly, 34
Duma River, 32, 210
Dunlap, Thomas, 4
Durkheim, Emile, 21

East Africa, 15, 33, 37, 45–46, 67, 71, 118, 127, 135, 139–40, 145, 156–57, 163, 165, 170, 180, 187, 195, 209, 216
East Coast fever, 37, 195
Eastern Sahelian, 46
East Nyanza Bantu, 46–49, 58, 77, 106, 115; speakers (*see* Bantu speakers: East Nyanza)
ebehwe, 105
ebenturu trees, 172
ebimenyo, 105
ebimoro, 95, 191
ebony, 74
ebyaro, 126. *See also* ekyaro
ecosystem, 9, 15, 30, 31, 34–35, 67; Serengeti-Mara, 1, 9–10, 24, 31–34
ecotone, 32, 46
ecotourism, 231
education, 11, 15, 192, 224, 226–28
eghise, 112, 121
Ehret, Christopher, 10
eka, 77, 80
ekeburu, 80
ekebuse, 71
ekehita, 78
ekehwe, 105
ekerisho, 79
ekeshoka, 78
eketoha, 72
ekimweso, 114, 119, 128. *See also* purification, ritual
ekinyariri, 36
ekitana, 127, 130
ekitando, 74
ekyaro, 74, 79–80, 82–83, 101–2, 104, 116, 118, 121–24, 126–31, 164
elder: Ikizu, 84, 112, 152, 190; Ikoma, 3–5, 25, 38, 52, 54–56, 63–64, 70, 74, 83–85,

125, 129, 140, 171, 187–88, 196; Ishenyi, 138–39, 164; Kikuyu, 129; Maasai, 129; Mugumu, 219; Nata, 59, 70, 104, 124, 126, 138, 158, 179, 182, 189–90, 193, 196, 202; Ngoreme, 123, 151, 155–56; Sonjo, 48–49; Tatoga, 7, 52, 107, 146; western Serengeti, 36–37, 54, 71, 81, 90, 116, 139, 141; Zanaki, 118
elephant, 34, 38, 55, 108–9, 141, 186–88, 205, 219, 225, 303n108
elephant-hunting association, 141–42
eleusine (finger) millet (oburwe), 41, 46, 50–51, 70–71, 260n74, 309n206. *See also* millet
El Niño, 139
Embagai Crater, 209–10
emergence, 28–29, 40, 42, 50, 53–54, 152; Ikizu, 57, 96, 152; Ikoma, 43, 89, 155; Nata, 30–31; sites, 42, 54, 118; stories, 40, 45, 47, 50, 63, 70, 101, 119, 128, 154–55, 233
emigiro, 91, 108
emisambwa, 44, 100, 101, 103, 106–14, 119, 123, 128, 130, 131, 235. *See also* ancestors; erisambwa
encirclement, 101, 103, 115, 118–19, 127–28, 130, 158, 233, 237
enemy, 39, 151, 290–91n70
Engaruka, 48
environment/environmental: authority, 16, 24, 229, 232, 236; change, 14–15, 34, 36; degradation, 15, 24, 139, 202, 210; history, 6, 14–15, 23, 40, 60–61, 75; interaction, 65, 199, 204, 232; management, 15, 38–39, 145, 172, 177, 185, 198–99; policies, 68, 170, 172, 181, 230, 237; science, 15; views of, 5–6, 14–17, 66
epidemic, 37, 28, 135–36, 138, 143–44, 178–79, 198, 218. *See also* disease
erisambwa, 106–14, 122, 128, 136, 150, 276n15, 291n75. *See also* ancestors; emisambwa
erosion, 68, 198, 205, 209, 218
esaiga, 117
esebe, 128
eseghero, 71
ethnic group, 7–9, 47, 68–69, 71, 77, 81–82, 84, 94–96, 114–19, 122, 191, 194, 235; settlement places, 101–2, 104
ethnic identity, 124–26, 135, 138, 142, 152, 154–55, 160–61, 184, 207, 213, 219; origin, 11, 22, 29–30, 40, 43–45, 60, 64–65, 86, 89–91, 233, 173–76
ethnicity, 14, 117, 164–65

Naabi Hill, 212
Narok area, 142
narrative, 2, 5, 11–12, 14–15, 18–21, 31, 44, 55,
   57–58, 64, 72, 74, 77, 85–86, 90, 113,
   124, 136, 140, 143, 169–70, 175, 202–3,
   218, 235–37, 246n26, 285n4
Nata: language, 10, 12, 30, 40, 55–56, 71, 82,
   86, 95, 105, 127–28, 156; people, 3, 7, 10,
   29–30, 41, 43, 55, 59, 64, 70, 78, 82–85,
   89, 91, 95, 104, 121, 124–26, 138, 142,
   146–48, 151, 156, 158, 160–61, 171–72,
   175, 179, 182, 184–85, 187, 189–91, 193,
   196–98, 202; site, 45, 54, 110–11; stories,
   19, 39–41, 43, 57, 64–65, 89, 102, 109, 137,
   148–49, 171, 175, 191; territory, 12–13, 41,
   45, 54, 68, 79, 84–86, 91, 118, 120–21,
   124–26, 137, 148–49, 158, 160–61, 164,
   177–78, 190–91, 197, 219, 223, 225, 231–32
nationalism, 4
National Parks Ordinance, 204–5, 207
native, 38, 51, 83, 141, 148, 170, 173–74,
   178–83, 188, 192, 194–95, 204–7;
   courts, 197
Natron, Lake, 166
Nature Conservancy, 211
Ndabaka, 148, 150, 160
Ndorobo, 53–56, 88, 125, 154, 208, 265,
   282n90, 290n62
network, 12, 14, 22, 40, 45, 63–66, 68–70,
   77, 86, 89, 91, 94, 96–98, 123, 138–39,
   151, 155, 163, 189, 191, 198, 213, 233–35
Neumann, Roderick, 17, 205
Ngirate, 121, 138, 158, 160, 163–64, 177, 184,
   196, 203, 219
NGOs, 24, 231–32
Ngoreme: language, 10, 71, 85, 126–27, 156;
   people, 3, 7, 12, 30, 40, 47, 82, 94, 106,
   123, 130, 150–51, 154–56, 174, 177, 184,
   188–89, 191, 196–97, 199, 210, 217, 219,
   222; site, 44, 54, 104, 106–7, 112, 150; stories, 39, 43–44, 47–49, 90, 147, 155, 158,
   163, 176; territory, 35, 40, 50, 75, 97,
   106–7, 118–19, 122, 124–26, 128, 140–42,
   145, 150, 156–58, 161–62, 166, 174, 178,
   184, 194, 219, 232
Ng'orisa, 109, 122
Ngorongoro, 116–17; Conservation Area,
   32, 211–22; Crater, 2, 31, 51, 53, 108, 140,
   142, 146, 209–11; highlands, 51, 140, 142,
   208, 210
North Mara. See Mara: North
northern extension of SNP, 44, 210, 216, 222
Nsoro, 112

ntemi, 48, 152
Nyabange, xii–iii, 170, 176
Nyabehu River, 74
Nyakinywa, 57–59, 70, 96, 151–52, 174
Nyakitono, 228, 230
Nyambeho, 177
Nyamunywa, 30, 32, 39–41, 45, 59, 61, 64,
   89
Nyamwezi, 141, 146
nyangi, 116–17, 173, 190–91, 232
Nyarabovo mountain range, 125
Nyasigonko, 30, 32, 40–41, 45, 61, 64
Nyerere, Edward Wanzagi, 131
Nyerere, Julius Kambarage, 88, 130, 212,
   217
Nyiberekira, 44, 105, 136–39, 143, 158
Nyichoka, 54, 73, 108, 228
Nyigoti, 74, 136–38, 158, 178

oath, 55, 83, 94–95, 109, 118, 130, 149–50,
   275n140
obosongo, 73
obugabho, 96, 152
obugo, 156
oburwe, 41, 50. See also eleusine millet
obutani, 186. See also orutani
ochre, 74, 147
Odhiambo, E. S. Atieno, 4, 16
Okiek speakers, 135
Olduvai Gorge, 32
omochama, 129. See also abachama
omokoro, 105
omorama, 43
omoreto, 128
omoroti, 108
omoshana, 119
omosimano, 83–84
omosimbe, 59, 85
omotangi, 123
omotoro, 76
omotware, 59, 85
omugambi, 105
omugongo, 122
omugongo wa mwensi, 123
omukina, 105
omunase, 119
omunibi, 97, 189
omunywa, 41
omusangura, 111
omwame, 97
omwerechi, 105
omwibororu, 83
Operation Mara, 218